International Law on Antarctic Mineral Resource Exploitation

Runyu Wang

International Law on Antarctic Mineral Resource Exploitation

PL ACADEMIC RESEARCH

Bibliographic Information published by the Deutsche Nationalbibliothek
The Deutsche Nationalbibliothek lists this publication in the Deutsche Nationalbibliografie; detailed bibliographic data is available in the internet at http://dnb.d-nb.de.

Zugl.: Hamburg, Univ., Diss., 2015

Library of Congress Cataloging-in-Publication Data
Names: Wang, Runyu, author.
Title: International law on Antarctic mineral resource exploitation.
Description: Frankfurt am Main ; New York : Peter Lang GmbH, 2017.
Identifiers: LCCN 2017013848 | ISBN 9783631724255
Subjects: LCSH: Mining law–Antarctica. | Convention on the Regulation of Antarctic Mineral Resource Activities (1988 November 25) | Antarctic Treaty system. | Antarctic Treaty (1959 December 1). Protocols, etc. (1991 October 4) | Mines and mineral resources–Environmental aspects–Antarctica. | Prospecting–Environmental aspects–Antarctica.
Classification: LCC KWX823.31988 .W36 2017 | DDC 341.4–dc23 LC record available at https://lccn.loc.gov/2017013848

D 18
ISBN 978-3-631-72425-5 (Print)
E-ISBN 978-3-631-72440-8 (E-PDF)
E-ISBN 978-3-631-72441-5 (EPUB)
E-ISBN 978-3-631-72442-2 (MOBI
DOI 10.3726/b11213

PL Academic Research is an Imprint of Peter Lang GmbH.

Peter Lang – Frankfurt am Main · Bern · Bruxelles · New York · Oxford · Warszawa · Wien

This publication has been peer reviewed.

www.peterlang.com

Contents

Abbreviations 11

Introduction 15

A. Antarctic Geography 15

B. Mineral resource 17

C. The climate 19

D. Life in the Antarctic 20

Part 1: The Antarctic Treaty 23

A. Sovereignty Claims of Seven States 24

- I. The United Kingdom 24
- II. New Zealand 28
- III. France 29
- IV. Australia 29
- V. Norway 31
- VI. Argentina and Chile 32
- VII. Non-claimant States' activities and attitudes 36

B. Territorial Acquisition in International Law 39

- I. Categories of Territorial Acquisition 39
- II. Occupation theory in the Antarctic Territorial Issue 41

C. Maritime Zones in the Antarctic 50

- I. Territorial Sea 51
- II. Exclusive Economic Zone (EEZ) 53
- III. High Seas and Article VI of the Antarctic Treaty 55
- IV. Antarctic Continental Shelf 60
- V. Conclusion 61

D. Options for a Potential Minerals Regime 65
I. National Regime 65
II. Open Use Regime 65
III. Condominium 66
IV. Moratorium 68
V. Common Heritage Model 70
E. The Antarctic Treaty and Antarctic Treaty System 76
I. The Antarctic Treaty 76
II. Antarctic Treaty System 82

Part 2: CRAMRA 87
A. Background 87
I. Preface to Mining Negotiations 87
II. Negotiations of Antarctic Mineral Resource Regime 92
B. General Provisions 98
I. Objectives and Principles 98
II. Area of Application 100
III. International Participation 102
IV. Compliance 102
V. Liability 103
VI. Inspection and Protected Areas 107
C. Institutions 108
I. The Antarctic Mineral Resources Commission 109
II. The Scientific, Technical and Environmental Advisory Committee 110
III. The Special Meeting of Parties 112
IV. The Regulatory Committees 112
V. The Secretariat 116
VI. Financial provisions 116
VII. Conclusion 116

D. Operational Provisions 117
I. Prospecting 118
II. Exploration 119
III. Development 122
E. Dispute Settlement 123
F. Interests Accommodated in the Negotiation Process 126
I. Internal Interests 126
II. External Interests 128
III. Protection of Environment and Mineral Resource Activities 130
G. Responses to CRAMRA 132
I. The UK's Antarctic Mineral Act 132
II. Attitude of the International Community toward Antarctica 136
III. Period Following Refusal of Australia and France to Sign CRAMRA 143

Part 3: CRAMRA – A Dead Convention? 151
A. Critiques of CRAMRA 151
I. CRAMRA, a Disaster or a Protector of the Environment? 151
II. Lack of Clear Terminology 154
III. UN-International Participation 156
IV. Inadequate Protection of Developing States 158
V. Disposition of Surplus 160
B. Merits of CRAMRA 161
I. CRAMRA and the Reconciliation of Internal Tensions 161
II. CRAMRA and Part XI of UNCLOS 163
III. CRAMRA and the International Community 164
C. Underlying Reasons 165
I. Political reasons 165
II. Economic and Technological Reasons 167
D. CRAMRA and the Environmental Protocol 167
I. Background 168

II. Mining Issue 169
III. Principles 170
IV. Environmental Impact Assessment 171
V. Responsibility & Liability 173
VI. Institutions, Inspection & Dispute Settlement 175

E. Influence of CRAMRA on the ISA 176
I. Impacts on Environmental Protection 176
II. Impacts on Institutions 182
III. Impacts on the Phases of mining activities 184
IV. Impacts on Confidential Nature of Date and Information 185

Part 4: Protection of the Antarctic Environment and Ecosystem 187

A. Management of Antarctic Living Resources before the Protocol 187
I. Antarctic Treaty 188
II. International Convention for the Regulation of Whaling 188
III. Agreed Measures for the Conservation of Antarctic Fauna and Flora 193
IV. Convention for the Conservation of Antarctic Seals (CCAS) 195
V. Convention on the Conservation of Antarctic Marine Living Resources 196

B. The Protocol on Environmental Protection to the Antarctic Treaty 202
I. General Rules 205
II. Prohibition of Mining Action 211
III. Environmental Impact Assessment (EIA) 213
IV. Implementation of the Protocol 217
V. Evaluation of the Protocol 224

C. Conclusion 226

Part 5: Exploitation of Antarctic Ice 229
A. Legal status of Antarctic Ice 230
B. Relevant Regimes for the Exploitation of Antarctic Ice 232
I. UNCLOS 232
II. Antarctic Treaty System 233
III. Common Heritage of Mankind 234
C. Exploitation of Antarctic Ice 235
I. Recommendation XV-21 235
II. CCAMLR 236
III. The Madrid Protocol 237
IV. The Precautionary Approach 241
D. Conclusion 242

Summary of Study 243

Literature 247

Table of Cases 267

International Treaties 269

UN Documents 271

ATS Documents 273

Abbreviations

A.J.I.L.	American Journal of International Law
A.F.D.I.	Annuaire Français de droit international
ASMA	Antarctic Specially Managed Areas
ASOC	Antarctic and Southern Ocean Coalition
ASPA	Antarctic Specially Protected Area
ATCM	Antarctic Treaty Consultative Meeting
ATCP	Antarctic Treaty Consultative Party
ATS	Antarctic Treaty System
AVR	Archiv des Völkerrechts
B.U. Int'l & Comp. L. Rev.	Boston College International and Comparative Law Review
BANZARE	British, Australian and New Zealand Antarctic Research Expedition
Cal. W. Int'l L. J.	California Western International Law Journal
CAN. Y.B. INT'L L.	Canadian Yearbook of International Law
Case W. Res. J. Int'l L.	Case Western Reserve Journal of International Law
CCAMLR	Convention for the Conservation of Antarctic Marine Living Resources
CCAS	Convention for the Conservation of Antarctic Seals
CEE	Comprehensive Environmental Evaluation
CLCS	Commission on the Limits of the Continental Shelf
COMNAP	Council of Managers of National Antarctic Programs
Cornell Int'l L. J.	Cornell International Law Journal
CRAMRA	Convention on the Regulation of Antarctic Mineral Resource Activities
Denv. J. Int'l L. & Pol'y	Denver Journal of International law and Policy
EEZ	Exclusive Economic Zone
EIA	Environmental Impact Assessment
Eur. J. Int'l L.	European Journal of International Law
Fla. Int'l L. J.	Florida International Law Journal
Geo. Int'l Envtl. L. Rev.	Georgetown International Environmental Law Review
IAATO	the International Association of Antarctica Tour Operators
ICJ	International Court of Justice

IEE	Initial Environmental Evaluation
IGY	International Geophysical Year
IHO	International Hydrographic Organization
ILM	International Law Materials
IMO	International Maritime Organization
Int'l & Comp. L.Q.,	International and Comparative Law Quarterly
IOC	Intergovernmental Oceanographic Commission
IOPC Funds	International Oil Pollution Compensation Funds
IPCC	Intergovernmental Panel on Climate Change
ISA	International Seabed Authority
IUCN	International Union for the Conservation of Nature
IWC	International Whaling Commission
Law & Pol'y Int'l Bus.	Law and Policy in International Business
NGO	Non-Government Organization
Ocean Dev. & Int'l L.	Ocean Development and International Law
OECD	Organisation for Economic Cooperation and Development
OPEC	Organization of Petroleum Exporting Countries
PCA	Permanent Court of Arbitration
PCIJ	Permanent Court of International Justice
RMP	Revised Management Procedure
RMS	Revised Management Scheme
San Diego L. Rev.	San Diego Law Review
SATCM	Special Antarctic Treaty Consultative Meeting
SCAR	Scientific Committee on Antarctic Research
SOS	Southern Ocean Sanctuary
SPA	Specially Protected Area
SPRI	Scott Polar Research Institute
SPSs	Specially Protected Species
SSSIs	Sites of Special Scientific Interest
Stan. J. Int'l L.	Stanford Journal of International Law
Tex. Int'l L. J.	Texas International Law Journal
UNCLOS	United Nations Convention on the Law of the Sea
UNDESA	United Nations Department of Economic and Social Affairs
UNEP	United Nations Environment Program
UNESCO	UN Educational Scientific and Cultural Organization
UNGA	United Nations General Assembly

Va. J. Int'l L.	Virginia Journal of International Law
Va. L. Rev.	Virginia Law Review
Vand. J. Transnat'l L.	Vanderbilt Journal of Transnational Law
Wis Int'l L. J.	Wisconsin International Law Journal
WMO	World Meteorological Organization
WTO	World Tourism Organization
WWF	World Wildlife Fund
Yale L. J.	Yale Law Journal
ZaöRv	Zeitschrift für ausländisches öffentliches Recht und Völkerrecht

Introduction

A. Antarctic Geography

Antarctica is a huge continent with an area of about 136,600,000 square kilometres[1], and a coastal line of approximately 14,000 miles.[2] The Antarctic consists of a polar continent surrounded by an ice-covered ring of ocean waters in which lie scattered islands group, some polar, others sub-polar in character.[3] Except for a small part of the peninsula and the archipelagoes immediately around it beyond 65°S, almost the entire Antarctic continent lies within the Antarctic Circle (John S. Cumpston said 66°30'S, Roberto Bargagli said 66°33'S[4]).

It was Nordenskjöld, in 1903, who first divided the Antarctic into West and East parts in order to facilitate research. According to his method, the Antarctic continent is divided by a line running along the western boundary of the Ross Ice Shelf, north of the Horlick Mountains and the Pensacola Range, and along the eastern edge of the Filchner Ice Shelf.[5] Generally speaking, the part south of Australia, Africa, and the Indian Ocean is East Antarctica, and the part of the continent lying south of the Americas constitutes what is referred to as West Antarctica.[6] East Antarctica is generally extended beyond the limits of the eastern hemisphere to include the geographic pole and a portion of land extending up to the great Antarctic Horst[7], and comprises the Trans-Antarctic Mountains, a large mountain range stretching for 3,500 kilometers from Victoria Land to the Pensacola Mountains, an extremely

---

1 At different times, there are different results for Antarctic data.5,058,356 sq. miles, or 13,101,154 sq. km., according to Kosack, H.P.: "Wie gross ist das Südpolargebiet?", 94 Petermanns Geographische Mitteilungen 202 (1950), citing Fairbridge, Rhodes W., "The Geology of the Antarctic", in Simpson, Frank A. (ed.), *The Antarctic Today: A Mid-Century Survey by the New Zealand Antarctic Society*, (Wellington: A. H. & A.W. Reed and New Zealand Antarctic Society, 1952) at 56.

2 Cumpston, John S. "Introduction the Antarctic", in *Simpson* (note 1), at 17.

3 Nordenskjöld, Otto and Mecking, Ludwig, *The Geography of the Polar Regions*, (New York: The Commonwealth Press, 1928) at 285.

4 Cumpston (note 2); Bargagli, Roberto, *Antarctic Ecosystems Environmental Contamination, Climate Change, and Human Impact*, (Berlinm / Heidelberg: Springer 2005), at 3.

5 Bentley, Charles R., "Glacial and Subglacial Geography of Antarctica", in Wexler, H. et al. (eds.), *Matthew Fontaine Maury Memorial Symposium on Antarctic Research*, (Richmond, Virginia: William Byrd Press, 1962), at 18; Nordenskjöld / Mecking (note 3), at 286.

6 Davies, William E. "Antarctic Stratigraphy and Structure", in Crary, A.P. et al. (eds.), *Antarctica in the International Geophysical Year: Based on a Symposium on the Antarctic*, (Baltimore, Maryland: Waverly Press, 1956), at 44.

7 Siple, Paul A., "Antarctic Geography", in *Crary* (note 6), at 13.

compact land mass that contains the centre of gravity of the continent[8]. In general, East Antarctica is characterized by narrow coastal strips and steep slopes rising sharply to the high Antarctic plateau. This area has an average elevation of about 2,300 meters. Western Antarctica comprises Marie Byrd Land, Ellsworth Land and the Antarctic Peninsula. The western part is only half as large as East Antarctica, and although its average elevation is about 850 meters, lower than its neighbour part, it does contain the highest peak in Antarctica, the Vinson Massif (4,897 meters). The Antarctica Peninsula constitutes a narrow north-south mountain range with an average width of 70 kilometres and a mean height of 1,500 meters.

The heavy dome of snow and ice that covers more than 98% of the Antarctic is its chief characteristic. The ice sheet is up to 4,500 km wide, with an area of about 12.09 million square kilometres, making it the greatest land-locked water mass on Earth. The ice and snow over East Antarctica rises toward the centre to above 3,962.4 meters in elevation, by contrast, the ice dome of West Antarctica is estimated at less than 3,048 meters in height.[9] The larger thicker East Antarctic Ice Sheet is resting on land which is mostly above sea level, giving it its terrestrial classification,[10] and the mean elevation of bedrock in East Antarctica is 15 meters[11]. On the other hand, the smaller, lower elevation West Antarctic Ice Sheet is considered a marine ice sheet because the mean elevation of the bedrock on which it rests[12] is 440 meters below sea level.[13]

Ice plays and has played an important role on our planet. It affects global climate and patterns of weather and ocean currents in the Southern Hemisphere, as well as contributing to the land and water masses as we currently know them. There are two huge bodies of water around the Antarctic continent, one is the Ross Sea, and the other is the lager Weddell Sea. The Ross Ice Shelf is the biggest flat-lying shelf ice about 500 miles wide east and west by 400 miles north and south. This sheet averages perhaps 600 and 1000 feet in thickness. It grows seaward at an average rate of about four feet per day and cyclically breaks back corresponding amounts once or twice a century.[14]

---

8 Nordenskjöld / Mecking, (note 3), at 286.

9 Siple (note 7), at 15. In the original paper the writer uses 13,000 feet instead of 3962.4 meters and 10,000 instead of 3,048 meters.

10 Keys, J. R. "Ice", in Glasby, G.P. (ed.), *Antarctic Sector of the Pacific*, (New York: Elsevier Science Publisher, 1990), at 96.

11 Drewry, D. J. / Jordan, S.R., "The bedrock surface of Antarctica", in, *Antarctica: Glaciological and Geophysical Folio*, (Cambridge: Scott Polar Research Institute, 1983).

12 Keys (note 10).

13 Drewry and Jordon (note 11).

14 Siple, (note 7), at 16.

B. Mineral resource

Frank Wild was the first person to discover coal resources close to the upper Beardmore Glacier during Schackleton's 1907–1909 expedition.[15] However, considering the long distances and high cost of exploiting this resource, compared to the relative availability of world mineral reserves, it was not until the 1970s and 1980s that the potential mineral resources of Antarctica began receiving more and more attention. Several oil companies carried out exploration during the 1970s[16] and claimed the mineral wealth of Antarctica. What and how many kinds of mineral resources does the Antarctic have? Unfortunately, this data is only speculative, based on superficial continental comparisons without adequate knowledge of geological environments and processes.[17] According to the hypothesis of tectonic plate theory, which emphasizes the relationship to formerly juxtaposed continents, the location of zones of economic resources on the once adjacent continents can be used in attempting to predict where similar zones might occur in Antarctica.[18]

It is therefore thought that **petroleum** occurs in thick sedimentary rock sequences, i.e. the Antarctic continental margins including the Ross Sea, Weddell Sea, and Prydz Bay adjacent shelves. There are 21 onshore and offshore sedimentary basins in Antarctica[19], at least 7 of these are considerable in size[20]. St John estimated the total potential to be 203 billion barrels of oil equivalent[21], however, these estimates are treated as meaningless[22]. Much additional research and drilling would be required to make reasonable assessment of potential petroleum resources.

---

15 David, T.W.E. / Priestley, R.E., *British Antarctic Expedition 1907–1909 Reports on the Scientific Investigations: Geology, Vol. 1 Glaciology, Physiography, Stratigraphy and Tectonic Geology of South Victoria Land*, (London: Heinemann, 1914), citing Hansom, James D. / Gordon, John E., *Antarctic Environments and Resources:a Geographical Perspective*, (New York: Addison Wesley Longman Limited, 1998), at 221.

16 Auburn, F.M., *Antarctica law and politics*, (London: Hurst & Company, 1982), at 243–246.

17 Hansom / Gordon (note 15), at 197 et seq.

18 Fifield, Richard, *International Research in the Antarctic*, (Oxford: Oxford University Press, 1987), at 55.

19 Ibidem, at 56.

20 Hansom and Gordon (note 15), at 222.

21 St John, B. "Antarctica—geology and hydrocarbon potential", in Halbouty, M.T. (ed.), *Future Petroleum Resources of the World: American Association of Petroleum Geologists*, (Tulsa, Oklahoma 1986), at 55–100, citing Hansom / Gordon (note 15), at 222.

22 Crockett, R.N. / Clarkson, P.D., " The exploitation of Antarctic minerals", 13 Environment International 121 (1987), at 121–132; Larminie, F.G., "Mineral Resources: Commercial Prospects for Antarctic Minerals", in Triggs, G.D. (ed.), *the Antarctic Treaty Regime: Law, Environment and Resource*, (Cambridge: Cambridge University Press, 1987), at 176–181; Behrendt, J.C., "Recent geophysical and geological research in Antarctica relate to the assessment of petroleum resource and potential environmental hazards to their development", in Splettstoesser, J.F. / Dreschhoff, G.A.M. (eds.),

Coal occurs throughout many locations of the Transantarctic Mountains as part of the Beacon Supergroup[23], and in the Prince Charles Mountains. Because of a high content of ash and moisture, the coal is normally thin, lenticular, and discontinuous. It has not yet proven economical to explore the coal in Antarctic.

Iron is thought to exist mainly in the Prince Charles Mountains, particularly at Mount Ruker, although the low iron content and distance from the coast has made exploration of these reserves uneconomical.

Copper has been found in the Antarctic Peninsula. In inland Pensacola Mountains, the Dufek Intrusion has been revealed to contain a variety of minerals, mainly platinum and chromium. It extends over an enormous area, perhaps 50,000 square kilometers.[24] Mineral nodules, also called manganese nodules, are scattered in the Pacific sector of the Southern Ocean and contain iron and manganese with some copper, nickel and cobalt[25]. These have also not yet been explored due to their low metal content.

Icebergs as a mineral are a source of fresh water for arid regions of the world. The estimate annual yield of icebergs in Antarctica is of the order of 1000 km^3.[26] The Scientific Committee on Antarctic Research (SCAR) report: Possible Environmental Effects of Mineral Exploitation in Antarctica points out: "...Theoretically, icebergs could be floated to any point accessible by water route with minimum depths of 200m. ..." However, there are still uncertainties and problems regarding the stability of towed icebergs, insulation and protection from melting and wave erosion, and processing at destination points.[27]

Mineral Resource Potential of Antarctica: American Geophysical Union, Antarctic Research Series 51, (Washington D.C.: 1990), at 163–174, citing Hansom / Gordon (note 15), at 222–223.

23 Splettstoesser, J.F., "Antarctic Geology and Mineral Resources", 1 Geology Today 41 (1985), at 41–45; Rose, G / McElroy, C.T., *Coal potential of Antarctica", Resource Report 2* (Canberra: Bureau of Mineral Resources, Geology and Geophysics. Australian Government Publishing Service, 1987); Coates, D.A. / Stricker, G.D. / Landis, E.R., "Coal geology, coal quality and coal resources in Permian rocks of the Beacon Supergroup, Transantarctic Mountains, Antarctica", in Splettstoesser, J.F. / Dreschhoff, G.A.M. (eds.), *Mineral Resource Potential of Antarctica: American Geophysical Union, Antarctic Research Series 51*, (Washington D.C.: 1990), at 133–162, citing Hansom / Gordon (note 15), at 222.

24 Fifield (note 18), at 57.

25 Frakes. L.A. / Moreton, D.L.E., "Manganese Nodule Provinces of the Southern Ocean", in Hansom / Gordon (note 15), at 222.

26 Fifield (note 18), 57.

27 Hansom / Gordon (note 15), at 222.

C. The climate

Antarctica is the coldest continent in the world.[28] The maximum surface air temperatures of the Antarctic is about -30℃ on the plateau and -4℃ at coastal locations, and the minimum happens in July or August -70℃ on the plateau and -25℃ at the coast.[29] In July 1983, at Vostok on the polar plateau, the world's coldest temperature was recorded with -89.5℃. Another feature of the climate is the short summer, only 6–8 weeks happening around the beginning of January. There are several reasons for the cold weather in the Antarctic. As the 23.3° inclination of the Earth's axis of rotation in relation to the plane of its orbital axis round the sun, there is a long period of polar night in Antarctica. Even during the polar day, it receives much less radiation than other parts of the world because of the curvature of the earth. Moreover, most of the solar radiation (80–90%) in Antarctic continent is reflected out into space by the mass mirror-like ice sheet, which means a large amount of radiation in Antarctica is not absorbed at all.[30] Finally, the high average altitude (2,300 meters) is another main contributing factor to Antarctica acting as one of the largest heat sinks on earth. The cold waters of Southern Ocean, the variable sea ice cover and the high continental ice sheet not only affect the climate of Antarctica, but also give Antarctica a very important role in the global climate system[31].

The Southern Ocean is characterized by two wind-driven currents, which result in the circulation of cold water around the continent. The mixing of this cold water with warmer sub-Antarctic water at the Polar Frontal Zone produces a zone of great temperature and salinity gradients in the ocean. The winter advance of the sea ice effectively doubles the ice-covered area of the Antarctic region and further reduces temperatures by increased albedo effects.[32]

Sea ice which could reflect radiation into space is one of the main factors determining variations in the Antarctic climate. Besides increasing albedo, sea ice reduces heat fluxes between the atmosphere and sea, exchanges salt with the ocean, and water evaporation during the winter.[33] Sea ice modifies the seasonal temperature circle, delaying cooling of the atmosphere in autumn by the release of "latent heat" as an increasing amount of water becomes ice, and having the reverse effect during ice-melt in spring.[34] At the end of summer, the formation of sea ice involves the release of dense, brine-rich waters, determining the formation of cold deep waters. These cold waters influence global oceanic circulation, because the generally

---

28 Bargagli (note 4), at 17–19; Fifield (note 18), at 82–91.

29 Bargagli (note 4), at 17.

30 Fifield (note 18), at 82; Hansom / Gordon, (note 15), at 35–36.

31 Phillpot, H.R., "Physical Geography—Climate", in Bonner, W. N. / Walton, D. W. H. (eds.), *Key Environments: Antarctica*, (Oxford: Pergamon Press, 1985), at 23–38.

32 Hansom / Gordon (note 15), at 36.

33 Parkinson, C.L., "Southern Ocean Sea Ice Distribution and Extents", (1992) *Philosohical Transactions: Biological Sciences*, Vol. 338, No. 1285, at 243–250.

34 Fifield (note 18), at 83.

divergent flow around Antarctica modifies both the concentration of salt and heat in the ocean by the net flow of ice towards the equator.[35]

In short, changes to the extent, thickness, persistence and annual distributions of sea ice affects the climate of Antarctica and consequently global atmospheric circulation, the exchanges of energy, the transfer of momentum between atmosphere and ocean, and the vertical mixing and salt balance in the ocean.

As to the effect of these ice sheets on global sea level – the maintenance of the ice sheets relies on the balance between snowfall over the continent and the rate of ice discharge in the Southern Ocean. This balance strictly depends on temperature, the extent of sea ice, and water evaporation in the Southern Ocean, which affects the transport of heat, air masses and precipitation from mid-latitudes. Changes of a few percent in the total volume of ice would affect the volume of the world's oceans. The West Antarctic ice sheet contains enough ice to raise global sea level by about 6 meters.[36] If the whole Antarctic ice sheet melted, sea level would rise by about 73 meters.[37]

D. Life in the Antarctic

In the early twentieth century, scientists believed that the vast area of inland ice and the extreme cold climate brought about uniformity of life in the Antarctic. They wrote that in the Antarctic "everywhere there are the same birds, especially penguins, everywhere there are the same few species of seals and whales in the sea and on the ice, moreover, in the Southern Ocean, great seasonal fluctuations in the ratio of open to ice-covered sea, with minute algae, krill and other invertebrates, fish, birds, seals, and whales forming relatively short food chains."[38] However, with the rapid development of research in the Antarctic, more and more species have been found in this "cold desert".

The Penguin family is widely distributed in the cooler waters of the southern oceans. Almost every Antarctic and sub-Antarctic island has more than one resident species. The largest concentrations and greatest numbers of species occur in the cold temperate, sub-Antarctic and Antarctic waters. There are seventeen different species of Penguins, four of these – the Emperor, Adelie, Chinstrap and Gentoo – breed on the Antarctic mainland. Most of the others live on rocky islands in the Antarctic Ocean. The principal species are the Emperor Penguin, the King Penguin, and the Adelie Penguin.[39] The Emperor Penguin, which prefer the cold, are found in South Victoria Land and East Antarctica, but not in the northern part of West

35 Bargagli (note 4), at 13; Fifield (note18), at 83.

36 IPCC, "Climate Change 2001," (Cambridge: Cambridge University Press, 2001).

37 Robin, G. de Q., "Changing the sea level", in B. Bolin et al. (eds.), *SCPOE 29 The Greenhouse Effect, Climate Change, and Ecosystems*, (Chichester: John Wiley & Sons, 1986), at 323–359.

38 Nordenskjöld / Mecking (note 3), at 285.

39 Nordenskjöld / Mecking (note 3), at 295–298.

Antarctica. It broods in mid-winter, in the coldest part of the Antarctic, and not on the land but on the sea ice. The habitat of the King Penguin is located north of the Antarctic Circle, and the Adelie penguin is most frequent along the border of the Antarctic Continent.

Besides the well-known penguins, there are many **other seabirds** in the Antarctic. The largest group of seabirds in the Antarctic is the petrels. The smallest is the storm petrel, weighing one to two ounces; it feeds at sea, folding its wings vertically, giving it the appearance of walking on water. The largest is the albatross, which can weigh up to 21 pounds and have a wing span of 12 feet. It can remain at sea for months as it sleeps on the water and drinks sea water. It can fly at speeds of 60 mph and with a favorable wind may reach speeds as great as 100 mph. The Arctic tern travels from the edge of the Arctic Ocean to the pack ice of Antarctica, a distance of some 22,000 miles, in order to flee the northern Arctic winter to enjoy a brief Antarctic summer. Additionally, there are gulls, cormorants, fulmars, and shearwaters and skuas.

Seals[40] normally live in pack ice. There are six species of seals in the Antarctic. Five of these species are true, or earless seals, without external pinnae. These are Elephant, Crabeater, Weddell, Ross, and the Leopard Seal. The sixth species is the Southern Fur Seal, which belongs to the sea lion family – the group of seals with external ears. The Elephant Seal seldom visits the Antarctic coasts and has already been almost exterminated by ruthless slaughter on the sub-Antarctic islands. The Crabeater Seal lives in the pack ice during winter and summer and occasionally reaches sub-Antarctic latitudes. It appears to be semi-migratory in relatively great numbers, living almost exclusively on pelagic shrimp. The Weddell Seal is more southern in distribution than the Crab-eater Seal, and its range is confined predominantly to the fast-ice along the coasts throughout the year. It feeds on fish and cephalopods. The circumpolar Ross Seal is the rarest species. It inhabits the pack-ice belt, is solitary in its habits, and appears to confine itself to the southerly parts of the area, being unknown in the northward of the pack ice. Its diet is probably fish and cephalopods. The Leopard Seal has a rather broad latitudinal range, and is found from the shores of the Antarctic Continent northward to parts of the temperate zone. It is semi-migratory, solitary in its habits, and is relatively widespread, but scarce. This aggressive seal preys on other seals and penguins, as well as on fish and cephalopods.

Whales are mammals who swim thousands of miles from other oceans to visit the Antarctic seas, for the abundance of krill and fish. There are two types of whales: baleen whales and toothed whales. Baleen whales live on krill which they "filter" through baleen plates in their mouths. The most common baleen whales found in the Antarctic are the Blue, Fin, Humpback, Minke, Sei, and Southern Right Whale. The toothed whales have small peg-like teeth and eat mainly fish. The Killer whale

40 See Eklund, Carl R. "Antarctic Fauna and Some of its Problems", in Crary (note 6), at 117–123.

and the Sperm whale are the only toothed whales found in the Antarctic region. The Killer whale eats small penguins which it hunts by tipping them off ice floes.

Approximately 100 species of **fish** live in the Southern Ocean south of the Antarctic Convergence, where the colder water meets the warmer water.

Besides varied animals, there are also different **plants** in the Antarctic.[41] Two species of flowering plants occur on the Antarctic Peninsula.[42] Not a single flowering plant has been found within the Antarctic continent. Grasses do not occur south of latitude 62°S[43]. The flora of continental Antarctica comprises 15 species of mosses[44], one species of liverwort[45] and at least 88 taxa of lichens[46]. Eight species of moss have been recorded from southern Victoria Land and Ross Island.[47] Algae color the snow; they also occur on damp ground and in water, especially in the sea. These number about four hundred species in the Antarctic and sub-Antarctic.

This list of data serves to provide an initial impression of the Antarctic. The huge ice layer and severely low temperatures have not impaired humankind's enthusiasm for exploring the Antarctic. The predicted resources to be found in the Antarctic is drawing international society's attention and interest particularly as current supplies of food, water and energy are limited, and there has been a great improvement in technology for exploitation in cold weather. Even though there are still some scientific and legal obstacles, it seems likely that Antarctic mineral resources will be exploited at some point in the future.

---

41 Skotnicki, M. L. / Selkirk, P. M., "Plant Biodiversity in an Extreme Environment – Genetic Studies of Origins, Diversity and Evolution in the Antarctic", in Bergstrom, D. M. et al. (eds), *Trends in Antarctic terrestrial and limnetic ecosystems*, (Dordrecht: Springer, 2006), at, 161–175.

42 Edwards, J. A. / Smith, R.I. Lewis, "Photosynthesis and Respiration of Colobanthus Quitensis and Deschampsia Antarctica from the Maritime Antarctic", (1988) *British Antarctic Survey Bulletin*, Vol. 81, at 43–63; Smith, R.I. Lewis, "The Enigma of Colobanthus Quitensis and Deschampsia Antarctica in Antarctica", in Huiskes, A.H.L. et al. (eds.), *Antarctic Biology in a Global Context*, (Leiden, Backhuys 2003), at 234–239.

43 Nordenskjöld / Mecking (note 3), at 295.

44 Smith, R.I. Lewis, "Terrestrial Plant Biology of the sub-Antarctic and Antarctic", in Laws, R. M. (ed.), *Antarctic Ecology*, (London: Academic Press, 1984), at 61–162.

45 Bednarek-Ochyra, H. / Bana, J. / Ochyra, R / Smith, R.I. Lewis, "The Liverwort Flora of Antarctica", (Cracow: Polish Academy of Sciences, Institute of Botany, 2000), pp. 238 et seq.

46 Øvstedal, D.O. / Smith, R.I. Lewis, "Lichens of Antarctica and South Georgia A Guide to their Identification and Ecology", (Cambridge: Cambridge University Press, 2001), pp. 411 et seq.

47 Seppelt, R.D. / Green, T.G.A., "A Bryophyte Flora for Southern Victoria Land, Antarctica", 36 New Zealand Journal of Botany 617 (1998), at 617–635.

Part 1: The Antarctic Treaty

There is still some debate as to who first discovered Antarctica, or even who first navigated Antarctic waters. In 1675, an English merchant Antonio de la Roché sighted a snow-covered island, which is believed to have been South Georgia. In 1772–1775, Captain James Cook circumnavigated the whole continent. On his third voyage in 1774, Cook reached the latitude of 71°10'S, 106°54'W, the furthest south of the 18th century. There have been some claims that Cook's achievement was to disprove of the theory of the southern continent, because Cook only met "little ice and no land".[48] Oppositely, others contribute the confirmation of the landmass in the South Hemisphere to Cook's exploration.[49]

In 1819–1821 Fabian Gottlieb von Bellingshausen, a Russian captain, sailed to the most southern point at longitude 92°19' W. and latitude 69° 53' S.[50] He also confirmed the existence of a continuous open sea south of 60°S. Moreover, he discovered two islands off the west coast of the Antarctic peninsula, and named them Peter I and Alexander Island.[51] The British confirm that Edward Bransfield, a British Captain, sighted the continent on January 20, 1820.[52] Americans contend that Bransfield sighted only an island, and Nathaniel B. Palmer was the first to see the mainland on November 17, 1820.[53] As to who was the first person to discover Antarctica, there are no consistent opinions. In 1841, Captain James Clark Ross, a British sailor, was chosen to lead an Antarctic expedition for magnetic measurements and arrived at 78°4'S, 173°20'W. He discovered and named the mountain chain of Victoria Land, the great volcanoes of Mounts Erebus and Terror and McMurdo Bay.[54] He gave his name to the Ross Sea but did not land on the continent because of the huge Ross Ice Shelf.[55]

In the 20th century, further expeditions and explorations went on in the Antarctic. There have been significant developments in techniques and equipment,

48 Glasby, G.P. (ed.), *Antarctic Sector of the Pacific: Elsevier Oceanography Series 51*, (Amsterdam: Elsevier Science Publishers B.V., 1990), at 6.

49 Dunnett, Harding McGregor, "Shackleton's Boat: The story of the James Caird", (Neville & Harding 1996), p. 2.

50 Cook, F. A., "Captain Fabian Gottlieb Von Bellingshausen, 1819–1821. The Discovery of Alexander I., Peter I., and Other Islands", 33 Bulletin of the American Geographical Society 36 (1901), at 38.

51 Glasby (note 48), at 7; Rubin, M.J., "Thaddeus Bellingshausen's scientific programme in the Southern Ocean 1818–1821", 21 Polar Record 215 (1982), at 215–229.

52 Sullivan, Walter, *Quest of a Continent*, (New York: McGraw-Hill, 1957), at 18; U.S. Antarctic Projects Officer, *An Introduction to Antarctica* (Washington: U.S. Antarctic Projects Officer, 1961), at 11–12.

53 U.S. Antarctic Projects Officer (note 52); Gould, Lawrence McKinley, *Antarctica in World Affairs Vol. 121*, (Foreign Policy Association, 1958), at 17–18.

54 Glasby (note 48), at 8.

55 Dunnett (note 49), at 3.

especially the aircraft and tractors used in Antarctica. The International Geophysical Year (IGY) of 1957–1958 marked the start of the most recent phase of Antarctic activity. More and more countries have begun to pay attention to this ice and snow covered continent.

A. Sovereignty Claims of Seven States

I. The United Kingdom

In July 1908, the United Kingdom first claimed territory in the Antarctic by founding the Falkland Islands Dependencies, which contains Graham Land[56] and islands in the South Atlantic – South Georgia, the South Orkney, South Shetland and South Sandwich Islands.[57] On March 28, 1917, Britain further redefined its territorial claim to the whole sector between 20°W and 80°W, including "all islands and territories whatsoever" as far as the South Pole.[58] Britain's claim was based on the discoveries and acts of annexation by British nationals in the period 1675–1843 and the display and exercise of sovereignty between 1843 and 1908.[59] All these territories are under the administration of the Falkland Islands. British claims to the territory of the Antarctic were based on three main grounds, historic discovery, display of sovereignty, and recognition of other States, which were clearly expressed in its application to the International Court of Justice against Argentina and Chile.[60]

According for marine zones of Antarctica, United Kingdom does not explicitly declare any zones relating to Antarctica;[61] however, for the Antarctic continental shelf, it presented the UN Commission on the Limits of the Continental Shelf (CLCS) with a "partial submission" that "will not include areas of the continental shelf areas appurtenant to Antarctica, for which a submission may be made later".[62] This

56 Graham Land is the Great Antarctic Peninsula, known to Americans as the Palmer Peninsula, to Argentines as Tierra San Martin, and to Chileans as Tierra O'Higgins.

57 Beck, Peter J., "British Antarctic Policy in the Early 20th Century", 21 Polar Record 475 (1983), at 477.

58 The Letters Patent of 28 March 1917 claimed the sector 20°W – 50°W south of 50°S and 50°W – 80°W south of 58°S, *Falkland Islands Gazette*, 2 July 1917, at: http://journals.cambridge.org/download.php?file=%2FPOL%2FPOL5_35-36%2FS0032247400003792Xa.pdf&code=021d2fce556eebbbb38b329bcd41b762.

59 Wilson, R.E., "National Interests and Claims in the Antarctic", 17 Arctic 15 (1964), at 19.

60 *Cases Antarctica 1955 (United Kingdom v. Argentina; United Kingdom v. Chile)*, ICJ Report 1956, at 12 and 15.

61 https://www.gov.uk/government/world/organisations/british-antarctic-territory.

62 The United Kingdom made two official documents about its outer continental shelf claims in the Antarctic. One is a Note to the UN Secretary General (Note no.168/08) on 9 May 2008 at:
http://www.un.org.Depts/los/clcs_new/submissions_files/gbr08.gbr_nv_9may2008.pdf, and the other is its Submission to the CLCS on 11 May 2009, at:

intention to claim the continental shelf of Antarctica has been formally objected by Netherlands and Japan.[63]

1. Historic Discoveries

Britain claims that British explorations of the Antarctica can be traced back to 17th century. It was the British merchant Anthony de la Roche who discovered the islands or lands of the Falkland Islands Dependencies by accident,[64] which may well have been those of *South Georgia* in 1675, that were rediscovered and named *South Georgia* by English Captain James Cook one hundred years later in 1775.[65] In the same year, Captain Cook also discovered the *South Sandwich Islands.*[66] In 1819, another English captain William Smith discovered some islands in 62°17'S., 60°12'W.,[67] named them the *South Shetland Islands* and claimed them for the British Crown.[68] In January 1820, Mr. Edward Bransfield landed on the largest island of *South Shetlands Islands*, and in February, he landed on the most easterly island of the group.[69] In 1821, the *South Orkney Islands* was discovered and claimed for Great Britain by a British sealing captain, George Powell.[70] Moreover, Britain insists that Edward Bransfield was the first person to sight the *Graham Land*. On January 6, 1843, Captain James C. Ross penetrated into a gulf on the eastern side of the peninsula and landed on a coastal island. He named the gulf *Erebus* and *Terror Gulf*, and the island *James Ross Island*, and claimed the island and its "contiguous lands" under the British Crown.[71]

Based on the names given to these areas, Britain is convinced that the first discoveries of South Georgia, the South Sandwich Islands, the South Orkneys, the South Shetlands and Graham Land were all made by British nationals. Furthermore,

---

http://www.un.org/Depts/los/clcs_new/submissions_files/submission_gbr_45_2009.htm.

63 http://www.un.org/Depts/los/clcs_new/submissions_files/submission_gbr.htm

64 Balch, Edwin Swift, *Antarctica*, (Philadelphia: Press of Allen, Lane & Scott, 1902), at 55.

65 Ibidem, at 71.

66 Ibidem.

67 Ibidem at 78; Murray, John, "The Renewal of Antarctic Exploration", 3The Geographical Journal 1 (1894), at 8.

68 Joyner, Christopher C., *Antarctica and the Law of the Sea*, (Dordrecht/Boston/London: Martinus Nijhoff Publishers, 1992), at 48.

69 Balch (note 64), at 80.

70 Joyner (note 68), at 48.

71 All these discoveries by English explorers were formally set out in defense of the British claim when the UK brought Argentina and Chile to the International Court of Justice in 1955 to adjudicate the question of sovereignty to their overlapping Antarctic claims. See Cases Antarctica (United Kingdom v. Argentina) (United Kingdom v. Chile) 1955, International Court of Justice, at:
http://www.icj-cij.org/docket/files/26/9065.pdf, at 11–12; and
http://www.icj-cij.org/docket/files/26/9065.pdf, at 51–52.

during the period from 1678 to 1843, acts of annexation in these five principal groups were only performed in the name of the British Crown; therefore, the United Kingdom claims that Britain "possessed... an original root of title to all the territories concerned".[72]

2. *Display of British Sovereignty*

a) Falkland Islands Dependencies in the Period 1843 – July 21, 1908

On June 23, 1843, Royal Letters Patent were issued making provisions for the government of the "Settlements in the Falkland Islands and their Dependencies". In 1904 an Argentina whaling company, the Compania Argentina de Pesca, applied to the British legation in Buenos Aires for a whaling license for South Georgia from the British Government in London. In 1905, a Chilean Company, the South Georgia Exploration Company, was granted a mining and grazing lease of South Georgia by the Governor of the Falkland Islands and proceeded to set up operations there. In the same year, the British Government replied to an inquiry from the Norwegian Government that the South Shetlands were not international but were British possessions, as were South Georgia, the South Orkneys and Graham Land, and that Norwegian whalers should apply to the governor of the Falkland Islands for any facilities that they might need.[73] The Falkland Islands Government promulgated a whaling ordinance in 1906 by which the taking of whales without licence was made unlawful. On July 21, 1908, South Georgia, the South Orkneys, the South Shetlands, the Sandwich Islands and the territory of Graham Land were by Royal Letters Patent formally constituted Dependencies of the Colony of the Falkland Islands and placed under its government. In 1917, The Britain issued an amended Letters Patent to define the lands comprised in the Falkland Islands, to explicitly include all islands and territories situated between longitudes 20° and 50° West, and south of latitude 50° South; and all islands and territories situated between longitudes 50° and 80° West, and south of latitude 58° South.

b) Dependencies in the Period July 21, 1908 – September 22, 1938

In 1908 the whaling Ordinance, together with the Regulations made under it, established a detailed and comprehensive code of whaling law for the Dependencies. These laws made it unlawful to take seals in the Dependencies without a licence, gave authority for the issue of licences, and provided for the creation of a closed season and of seal reserves. British sovereignty had also been displayed and exercised in the Dependencies through magistrates commissioned by the Falkland Islands Government. In addition, a large-scale expedition, the British Graham Land Expedition, visited the southern parts of the Falkland Islands Dependencies in 1934–37 to make land investigations. Reaching Deception Island late in 1934, the expedition

72 Cases Antarctica (note 71).
73 Antarctic case, ICJ Application, at 15.

proceeded to Port Lockroy in the Palmer Archipelago in January 1935. Shortly afterwards, a base was established and occupied further south, and from there various sledge and plane journeys were made, revealing new information on the geography of the area. In February, 1936, the expedition moved still further south to Marguerite Bay and established a base on the Debenham Islands. From there, important sledge journeys were made across Graham Land to the east coast and far southwards into King George VI Sound. Numerous reconnaissance flights were carried out, and the expedition ascertained conclusively for the first time that Graham Land is attached to the Antarctic mainland.

c) *Recognition of British Claims by Norway, Argentina and Chile After the Issue of the Letters Patent of July 21, 1908*

As mentioned, the Norwegian Government had addressed enquiries to Great Britain in 1905 and in 1907 concerning the sovereignty of the territories situated between 35° and 80° west. The United Kingdom reasserted her claim. Norway did not make any protest or reservation against the assertion and exercise of British sovereignty over the Dependencies, neither then nor after the issue of the Letters Patent of 1908. Moreover, numerous Norwegian whaling companies applied for British licences and complied with the laws of the Falkland Islands Government. From Britain's point of view, these facts proved implicitly Norway's recognition of British sovereignty over the Dependencies in or about 1908.[74] Furthermore, the western boundary of Norway's own Antarctic claim on January 14, 1939 is a line coinciding with the eastern boundary of the Falkland Islands Dependencies. The British treat Norway's territorial claim in the Antarctic as a complete confirmation of Britain's sovereignty.[75]

Moreover, an Argentine company took out a British lease of land in South Georgia for 21 years. Regarding to British Letters Patent of 1908, the British Minister commented that he concluded from its terms that the "Argentine Government do not dispute the rights of Great Britain over the South Orkneys".[76] Again, Argentina made no protests or reservations against the issue of the British Letters Patent of 1917, establishing beyond question, at least from a British perspective, that Argentina recognised British sovereignty over the Dependencies.[77]

In 1905, a Chilean company applied for a British lease of land in South Georgia from the Falkland Islands Government. In 1907, another Chilean company, the Sociedad Ballenera de Magallanes of Punta Arenas, took out a British whaling licence of the South Shetlands and Graham Land. Neither then nor after the issue of the Letters Patent of 1908 or of 1917 did Chile make any protest or reservation against the assertion and exercise of British sovereignty over the Dependencies. These facts,

---

74 ICJ, Antarctic UK v Argentina, Application, p22.
75 Ibidem.
76 Ibidem, at 23.
77 Ibidem, at 23.

from British point of view, provide evidence that at this period, Chile also recognised British sovereignty over the Dependencies.[78]

II. New Zealand

In 1923 the United Kingdom claimed the Ross Dependency by an Order in Council, lying between 160°E and 150°W, and formally transferred this area to New Zealand in 1933.[79] In the beginning, New Zealand's interest in the Antarctic was following Britain's lead.[80] In 1923 New Zealand staked a claim to the territory known as the Ross Dependency under the "direct suggestion of London"[81]: It was Ross who discovered that area covering the Ross Sea, the Ross Ice Shelf, and McMurdo Sound. New Zealand's official response was mainly based on her historical dependence on the United Kingdom.

Prior to the International Geophysical Year, New Zealand government made a reluctant commitment of the resources required to assist in the trans-Antarctic expedition, which was also being pushed by the British.[82] As part of this expedition, a base for scientific researches was established at McMurdo Sound and named Scott Base.

During the negotiation of the Antarctic Treaty, New Zealand attempted to straddle the claimant and non-claimant division, even volunteering to relinquish its claim if this would lead other claimants to give up their positions and thus help in the establishment of a workable international regime. A motivating factor for Prime Minister Walter Nash in making this offer was the desire to strengthen the peaceful, non-military component of the treaty that finally emerged. As he told parliament in 1960: "What we as a Government desired on behalf of the people of New Zealand was the abandonment in its entirety of claims by nations having or

---

78 Ibidem, at 24.

79 "Order in Council under the British Settlements Act, 1887 (50 &51 Vict. C. 54) Providing for the Government of the Ross Dependency, (30 July 1923)", which is renamed as "Ross Dependency Boundaries and Government Order in council 1923 (SR 974)", (Wellington, 14 August 1923), at: http://legislation.govt.nz/regulation/imperial/1923/0974/latest/DLM1195.html?search=ta_regulation_O_ac%40bc%40rc%40ainf%40anif%40bcur%40rinf%40rnifan%40bn%40rn_resel_50_y&p=1&bt=bt; Joyner (note 68), at 47.

80 Alley, Roderic, "New Zealand and Antarctica", 39 International Journal 911 (1984), at 914.

81 Ibidem, at 913.

82 Hatherton, Trevor, "Preferred futures: a scientific view", in New Zealand Institute of International Affairs, *The Antarctic: Preferred Futures, Constraints and Choices: Proceedings of a Seminar Held by the New Zealand Institute of International Affairs at Wellington on 17 and 18 June 1983,* (Wellington: The Institute, 1983), 63 at 64.

asserting sovereignty over any area, thus making it a completely international area without any possibility whatever of war."[83]

In 1977, New Zealand proclaimed a 200 nautical miles Exclusive Economic Zone (EEZ) intended to apply to the Ross Dependency at any time when the Governor-General, by Order in Council, so declared.[84] Moreover, in April 2006, New Zealand also reserved the right to claim the continental shelf to Antarctica, and submitted the CLCS a partial submission, "not including areas of continental shelf appurtenant to Antarctica", which has been formally rejected by Japan and Netherlands.[85]

III. France

France had claimed 136° E to 142°E, an area named as Terre Adélie. The smallest claim in Antarctica is based on discoveries by Dumont d'Urville in 1840.[86] On November 21, 1924, Adélie Land (Terre Adélie) in Antarctica was officially claimed by France. In January 1950, the first French Antarctic station was established. France has not proclaimed an EEZ off Antarctica; however, in February 2009 it also made a "partial submission" to the UN CLCS which noted that the outer limit of its Antarctic continental shelf claim may be made later.[87] This intended submission has been formally rejected by Netherlands and Japan.[88]

IV. Australia

Australia claims two segments of Antarctica: 45°E to 136°E and 142°E to 160°E. The Australian Antarctic Territory covers nearly 5.9 million square kilometers, about 42% of Antarctica and nearly 80% of the total area of Australia itself. This is the largest claim on the continent. Australia issued the Australian Antarctic Territory Acceptance Act, in 1933, to claim her sovereignty over the Australian Antarctic Territories. Her claim is mainly based on discovery and expedition.

The Australian territorial claim in the Antarctic is partly supported by the British discovery of the Antarctic from 18th to early 20th centuries and is based on a long

---

83 Second Reading, Antarctic Bill, 12 October 1960, *Parliamentary Debates (Hansard), Fourth Session, Thirty-second Parliament, House of Representatives,* Vol. 325, (1960), pp. 2979–2995, at 2979

84 Article 9(4), Territorial Sea, Contiguous Zone, and Exclusive Economic Zone Act 1977 (1977 No 28), New Zealand, whose name is amended on 1 August 1996 by section 5 (1) of the Territorial Sea and Exclusive Economic Zone Amendment Act 1996 (1996 No 74), at: http://www.legislation.govt.nz/act/public/1977/0028/latest/DLM442579.html

85 http://www.un.org/Depts/los/clcs_new/submissions_files/submission_nzl.htm

86 Balch (note 64), at 127–135.

87 http://www.un.org/depts/los/clcs_new/submissions_files/submission_fra1.htm; http://www.un.org/depts/los/clcs_new/submissions_files/fra09/SUMREC_FRA1_19_04_2012.pdf, at 1.

88 http://www.un.org/Depts/los/clcs_new/submissions_files/submission_fra1.htm

historical association with this part of Antarctica. Australia's Douglas Mawson led a group of Australians and New Zealanders in the 1911 to 1914 Australasian Antarctic Expedition, which had bases at Commonwealth Bay, south of Tasmania, and the Shackleton Ice Shelf south of Perth. This expedition explored extensively along the coast near the bases. Mawson also led the British, Australian and New Zealand Antarctic Research Expedition (BANZARE) of 1929 to 1931. During this expedition Mawson claimed what is now Australian Antarctic Territory as British sovereign territory. In 1933, Britain asserted sovereign rights over the claimed territory and placed the territory under the authority of the Commonwealth of Australia.

In 1933, British Government issued an Order in Council asserting British sovereignty rights over the Australian Antarctic Territory and placed the territory under the authority of the Commonwealth of Australia. The transfer of sovereignty was affected by the Australian *Antarctic Territory Acceptance Act 1933* which came into effect in 1936. It stated: "That part of the Territory in the Antarctic seas which comprises all the islands and territories, other than Adelie Land, situated south of the 60th degree south latitude and lying between the 160th degree east longitude and the 45th degree east longitude, is hereby declared to be accepted by the Commonwealth as a Territory under the authority of the Commonwealth, by the name of the Australian Antarctic Territory."[89]

In 1954, *Australian Antarctic Territory Act 1954* provided for the application of Commonwealth legislation to the Territory. Australia also claimed maritime zones in Antarctica. According to the Sea and Submerged Lands Act 1973, Australian "sovereignty" extended the "belt of sea adjacent to the coast of Antarctica, known as the territorial sea, ... and the bed and subsoil of the territorial sea",[90] which did not separate Antarctica, thereby in 1970s a three nautical miles territorial area[91] appurtenant to all its "territory". In 1990, Australia extended territorial sea from three to twelve nautical miles;[92] and in 1991, it sought to establish a contiguous zone and an EEZ and redefine the continental shelf.[93] Moreover, Australia was the first to clearly state their claims to Antarctic continental shelf in its submission to the

---

89 http://www.aad.gov.au/default.asp?casid=78

90 Act No. 16 of 1973 as amended, at: http://www.comlaw.gov.au/Details/C2004C00306

91 In 1970s, the territorial sea of Australia was 3 nautical miles. See Harry, Ralph L., "Antarctic Regime and the Law of the Sea Convention: An Australian View", 21 Va. J. Int'l L. 727 (1980–1981), at 730.

92 Evans, G. / Duffy, M., "Australia Extends Territorial Sea", 61 Australian Foreign Affairs and Trade 816 (November 1990), at 816.

93 Blewett, N. / Duffy, M. / Crean S., "Australia Enhances its Power Over Marine Conservation and Resources", (News Release by Minister for Foreign Affairs and Trade, Canberra, September 24, 1991), citing Opeskin, Brian R. / Rothwell, Donald R., "Australia's Territorial Sea: International and Federal Implications of Its Extension to 12 Miles", 22Ocean Dev. & Int'l L. 395 (1991), at 395–396.

CLCS in November 2004.[94] However, its claims to the Antarctic continental shelf have been formally rejected by the United States, Russian, Japan, the Netherlands, Germany and India.[95]

V. Norway

Norway laid claim to Queen Maud Land, the largest remaining slice of the pie, known in Norwegian as Dronning Maud Land, its total area extends from 20°W to 45°E and exceeds two and half million km². As the British extended their claims in Antarctica,[96] Norway was concerned about the exclusion of Norwegian whalers in the Southern Ocean.[97] Norway claimed Bouvet Island (54°26'S, 3°24'E) in 1928,[98] and Peter I Island (69°00'S, 91°00'W) in 1931.[99] In 1939 a German expedition was being sent to explore areas near 35°W.[100] In order to prevent a claim being made by Nazi Germany over an area explored and used by Norwegians, Norway claimed the territory between 20°W and 45°E in the Antarctic in January 1939.[101]

Germany concentrated their activities on aerial reconnaissance and mapping between 20°W and 45°E in the season of 1938–39. A more substantial expedition

---

94 Commission on the Limits of the Continental Shelf, Outer Limits of the Continental Shelf beyond 200 Nautical Miles from the Baselines: Submission to the Commission: Submission by Australia, at: http://www.un.org/Depts/los/clcs_new/submissions_files/aus04/clcs_03_2004_los.pdf.

95 http://www.un.org/Depts/los/clcs_new/submissions_files/submission_aus.htm.

96 In 1920, the British government expressed that "it is desirable that the whole of the Antarctic should ultimately be included within the British Empire." See "Letter from the Colonial secretary to the Governor-General of Australia", 6 February 1920, text in Bush, W.M., *Antarctica and International Law: a Collection of the Interstate and National Documents II, (New York: Oceana Publications, 1982)*, at 104 (hereafter as Bush, *Antarctic and International Law*). Then, in 1923, Britain claimed 160°E and 150°W between 60°S and the South Pole, which was transferred to New Zealand later. In 1926, it claimed 20°W to 80°W between the 60°S and the South Pole, see Costa, Jean Fitch Da, *Souverainete sur l'Antarctique*, (Paris: Librairie Generale de Droit et de Jurisprudence, 1958), at 14,

97 Memorandum of the Norwegian Minister to the British Foreign Office, 26 January 1934, text in Bush, *Antarctic and International Law* (note 96), at 150; and Isachsen, Fridtjov, "The New Norwegian Dependency in the Antarctic", 1 Le Nord 67 (1939), at 73.

98 Order-in-Council of 23 January 1928, text in Norway, *Collection of Laws*, 1926–1930, at 33

99 Order-in Council of 4 May 1931, text in *British and Foreign State Papers*, Vol. 134, at 1010.

100 Peterson, M.J., *Managing the Frozen South-The Creation and Evolution of the Antarctic Treaty System: Studies in International Political Economy Vol. 20*, (Berkeley, Los Angeles, London: University of California Press, April 1, 1988), at 34.

101 Jessup, Philip C. / Taubenfeld, Howard J., *Controls for Outer Space and the Antarctic Analogy*, (New York: Columbia University Press, 1959), at 152.

was carried out the next year. This operation extended up to 300 miles inland, and resulted in the discovery of mountains over 13,000 feet high. The Germans named this area Neu Schwabenland.[102] On January 14, 1939, Norway announced claim to the area between 20°W and 45°E. However, after the outbreak of the World War II German vessels in Antarctic waters seized a number of Norwegian whalers and factory ships and operated them for Germany.

From the Norwegian point of view, the claim is based on discovery and exploration in Antarctic. Roald Engelbregt Gravning Amundsen, a Norwegian explorer of Polar Regions, led the first Antarctic expedition to reach the South Pole between 1910 and 1912.[103] Amundsen with Douglas Mawson, Robert Falcon Scott, and Ernest Shackleton, was a key expedition leader during the Heroic Age of Antarctic Exploration. In 1928, Bouvet Island was formally claimed by Norway, Peter I Island was claimed formally in 1930, and 14 Dronning Maud Land claimed in 1939. In 1948, Norwegian Polar Institute, which is a part of the Ministry of the Environment, was assigned to administer Queen Maude Land. In 1957, Dronning Maud Land was declared subject to Norwegian sovereignty as a dependency. Norway has not proclaimed any marine zones off Antarctica, despite having claimed continental shelf of Dornning Maud Land, in May 2009, it requested the CLCS not to take any action on that portion of its submission relating to areas of the seabed and subsoil adjacent to Antarctica,[104] which obtained support from the United States, Russia, India, Netherland and Japan.[105]

VI. Argentina and Chile

Argentina has asserted that the area from 74° W to 25°W has been part of its territory since Argentina became a republic. Chile has made a formal claim to the area between 90°W and 53°W in 1940. Their Antarctic territorial claims are based on similar theories, inheritance, succession to Spanish rights, discovery and early exploration, geographical proximity and geological affinity, the sector principle, and effective occupation.

Under the principle of *uti possidetis, ita possideatis*, Argentina and Chile trace their historical rights to Antarctica to the Papal Bull *Inter Caetera* of 1493, and the

---

102 Wilson (note 59), at 21.

103 See more details from: http://www.historynet.com/roald-amundsen-and-the-1925-north-pole-expedition.htm and http://www.pbs.org/wgbh/amex/ice/peopleevents/pandeAMEX87.html.

104 Continental Shelf Submission of Norway regarding to Bouvetoya and Dronning Maud land, Executive Summary, http://www.un.org/Depts/los/clcs_new/submissions_files/nor30_09/nor2009_executivesummary.pdf.

105 At: http://www.un.org/Depts/los/clcs_new/submissions_files/submission_nor_30_2009.htm.

Treaty of Tordesillas 1494, which gave lands west of the 46th meridian to Spain.[106] In 1539 Charles V of Spain is said to have granted Pedro Sánchez de Hoz all the territory from the Strait of Magellan to the South Pole west of 40° W.[107] Argentina asserts it is the legal heir to the possessions of the King of Spain in the Viceroyalty of the River Plate.[108] Chile can also point to a long series of Spanish decrees, notes, and references that reinforce that Spain envisioned the Captaincy General of Chile as being the administrative center of its domains to the south.[109] In 1831, after independence, Chile's first president wrote a letter stating that the South Shetland Islands, which had recently been visited by British, American, and Russian navigators, were Chilean territory.

According to discovery theory, Argentina argues that sealers based in Buenos Aires were operating in Antarctic waters at least two years before the discoveries made by the British, Americans and Russians in and around 1820. Chile contends that sealers and sailors departing from ports in what is today Chile were among the earliest men in Antarctica, e.g. Admiral Gabriel de Castilla in 1603, who reached the Antarctic Peninsula.

Geographical factors are another argument which supports Argentina and Chile's territorial claim in the Antarctic. Argentina and Chile are the southernmost nations in the world, i.e. the closest nations to the Antarctic. Moreover, from geological perspective, the mainland Andes Mountains is related to the Antarctic Peninsula, which is a continuation of the Andean Cordillera as indicated by a submarine ridge that connects in a wide curve to the east the rip of Tierra del Fuego and the Antarctic Peninsula.[110]

Some scholars argue that the claims of permanent Argentine occupation of Antarctic territory date to 1904, when Argentine whalers first undertook the management of the weather station in the South Orkneys.[111] In 1925, Argentina erected a wireless station at the Laurie Island weather observatory without asking the British for a license. When Britain called this to Argentina's attention, the answer was that the station was on Argentina's territory.[112] In 1940 Argentina and Chile issued a proposal that their governments would cooperatively proclaim certain hemispheric

106 Shusterich, Kurt M, "The Antarctic Treaty System: History, Substance, and Speculation", 39 International Journal 800 (1984), at 801; Joyner (note 68), at 44.

107 Wilson (note 59), at 17.

108 Child, Jack, *Antarctica and South American Geopolitics: Frozen Lebensraum*, (New York: Praeger Publishers, 1988), at 68.

109 Ibidem, at 108.

110 Wilson (note 59), at 23; and Child (note 108), at 69.

111 Joyner, Christopher C. / Ewing, Blair G. Jr, "Antarctica and the Latin American States: the interplay of law, geopolitics and environmental priorities", 4 Geo. Int'l Envtl. L. Rev. 1 (1991–1992), at 4.

112 *Application of the Antarctica Cases (UK v Argentina)* ICJ Press release at: http://www.icj-cij.org/docket/files/26/9065.pdf at 25.

rights extending southward to Antarctica.[113] In 1940, Argentina established the National Commission on the Antarctic and offered to host an international Antarctic conference. In 1942 and 1943 Argentina officially stated her territorial claim to the sector lying between 25° to 68°34′W (with 68°34'W being the boundary of Tierra del Fuego) south of 60°S. In 1946 an Argentina postage stamp was issued showing the western limit of Argentina's Antarctic Territory as the 74th meridian, which is approximately the westernmost point of Argentina's mainland. That has been the boundary shown on maps and official publications ever since. Furthermore, in addition to the Argentine meteorological station on Laurie Island and the next oldest permanent settlement at British Base B, Argentina established a second station on Melchior Island in 1947.

In 1906, Chile granted a fishing concession to two Chilean citizens, and established a base of operations on Deception Island in the South Shetland Islands. The based was used until 1911 and then abandoned. In 1907, Chile proposed to Argentina that the two countries should reach an agreement as to the division of the land within the spirit of the Treaty of 1881, but Argentina refused, saying "England claimed all these islands"[114]. On November 6, 1940 Chile's president, Pedro Aguirre Cerda issued a decree defining Antarctic territory as "all the lands, islands, islets, reefs, glaciers (ice pack) already known or to be discovered and their respective territorial waters in the sector constituted by the meridians 53° and 90° west of Greenwich".[115] In 1947, Chile established her first permanent settlement, Prat Base, at Isla González Videla, and from that time on the Maritime Governor of Chilean Antarctica and an officer in charge of the postal agency have resided there. One year later, the second Chilean base was inaugurated in 1948 by President Gavriel Gonzalez Videla, the only chief of state of any country over to set foot on Antarctic soil. This base was maintained by the army at the extreme tip of the Palmer Peninsula, 63°19′S, 57°54′W. By 1961, Chile had three other permanent Antarctic stations.

These claims are based on several legal principles: discovery, occupation, sector principle, inherited rights, and the contiguity. Only five claimants, Australia, France, New Zealand, Norway, and the United Kingdom, mutually recognize each other's

---

113 Joyner /Ewing (note 111), at 5.

114 *Application of the Antarctica Cases* (note 112), at 22.

115 World War II greatly intensified States' interests and activities in the Antarctic. The U.S.A. feared that German activities based in the Antarctic might endanger American and allied shipping, moreover, that Germany might seize the British-claimed Antarctic territory and thus gain a foothold close to the South American continent. On May 24, 1940 Secretary of State Cordell Hull said "Considerations of continental defence make it vitally important to keep for the 21 American republics a clearer title to the Antarctic continent south of America than is claimed by any non-American country". This statement seems encourage both Chile and Argentina to press their territorial claims. More information see: Wilson (note 59); Joyner /Ewing (note 111).

claims. Most non-claim States do not recognize these seven territorial claims in the Antarctic. The following analysis from an international law perspective will examine whether the theories supporting these claims are tenable or not.

In 1966 Argentina promulgated Law No. 17094, which claimed that: "[t]he sovereignty of the Argentine Nation extends to the sea adjacent to its territory up to a distance of two hundred nautical miles..." and "[t]he sovereignty of the Argentine Nation likewise extends to the seabed and the subsoil of the submarine areas adjacent to its territory to a depth of two hundred meters or ...to where the depth of the waters allow the exploitation of the national resources of such zones."[116] In 1991 Argentina redefined its territorial sea to a distance of 12 nautical miles from baselines, and a 24 nautical miles contiguous zone and 200 nautical miles EEZ,[117] which could be applied to Antarctica. Moreover, according to Law No. 23.968 Argentina proclaimed Antarctic continental shelf up to the outer edge of the continental margin or up to 200 nautical miles when the out edge was below these limits.[118] Argentina's claim to the Antarctic continental shelf has been formally objected by the United Kingdom, the United States, Russia, India, Netherlands and Japan.[119]

Chile claimed continental shelf before the entry of force of the Antarctic Treaty in 1947, which confirmed sovereignty "over all the continental shelf adjacent to the continental and island coasts of its national territory, whatever maybe their depth below the sea, and claims by consequence all the natural riches which exist on the said shelf, both in the under it, known or be to discovered."[120] Later, Chile declared that "national sovereignty and jurisdiction can be extended to the submerged continental... shelf... whatever the depth and extension of this shelf may be."[121] So far, Chile has only submitted a "Preliminary Information" of the outer limits of the

116 Article 1 and Article 2 of Law No. 17904, "The President of the Argentine Nation Sanctions and Promulgates with the Force of the Law, Argentina, on December 29, 1966, cited from The Geographer Office of the Geographer Bureau of Intelligence and Research, "International Boundary Study Series A: Limits in the Sea, No. 44 Straight Baselines Argentina", August 10, 1972, at http://www.state.gov/documents/organization/61535.pdf.

117 Article 3, 4 and 5, Argentina Act No. 23.968 of 14 August 1991, at: faolex.fao.org/docs/texts/arg1223.doc.

118 Continental Shelf Submission, Argentine Submission, Executive Summary, http://www.un.org/Depts/los/clcs_new/submissions_files/arg25-09/arg2009e_summary-eng.pdf. Argentina's submission includes a map and geographical coordinating for outer continental shelf limits overlapping the Antarctic Peninsula.

119 http://www.un.org/Depts/los/clcs_new/submissions_files/submission_arg_25_2009.htm.

120 Presidential Declaration Concerning Continental Shelf of June 23, 1947, *El Mercurio* (Santiago de Chile), June 29, 1947, p. 27, translation from (1948) *Int'l & Comp. L. Q.*, Vol. 2, at 135; and Presidential Decree No. 781 of August 1, 1947, *El Peruano: Diario Oficial*, Vol. 107, No. 1983, 11 August 1947.

121 Presidential Decree No. 781 (note 120).

continental shelf to the CLCS, it also maintains that a submission about Antarctic continental shelf "may be made later".[122]

VII. Non-claimant States' activities and attitudes

In the early stages of exploration of Antarctica, two States outside of the UK, that is, the US and Russia, made remarkable contributions to Antarctic exploration, indeed with the continent and its offshore islands being seen by their citizens during the nineteenth century.[123] However, neither of these two States officially claimed any territory in Antarctica. It was not until the first territorial claim in Antarctica, which was followed by other claims that turned the American mind to political considerations of the faraway ice-covered continent. The Soviet Union, as a successor of Russia, showed its concern for the Antarctica much later only when signing the International Whaling Convention and began sending whaling flotillas into the South Seas in 1946.[124] Although there were several privately-financed Byrd expeditions of 1929–1931[125] and 1933–1935, as well as Lincoln Ellsworth's expedition in 1935–1936,[126] which might advance US' territorial claims in Antarctica,[127] the American policy on Antarctica had been the same as the opinion expressed in 1924 by the then Secretary of US, Charles Evans Hughes, which states that: "discovery of lands

---

122 "Preliminary Information Indicative of the outer limits of the Continental Shelf and description of the status of preparation and intended date of making a submission to the Commission on the Limits of the Continental Shelf", at: http://www.un.org/Depts/los/clcs_new/submissions_files/preliminary/chl2009premilinaryinformation.pdf ,p. 19.

123 Admiral Bellingshausen, a Russian, circumnavigated Antarctica in 1820–1821, see Gould, Laurence McKinley, *The Polar Regions in Their Relation to Human Affairs*, (New York: American Geographical Society, 1958), at 15. Nathaniel Palmer, an American sealing captain, claimed to see the mainland late in 1820, see Bertand, Kenneth J., *Americans in Antarctica, 1775–1948*, (New York: American Geographical Society, 1971), at 72.

124 Toma, Peter A., "Soviet Attitude towards the Acquisition of Territorial Sovereignty in the Antarctic", 50 A.J.I.L. 611 (July 1956), at 611 footnote 2.

125 During the first expedition, Admiral Byrd claimed a large part of Marie Byrd Land for the US, see Walter, Sullivan, "Antarctica in a Two-Power World", 36 Foreign Affairs 154 (October 1957), at 156.

126 Lincoln Ellsworth made the first trans-Antarctic flight and surveyed in the air the section known as the American Highland. Wilson (note 59), at 20.

127 Whiteman, Marjorie M., *Digest of International Law Vol. II*, (Washington, DC: Government Printing Office, 1963), p. 1245. These expeditions were to explore the area between 78W and 148 W, where US might claim a potential claim. Moreover, President Franklin D. Roosevelt required Amiral Byrd to "take appropriate steps such as the dropping of written claims from airplanes, depositing such writing in cairns, etc., which might assist in supporting a sovereignty claim by the United States Government." See Sullivan, Walter, *Quest for A Continent*, (McGraw-Hill, New York, 1957), p. 139; and Sullivan (note 125), at 156.

unknown to civilization, even when coupled with the formal taking of possession, does not support a valid claim to sovereignty unless the discovery is followed by actual settlement of the discovered territory"[128]. This official policy restrained the US' opportunity to declare a territory in Antarctica and also denied the legality of other states' sovereignty claims in Antarctica.

Although German expedition in Antarctica can trace its history to 1873–1874,[129] its third expedition in the season 1938–1939 to an area unclaimed at that time, between 20°W and 45°E, drew more international attention, and compelled Norway to claim the same area as its territory in 1939. German activities also threatened American interests in Antarctica, while the latter worried that if Germany conquered British claimed area, it would provide a foothold close to the South American continent. Therefore, on May 24, 1940, the US declared that "considerations of continental defence make it vitally important to keep for the 21 American republics a clearer title to the Antarctic continent south of America than is claimed by any non-American country."[130] This pronouncement encouraged Argentina and Chilean territorial claims in Antarctica, and they also cited the 1947 Inter-American Treaty of Reciprocal Assistance to support their claims.[131] This treaty states that an armed attack by any State against an American State in the sector between 24°W and 90 °W, which covered the whole claimed area, would be considered as an attack upon all American States. This led to an escalation of tension among the three rival governments for the same area, Argentina, Chile and UK, and almost caused a physical clash in 1948.[132] In order to avoid potential embarrassing damage to the relationship between Argentina, Chile and UK, and to safeguard its interests and security from potential increasing Russian activities in Antarctica, US made the first attempt to bring the Antarctic sovereign questions in an international forum including all seven claimed States in 1948.[133] The US intended to set up an eight-state condominium in Antarctica with a full legislative and executive Commission and with some kind of cooperation with the UN and other international organizations. However, this proposal did not receive much support, as no claimant would accept relinquishing its national "sovereignty" in Antarctica.[134] As the Soviet Union

128 "Letter to A.W. Prescott in G.H. Hackworth", in Hackworth, Green Haywood, *Digest of International Law Vol. I.* (Washington, DC: Government Printing Office, 1940), p. 399.

129 Stäblein, Gerhand, "Historische Aspekte der deutschen geowissenschaftlichen Polarforschung", 52 Polarforschung 219 (1981), at 221.

130 Sullivan (note 127), at 138.

131 Hayton, Robert D., "The 'American' Antarctic", 50 A.J.I.L. 583 (1956), at 593.

132 Hayton, ibidem, at 592; Peterson (note 100), at 37.

133 *Department of State Bulletin*, Vol 19, 1948, p. 301; *Documents on American Foreign Relations*, Jan. 1-Dece. 31, 1948, Vol. 10, p. 544; Hayton (note 131), at 583–610, at p. 584.

134 Myhre, Jeffrey D., *The Antarctic Treaty System: Politics, Law and Diplomacy*, (Boulder: Westview Press, 1986), at 26–30; Wilson (note 59), at 24–25; Sullivan (note 125), at

did not claim territory, it was not included in the US proposal. However, based on early Russian exploration and the increasing whaling interest in Antarctica, Soviet Union showed its desire to get involved in Antarctica and claimed that "[n]o solution of the problem of a regime for the Antarctic without the participation of the Soviet Union can have legal force, and the USSR has every reason not to recognize any such solution".[135] In 1950, Soviet Union officially declared its position in Antarctic problem: "that insofar as the fate of the Antarctic is of interest to many countries no, it would be expedient to discuss the question of the Antarctic regime internationally in order to reach an agreement in accordance with the lawful interests of all the states concerned."[136] Moreover, in October 1948 Belgian and the South African governments notified the US that they should be included in the discussion of the question of Antarctica, as Belgian had been exploring in this area for a long time since the Gerlache Expedition of 1899, and because of South Africa's geo-proximity.[137]

The substantial commencement of the international cooperation of Antarctica began with scientific cooperation in 1950s. In 1951 the International Council of Scientific Unions decided to sponsor the International Geophysical Year (IGY) in 1957–1958, which was planned as a world-wide research program by individual scientists and national scientific academies.[138] All claimants, the US, USSR, as well as Belgium, Japan and South Africa established IGY Stations in Antarctica or Sub-Antarctica. Antarctic IGY program was a success in that it showed cooperation, better understanding and harmony among the participants, and encouraged them to attempt a new management regime for Antarctica. In October 1958, the US invited all seven claimants (Argentina, Australia, Chile, France, New Zealand, Norway, and the UK) as well as other four states who built stations in the IGY program (Belgium, Japan, South Africa and USSR) to a conference in Washington to negotiate an Antarctic regime, including a Treaty in which the conflicting views on sovereignty questions could be accommodated. The US and the Soviet Union highly recommended an international administration in Antarctica, which got approval from Belgian and Japan.[139] Although there were some obstacles during the negotiating of

160; Peterson (note 100), at 36–37; Hayton, Robert D., "The Antarctic Settlement of 1959", 54 A.J.I.L. 348 (April 1960), at 352.

135 "Memorandum of the Soviet Government on the Antarctic Regime Question", in 10 USSR Information Bulletin (23 June 1950), at 380.

136 Memorandum of the Soviet Government on the Question of the Regime of the Antarctic, June 7, 1950, translated in Toma (note 124), at 612, Appendix I p. 624; the other translation edition in Whiteman, Marjorie M., *Digest of International Law Vol II* (note 127), at 1255; see also Boczek, Boleslaw A., "The Soviet Union and the Antarctic Regime", 78 A.J.I.L. 834 (1984), at 837.

137 Peterson (note 100), at 37–38.

138 Peterson (note 100), at 38–39; Myhre (note 134), at 31.

139 The US statements can be found in *Department of State Bulletin* Vol. 38, 1958, p. 910; Hayton (note 134), at 354. Soviet Union agreed with international administration

the treaty, with Argentina and Chile in particular trying to avoid any discussion of Sovereignty,[140] the final treaty was signed six week later on 1 December 1959, and entered into force on 23 June, 1961.

B. Territorial Acquisition in International Law

I. Categories of Territorial Acquisition

When Hugo Grotius[141] created the science of the Law of Nations, State territory used to be more or less identified with the private property of the monarch of the State. Grotius and his follows applied the rules of Roman law concerning the acquisition of private property to the acquisition of territory by States.[142] By 20th century, however, Roman law concerning the acquisition of private property could no longer be applied, as the acquisition of territory by a State could mean nothing other than the acquisition of sovereignty over that territory.[143] Nowadays, According to classical international law theories, an existing State could acquire territory through five modes namely: cession, accretion, conquest, prescription and occupation in International Law.[144]

---

but reiterated its reservations concerning the territorial claims which it had stated in its Memorandum of 1950. See Note of the Soviet Government to the United States, June 2, 1958, in Whiteman (note 127), at 1254–1255. Belgian's attitude can be found in Bush, *Antarctica and International Law* (note 96), at 268. Japanese opinion is in Conference on Antarctica (Washington), Doc. No. 9, p. 2. (Oct. 15, 1959, Mimeo), cited from Hayton (note 134), at 356–357.

140 Argentine opinion can be found in Conference on Antarctica (Washington), Doc. No. 13, pp. 1–2, cited from Hayton (note 134), at 355; Chilean opinion can be found in Bush, *Antarctica and International Law* (note 96), at 418. See also Hanessian, John, "The Antarctic Treaty 1959", 9 Int'l & Comp. L. Q. 436 (July 1960), at 462.

141 Hugo Grotius (1583 –1645) laid the foundations for international law. In one of his famous works *The Free Sea* (*Mare Liberum*, published 1609), he claimed that the sea was international territory and all nations were free to use it for seafaring trade. The free sea principle established the foundation for modern law of the sea.

142 Oppenheim, L., *International Law*, (7th ed., London, New York, Toronto: Longmans, 1948), at 497.

143 Oppenheim (note 142), at 497.

144 Oppenheim (note 142), at 498; Adjudication is sometimes listed as a mode of acquisition. See: Akehurst, Michael, *A Modern Introduction to International Law*, (3rd ed., London: George Allen and Unwin Ltd., 1977), at 140–146; Malanczuk, Peter, *Akehurst's modern introduction to International Law*, (7th ed., London and New York: Routledge, 1997), at 147–154. However, Brownlie observes that "the whole concept of modes of acquisition is unsound in principle and make the task of understanding the true position much more difficult", see: Brownlie, Ian, *Principles of Public International Law*, (4th ed., Oxford: Oxford University Press, 1990), at 131.

Cession is the transfer of sovereignty over State territory by one owner-State to another.[145] Accretion is the name for the increase of Land though new formation.[146] For instance, an island rises within a river, or the formation of dry land as a result of the change of flow of a river, which is totally within the territory of one and the same State. Accretion in Antarctica is referred to as calving, and only occurs in a related and reverse form.[147] Conquest[148] normally means that a State defeated in a war ceded territory to the victor by treaty. But conquest without a treaty could also confer title on the victor under traditional law.[149] Prescription may be defined as the acquisition of sovereignty over a territory though continuous and undisturbed exercise of sovereignty over it during such a period as is necessary to create, through historical development, the general conviction that the present condition of things is in conformity with international order.[150] Occupation is the act of appropriation by a State through which it intentionally acquires sovereignty over such territory as is at the time not under the sovereignty of another State.[151]

Not all these five theories can be applied to the issue of the territorial sovereignty of the Antarctic. Only two of them, cession and occupation, might be applicable with the Antarctic. However, cession has not occurred in Antarctica and need not be discussed.[152] Thus occupation, coupled with discovery is the primary mode of acquiring sovereignty in Antarctica.[153]

---

145 Oppenheim (note 142), at 499.

146 Oppenheim (note 142), at 515.

147 Bernhardt, J. Peter A, "Sovereignty in Antarctica", 5 Cal. W. Int'l L. J. 297 (1974–1975), at 317.

148 This appears as "Subjugation" in Oppenheim (note 142), at 518–525. Oppenheim emphasizes that conquest alone does not *ipso facto* grant the conquering State sovereignty over the conquered territory, although such territory comes through conquest for the time under the control of the conqueror. Conquest is the taking procession of enemy territory through military force in war time. Subjugation, from Oppenheim's view, is the situation in which the conqueror formally annexes the territory of the enemy State, and thereby brings the war to an end. See Oppenheim (note 142), at 518–525.

149 Akehurst (note 144), at 144.

150 Oppenheim (note 142), at 527.

151 Oppenheim (note 142), at 507.

152 Bernhardt (note 147), at 317; Parriott, Todd Jay, "Territorial Claims in Antarctica: Will the United States Be Left Out the Cold?", 22 Stan. J. Int'l L. 68 (1986), at 77.

153 Bernhardt (note 147), at 317; Smedal, Gustav, *Acquisition of sovereignty over polar areas*, (J. Dybwad Publisher, 1931), at 64; Simsarian, James, "The Acquisition of Legal Title to Terra Nullius", 53 Political Science Quarterly 111 (March 1938), at 111; Parriott (note 152), at 77. Other point also see: Honnold, Edward, "Thaw in International Law? Rights in Antarctica under the Law of Common Spaces", 87 Yale L. J., 804(March 1978) at 804–859 (hereafter cited as Thaw); Toma (note 124), at 612–622.

II. Occupation theory in the Antarctic Territorial Issue

1. *Discovery*

Although not considered sufficient to vest legal title in any of the claimants, discovery may be accompanied by symbolic acts, and was extensively employed and clearly sufficient to constitute title to *terra nullius* before 1700.[154] For instance Spain, England, France and the Netherlands, during 17th century, considered discovery with symbolic taking of possession an adequate basis for legal title to *terra nullius* in North America.[155] As has been mentioned previously, almost all of the seven States claiming territory, Australia, France, Norway and the United Kingdom, used discovery as one source of title or potential title. However, in the theory and practice of international law it is clear that discovery alone of a region in Antarctica does not adequately legitimize a claim.[156] Grotius wrote: "No one is a sovereign of a thing which he himself has never possessed and which no one else has ever held in his name.... To discover a thing is not only to capture it with the eyes but to take real possession thereof.... The act of discovery is sufficient to give a clear title of sovereignty only when it is accompanied by actual possession."[157] In the *Island of Palmas Case*, the United States, as successor to Spain, based its title on discovery in the first instance. Mr. Huber, as the arbitrator, believed that International Law since the middle of the 18th century had developed a process and laid down the principle that occupation, to constitute a claim to territorial sovereignty, must be effective and the existence of right shall follow the conditions required by the evolution of law.[158] He asserted that: "..., discovery alone, without any subsequent act, cannot at the present time suffice to prove sovereignty over the Island of Palmas. ...discovery does not create a definitive title of sovereignty, but only an 'inchoate' title, such a title exists, it is true, without external manifestation."[159]

The view has come to be widely accepted in modern law to provide that discovery only gives an inchoate title, an option, which acts as a temporary bar against

---

154 Simsarian (note 153), at 112.

155 Simsarian (note 153), at 111.

156 See: Bergin, Patrick T, "Antarctica, the Antarctic Treaty regime, and Legal and Geopolitical Implications of Natural Resource Exploration and Exploitation", 4 Flo. Int'l L. J. 1 (1988), at 1–40; Simsarian (note 153); Keller, Arthur Schopenhauer / Lissitzyn, Oliver James / Mann, Frederick Justin, *Creation of Rights of Sovereignty through symbolic Acts, 1400–1800*, (New York: Columbia University Press, 1938), at 148.

157 Grotius, Hugo, *The freedom of the Seas* or *The right which belongs to the Dutch to take part in the east Indian trade*, Magoffin, Ralph Van Deman, translated, (New York: Oxford University Press, 1916), at 11.

158 *Island of Palma case (United States of America v the Netherland)* PCA Awards of 4 April 1928, 2 Reports of International Arbitral Awards 829, at 845; Simsarian (note 153), at 128.

159 Ibidem, at 846.

another State.[160] If, during this period, the discovering State does not take any effective occupation to turn its inchoate title into a real title of occupation, the inchoate title would perish.[161] As international law developed, theory and practice came to agree that discovery only makes sense if it is placed firmly in the context of effective occupation. Mr. Huber further stated: "...according to the view that has prevailed at any rate since the 19th century, an inchoate title of discovery must be completed within a reasonable period by the effective occupation of the region claimed to be discovered."[162] However, international law provides no standard reasonable period during which the right of inchoate title must be exercised.[163] Therefore, discovery alone cannot provide legal grounds for a State to acquire sovereignty over the land, and more detailed requirements needed to be gradually elaborated and finally developed into the doctrine of "effective occupation" by end of the eighteenth century.[164]

Historical and physical conditions also deeply impair application of discovery theory in Antarctic. Firstly, there is still a lack of consensus about who actually discovered what region and when.[165] Furthermore, the harsh climate in Antarctica can be deceptive for a discoverer.[166] In summary, it is undoubted that claims in Antarctic based on discovery are invalid.

---

160 Oppenheim (note 142), at 510; Brownlie (note 144), at 139; O'Connell, Daniel Patrick, *International Law*, (2nd ed., London: Stevens, 1970), at 510; Brierly, James Leslie, *The Law of nations*, (Oxford: The Clarendon Press, 1928), at 93.

161 Von der Heydte, Friedrich August Freiherr, "Discovery, Symbolic Acquisition and Virtual Effectiveness in International Law", 29 A.J.I.L. 448 (1935), 448–471; Oppenheim (note 142), at 510–511; Hall, William Edward, *A Treatise on International Law*, Higgins, A. Pearce (ed.) (8th ed., Oxford: Clarendon Press, 1924), at 126; *Island of Palmas Case* (note 158), at 884; Auburn, F.M., *Antarctic law and politics*, (London: C. Hurst & Company, 1982), at 9; Brierly (note 160), at 93.

162 *Island of Palmas Case* (note 158), at 846; also see Von der Heydte (note 161), at 452–462; Smedal, (note 153), at 48.

163 Paul Fauchille suggested one year would be sufficient, see his *Traité de Droit International Public*, Vol. 1 (Paris: A. Rousseau and Co., 1925), pp. 720–721. Semdal suggested a maximum 25 years, in his *Acquisition of Sovereignty over Polar Areas* (note 153), at p. 49. Hayton argued for 20 years, in his "The 'American' Antarctic" (note 131), at 602.

164 Hill, Norman, *Claims to Territory in International Law and Relations*, (London, New York, Toronto: Oxford University Press, 1945), at 145–158; Oppenheim (note 142), at 558.

165 Bergin (note 156), at 16.

166 Auburn (note 161), at 17; Dufek, George J, *Operation Deepfreeze*, (New York: Harcourt, Brace and Company, 1957), at 137.

2. *Occupation*

Occupation is the acquisition of *terra nullius*, i.e. only such territory which, immediately before acquisition, belonged to no state.[167] The object of occupation may never have belonged to any State or it has been abandoned by a previous State. The concept of "occupation" drives from Roman private law right to acquire *res nullius* by *occupatio*,[168] the taking of possession of a *res nullius* with the intention of becoming its owner. According to occupation of territory, the Permanent Court has observed clearly in the *Eastern Greenland Case*: "...a claim to sovereignty based not upon some particular act or title such as a treaty of cession but merely upon continued display of authority, involves two elements each of which must be shown to exist: the intention and will to act as sovereign, and some actual exercise or display of such authority."[169] What is important is State activity[170], especially acts of administration; moreover the State acts of occupation must be of a public nature.[171] The underlying principle is that territory is occupied when it is placed under effective control.[172] Theory and practice conclude that occupation is effected through taking possession of, and establishing an administration over, territory in the name of the acquiring State.[173] In other words, the essential requirements of effective control are *possession* and *administration*.[174]

The territory must be actively taken into possession by the occupying State, which means that the first step for possession is to formally announce possession of the

---

167 Guggenheim, P., *Lehrbuch des Völkerrechts*, Bd. 1, (Basel, 1948), at 399 et seq.; Verdross, A., *Völkerricht*, (4 ed., Wien 1959), at 222; Boetzinger, J., *Völkerrecht der Antarktis*, (diss. Göttingen 1958) (Maschinenschrift), at 56; de Vattel, E., *Le droit des gens ou principes delaloi naturelle, Klassiker des Völkerrecht*, Bd.3, (Tübingen 1959), at 224; Simsarian (note 153); Akehurst (note 144), at 141–143; Moelwyn-Hughes, Ronw ed., *International Law*, (6th ed., London & Toronto: J.M. Dent & Sons Ltd., 1927), at 93; Oppenheim (note 142), at 507.

168 Moelwyn-Hughes (note 167), at 91; Brownlie (note 144), at 142.

169 *The Legal Status of Eastern Greenland Case (Norway v Denmark)* (Judgment) PCIJ Rep Series A/B No 53, at 45–46.

170 Brownlie (note 144), at 138.

171 Moelwyn-Hughes (note 167), at 91.

172 Different scholars named it differently. "Effective control" is used by Akehurst and Malanczuk, see: Akehurst (note 144), at 142; Malanczuk (note 144), at 149; Lowe, Vaughan, *International Law*, (Oxford: Oxford University Press, 2007), at 142. Other writers use "effective occupation" in contradistinction to *fictitious* occupation, such as: Oppenheim (note 142), at 509; Brownlie (note 144), at 138–145. In this paper, I prefer to use "effective control", in order to avoid confusion between the occupation as an acquisition method and effective occupation as a base condition of a real occupation. However, there is no essential distinction in meaning between them.

173 Oppenheim (note 142), at 509; Brownlie (note 144), at 138–145.

174 Effective control is a relative concept; it varies according to the nature of the territory concerned. See: Akehurst (note 144), at 142. It is undoubted that there are some requirements which are widely accepted by most International Law scholars, such as: Oppenheim (note 142), at 509; Brownlie (note 144), at 138.

territory and to fulfil responsibilities of sovereignty over the land.[175] The activity must be *à titre de souverain* in the sense that it must be actions of a State and not of unauthorized natural or legal persons.[176] Some kind of administration, or "effective occupation"[177] must be established by the possessor after possession of a territory has been taken, the main evidence of which includes administration, exercising or displaying sovereign authority though performance of government over people, settlements and activities in the territory.[178] This kind of activity shows that the territory is really governed by the possessor. If the possessor does not establish some responsible administration within reasonable time, there is no effective occupation as no sovereignty is exercised by any State over the territory.[179] Assuming that Antarctica is *terra nullius*, occupation is the only applicable method of acquiring sovereignty in Antarctica.[180] The claimant States attempt to satisfy the standard of effective occupation by the establishment of scientific bases and research stations, but these can hardly be recognized as sufficient evidence to support their claims, as some non-claimant States also established similar kind of bases and stations in Antarctica.[181]

It has been contended that inhospitable conditions provide an exception to the application of *occupatio* doctrine to Antarctica,[182] which might be referred to as "broad occupation theory"[183] which claims that effective control of Antarctic could only be continuously carried on in the area in question, but not the whole of Antarctica.[184] As one academic voice among many, Joyner, insists that international law should be applied all over the world no matter the latitude and longitude, and "effective possession must be properly required as a condition for occupation in polar region".[185] However, other international scholars cite analysis of three cases[186],

---

175 Oppenheim (note 142), at 509; Smedal, (note 153), at 39–42; Lindley, Mark Frank, *The Acquisition and Government of Backward Territory in International Law: Being a Treatise on the Law and Practice Relating to Colonial Expansion*, (Longmans, Green and Company, 1926), at 293–296; Joyner (note 68), at 52.

176 Oppenheim (note 142), at 507; Brownlie (note 144), at 140.

177 Oppenheim (note 142), at 510; Smedal (note 153), at 38.

178 Joyner (note 68), at 52.

179 Oppenheim (note 142), at 510.

180 Smedal, (note 153), at 64; Bernhardt (note 147), at 318; Simsarian (note 153), at 111.

181 Wolfrum, R., *Die Internationalisierung staatsfreier Räume*, (Springer-Verlag, Berlin Heidelberg New York Tokyo, 1984), p. 46; Conforti, Benedetto, "Territorial Claims in Antarctica: A Modern Way to Deal with an old Problem", 9 Cornell Int'l L. J. 249 (1986), at 255–256; Joyner (note 68), at 54.

182 Costa (note 96), at 31; Verdross (note 167), at 222; Bernhardt (note 147), at 319.

183 This term was originally expressed in German as "weite Okkupationstheorie", in Wolfrum (note 181), at 45.

184 Costa (note 96), at 31; Verdross (note 167), at 222; Von der Heydte (note 161), at 462 et seq.

185 Joyner (note 68), at 54.

186 The three cases are: Eastern Greenland Case, Island of Palmas Case (note 158) and Clipperton Island Case.

as grounds for a relaxation of the requirements for effective control in inhospitable land.[187] Waldock states: "The three cases ... mark a change in the concept of effective occupation.... The emphasis has shifted from the taking of physical possession of the land and the exclusion of others to the manifestation and exercise of the functions of government over the territory. This change is a natural consequence of the recognition that in modern international law occupation is the acquisition of sovereignty rather than of property."[188] The most difficult question concerning Antarctic sovereignty is whether or not exceptions should be made for such an inhabitable area and if so, the form of and degree to which they should be articulated.[189]

It is beyond doubt that when the issue of sovereign claims in Antarctica was first presented, the climatic conditions in the continent precluded the possibility of any permanent Antarctic settlement. In *Island of Palmas Case*, Mr. Huber states: "Manifestations of territorial sovereignty assume, it is true, different forms, according to conditions of time and place. Although continuous in principle, sovereignty cannot be exercised in fact at every moment on every point of a territory. The intermittence and discontinuity compatible with the maintenance of the right necessarily differ according as inhabited or uninhabited regions are involved, or regions enclosed within territories in which sovereignty is incontestably displayed or again regions accessible from, for instance, the high seas....the actual continuous and peaceful display of State functions is in case of dispute the sound and natural criterion of territorial sovereignty."[190] In Huber's opinion, the minimum degree of effective control would be the actual continuous and peaceful display of State functions. The requirements for effective occupation were relaxed again in *Clipperton Island* Arbitration award as: "... Thus, if a territory, by virtue of the fact that it was completely uninhabited, is, from the first moment when the occupying state makes its appearance there, at the absolute and undisputed disposition of that state, from that moment the taking of possession must be considered as accomplished, and the occupation is thereby completed."[191]

Both of these cases involved small islands with either no native population or a small native population, and with no resources worth exploiting.[192] However, neither of these two judicial criteria could support any sovereign claim in Antarctica, because no State was able to assert a continuous display of authority in Antarctica until the Antarctic Treaty entry into force, and by the time effective year-long occupation was achieved, its concomitant, peaceful and uncontested display, had been ruled out by competing claims of other states.[193] Moreover, the major factual difference between *Clipperton Island* and Antarctica is that the former is merely a

---

187 Parriott (note 152), at 81–86; Bernhardt (note 147), at 326–332.

188 Waldock, C.H.M., "Disputed sovereignty in the Falkland Islands Dependencies", 25 British Yearbook of International Law 310 (1948), at 317.

189 Bernhardt (note 147), at p. 322.

190 Island of Palmas Case (note 158), p. 840.

191 *Clipperton Island Case, (Mexico v France)*, 26 A.J.I.L. 390 (1932), at 394.

192 Parriott (note 152), at p. 84.

193 Bernhardt (note 147), at 330; Parriott (note 152), at p. 84.

small island, as compared with the continent that is Antarctica. Antarctica was never "at the absolute and undisputed disposition" of any single country. Therefore, the *Palmas* and *Clipperton Island* decisions fail to apply.[194]

The *Legal Status of Eastern Greenland* case, which dealt with sovereignty over a large uninhabited territory, enunciates that the degree of administration which could be found sufficient to perfect title is a varying standard. In largely uninhabited areas it is not reasonable to require possession of the territory; however, to establish effective control in an uninhabited land, all that is required is the effective administration of such territory.[195] If a territory does not have a human population, sovereignty can still arise through control over the land's resources.[196] Take the other point of view, Von Der Heydte insisted that "[e]ffectiveness then seems to be illustrated by actual display of sovereign rights, the maintenance of order, and protection"; that a "sovereign right can be exercised only over human beings, in inhabited land"; and "it would be a misconstruction of the doctrine of effectiveness to say that sovereignty over completely uninhabited land presupposes in every case actual occupation."[197] Although effective administration is not a satisfactory and easily applicable standard, it is hard to find that any State who claimed territorial sovereignty in Antarctica fulfilled even this relaxed requirement of effective control.[198] Moreover, the nature of Antarctica either as *terra nullius* or as comment-wealth plays a critical role when considering the application of occupation theory to Antarctica.

3. Nontraditional Theories of Territorial Acquisition

a) Geographical Doctrines

Geographical doctrines, i.e. principles of continuity and contiguity, are not the roots of title but act as evidence in borderline cases derived from some orthodox source of title such as effective occupation.[199] In other words, the principles are simply a part of judicial reasoning for effective occupation, having no intrinsic merit *per se*.[200]

---

194 Bernhardt (note 147), at 331; Parriott (note 152), at p. 84.

195 Parriott (note 152), at 82; O'Connell (note 160), at 38; Bernhardt (note 147), at 331–332.

196 Parriott (note 152), at 82; von der Heydte (note 161), at 463.

197 von der Heydte (note 161), at 463.

198 Boetzinger (note 167), at 74–75; von Münch, Ingo, "Völkerrechtsfragen der Antarktis", 7 AVR 225 (1958/59), at 248–249; Reeves, J.S., "Antarctic Sector", 33 A.J.I.L. 519 (1939), at 519; Dupuy, René-Jean, "Le statut de l'Antarctique", 4 Annuaire français de droit international 196 (1958), at 209–210; Balch, Thomas Willing, "Arctic and Antarctic Regions and the Law of Nations", 4 A.J.I.L. 265 (1910), at 267–268.

199 Wolfrum (note 181), at 46–47; Brownlie (note 144), at 149; Akehurst (note 144), at 150–151.

200 Brownlie (note 144), at 150; Bernhardt (note 147), at 341.

The contiguity principle has been invoked by coastal States in order to extend sovereignty over adjacent coastal islands[201], although this requires a natural connection between the State and territory under claim[202] – that islands are within a reasonable distance from the coast, and have a socio-economic relationship with the contiguous mainland.[203] However, even the closest States to the Antarctica, Chile and Argentina, are separated by great distances and there is a lack of economic links to the Antarctica.[204] Furthermore, it would be inapposite for a State to apply the contiguity principle to Antarctica, because the principle has embraced islands with continental coastlines, but the case of Antarctica deals with a projection of a continental coast to a continent or a projection from an island or islands to a continent.[205] Even were it not to founder for the above reasons, "the title of contiguity, understood as a basis of territorial sovereignty, has no foundation in international law",[206] and is little more than a technique in the application of the normal principles of effective occupation.[207]

Continuity maintains that when a State took possession of a territory, it obtained sovereignty over the adjacent area.[208] This principle has not been accepted by international law, and has been rejected in the case of Antarctica.[209]

b) *Sector Principle*

The sector principle was first introduced in the Arctic region by Canadian Senator Pascal Poirier.[210] This "principle" is a modification of contiguity principle.[211] Under this theory, Antarctic territorial boundaries are defined according to lines of longitude converging on the South Pole to coast to form triangular territories.[212] The baselines are composed of: mainland boundaries of the claimant state, or a length of the Antarctic coast "discovered" or "occupied" by the claimant state.[213]

---

201 Quigg, Philip W., *A pole Apart: The Emerging Issue of Antarctica: A Twenty Century Fund Report,* (New York: New Press McGraw-Hill Book Company, 1983), at 115; Thaw (note 153), at 815; Bernhardt (note 147), at 339–340.

202 Verdross (note 167), at 222; von Münch (note 198), at 236; Guggenheim (note 167), at 402; Schenk, D., *Kontiguität als Erwerbstitel im Völkerrecht,* (Ebelsbach 1978), at 34 et seq.

203 Bernhardt (note 147), at 340–341.

204 Parriott (note 152), at 86; Bernhardt (note 147), at 341.

205 Bernhardt (note 147), at 342.

206 Island of Palmas Case (note 158), at 867.

207 Brownlie (note 144), at p. 150.

208 Smedal, (note 153), at 43.

209 Bernhardt (note 147), at p. 343–345.

210 Lakhtine, W., "Rights over the Arctic", 24 A.J.I.L. 703 (1930), at 708.

211 von Münch (note 198), at 237; Verdross (note 167), at 223; Taubenfeld, H.J., "A Treaty for Antarctica", 531 International Conciliation 245 (1961), at 253.

212 Wolfrum (note 181), at p. 47.

213 Thaw (note 153), at 822–823.

The sector principle is supported by all claimant States except Norway;[214] however, it is misapplied for Antarctica.[215]

Firstly, the two approaches of baselines are theoretically incompatible, as the application of either one could slice the entire Antarctic pie. Both approaches have been applied in Antarctica, leading to overlapping and competing claims.[216] Secondly, its application should be based on a perfect reflection of the varying degrees of occupation or discovery mode. However, it has been found that neither of these two modes is satisfied for claims.[217] Thirdly, "'sector theory' is not a theory at all, but merely a convenient manner of delimiting claims founded on other theories."[218] It is based on contiguity; however the latter mode is rejected as valid method of territory acquisition.[219] Thus there are no grounds for Southern Hemisphere States to claim sector lines from their territories to Antarctica. Similar to Northern Hemisphere States, without support from discovery and occupation theories, they cannot establish the Antarctic coastal rights needed as a base from which to draw their sector lines.[220] Finally, the sector theory is controversial even in the north-polar area, in where it is originally applied,[221] and has never been recognized by international law in practice.[222] Therefore, it is clear that sector principle cannot be applied to Antarctica.[223]

c) *Uti Possidetis*

Uti possidetis has been defined as "a phrase used to signify that the parties to a treaty are to retain possession of what they have acquired by force during the war."[224] Argentina and Chile argue that their sovereignties in Antarctica are mainly

214 Bergin (note 156), at 12.
215 Bernhardt (note 147), at 332; Parriott (note 152), at 87–88.
216 Thaw (note 153) at 823; Parriott (note 152), at 87–88.
217 Bernhardt (note 147), at 332; Parriott (note 152), at 88.
218 Thaw (note 153), at 823.
219 Bernhardt (note 147), at 332.
220 Thaw (note 153), at 823.
221 Wolfrum (note 181), at 47. Detailed analyses see: Hayton (note 131), at 603–607; von Glahn, G., *Law among Nations: An Introduction to Public International Law*, (2 ed. London, 1970), at 283. Unfavorable papers are: Schmitz, E. / Friede, W., "Souveränitätsrechte in der Antarktis", 9 ZaöRv 219 (1939–1940), at 256 et seq.; Verdross (note 167), at 223; Thaw (note 153), at 823; Joyner (note 68), at 56–58.
222 Auburn (note 161), at 31; Bergin (note 156), at 13.
223 Thaw (note 153), at 823; Auburn (note 161), at 31; Bergin (note 156), at 13; Smedal, (note 153), at 64; Bernhardt (note 147), at 334.
224 Bernhardt (note 147), at 346; Parriott (note 152), footnote 139 at 87. More detailed analyses see, Fischer, F.C., "The Arbitration of the Guatemalan-Honduras Boundary Dispute", 27 A.J.I.L. 403 (1933), at 415–416.

based on historical right, or *uti possidetis*,[225] because in 1493 the Pope Alexander VI of Spain issued papal bulls which granted to Spain: "All island and mainland found and to be found, discovered and to be discovered towards the west and south, by drawing and establishing a line from the Arctic pole, namely the north, to the Antarctic pole, namely the south, no matter whether the said mainlands and islands are found and to be found in the direction of India or towards any other quarter, the said line to be distant one hundred leagues towards the west and south from any of the islands commonly known as the Azores and Cape Verde."[226] Thus Argentina and Chile maintain that they succeeded to the rights to Antarctica, which were vested in Spain, by right of "*uti possidetis* of 1810", at the time of the break with Spain.[227] However, the dominant view came to be that the papal grant was not truly a source of legal title,[228] and the view of *uti possidetis* has been considered highly problematic by international scholars.[229]

Firstly, it was uncertain, to Spain in 15th century, whether Antarctica existed or not, moreover, at the time of the break with Spain, Antarctica was still unknown as it was first sighted ten years later. Thus Argentina and Chile cannot lay claim to any part of Antarctica *vi et armis* as required.[230] Secondly, Spain never possessed Antarctica, not to mention establishing a "continuous and peaceful display" of territorial sovereignty in Antarctica, so there is no presumption in favor of Spanish sovereignty in Antarctica based in international law. Therefore it is impossible for Argentina and Chile to succeed to territories that did not once belong to Spain.[231] Third, assuming that Spain had inchoate title to Antarctica and passed it automatically, Antarctica still as *terra nullius*, Argentina and Chile had to use the prevailing method of discovery and occupation to acquire territory in Antarctica.[232] In brief, although both Argentina and Chile rely heavily on the *uti possidetis* principle, neither of their territorial claims can be established based on this theory.

---

225 Quigg (note 201), at 114; Bernhardt (note 147), at 345–347; Hayton (note 131) pp. 585–587; See also: Fisher (note 224).

226 Lowe (note 172), at 140.

227 Hayton (note 131), at 585; Quigg (note 201), at 114; Bernhardt (note 147), at 345–346; Parriott (note 152), at 87; Bergin (note 156), at 10; See also: O'Connell (note 160), at 426.

228 Lowe (note 172), at 141; Brownlie (note 144), at 139; Malanczuk (note 144), at 149.

229 Quigg (note 201), at 114.

230 Bernhardt (note 147), at 346; Parriott (note 152), footnote 143 at 87.

231 Hayton (note 131), at 601; Bernhardt (note 147), at 346; Parriott (note 152), footnote 143 at p. 87.

232 Hayton (note 131), at 601; Bernhardt (note 147), at 346; Parriott (note 152), footnote 143 at p. 87.

C. Maritime Zones in the Antarctic

After almost ten years of tough negotiations, the UNCLOS was signed on December 10, 1982 at the Third United Nations Conference on the Law of the Sea. The result of that vote, unprecedented in the history of treaty law (130 delegations voted in favour, four against with 17 abstentions) represented an enormous success of a new international order for the seas.[233] UNCLOS came into force on November 16, 1994, and to date, 158 countries and the European Community have joined in the Convention. Although UNCLOS, as the Charter of the Sea, covered almost all aspects of the issues of the sea, it failed to address the issue of sovereignty in the Antarctic,[234] and some scholars even argue that the issue of sovereignty has been further complicated by UNCLOS.[235]

Under UNCLOS, the sea is composed of various zones: internal waters, territorial seas, contiguous zones, EEZ, continental shelves, and high seas. The legal status of Antarctica affects whether States *ipso facto* have sovereign rights over any adjacent continental shelf and territorial sea up to 12 nautical miles.[236] Although the Antarctic Treaty freezes sovereignty claims, the issue is still unresolved and non-Treaty States could argue that they are not bound by the Treaty. It has been suggested that the 'common heritage of mankind' principle regulated in UNCLOS and accepted by the majority of the world, might provide the best way to deal with the critical issue and the theoretical basis for further analysis of mineral exploitation in the Antarctic.

The high seas were defined in 1958, as "all parts of the sea that are not included in the territorial sea or in the internal waters of a State".[237] Since the establishment and broad acceptance of the concept of the EEZ in UNCLOS, the geographical coverage of the high seas has been modified to "all parts of the sea that are not included in the EEZ, in the territorial sea or in the internal water of a State, or in the archipelagic waters of an archipelagic State."[238] The application of the UNCLOS definition of the high seas has been placed beyond dispute, as it has been ratified by many more parties than the High Seas Convention.[239] Thus, the question of

233 Oxman, Bernard H., et al., "Law of the Sea", **77** Proceedings of the Annual Meeting (American Society of International Law), 150 (April 14–16, 1983), at 151.

234 Auburn (note 161), at 126.

235 Joyner, "Antarctica and the Law of the Sea: Rethinking the Current Legal Dilemmas", 18 San Diego L. Rev. 415 (1981), at 425; Parriott (note 152), at 93; Joyner (note 68), at 75–76.

236 Triggs, Gillian Doreen, "The Antarctic Treaty System: Some Jurisdictional Problems", in Triggs, Gillian Doreen (ed.), *The Antarctic Treaty regime: Law, Environment and Resources*, (Cambridge: Cambridge University Press, 1987), 88–109 at 90; UNCLOS, Articles 3, 76 and 77.

237 Article 1 Convention on High Seas.

238 Article 86 UNCOLS.

239 According to the Vienna Convention of on the Law of Treaties, Article 59 (1) (a), which regulates: "A treaty shall be considered as terminated if all the parties to it conclude a later treaty relating to the same subject matter and: (a) it appears

whether waters surrounding the Antarctic[240] qualify as high seas can be determined by a process of elimination, i.e. if the waters are not territorial sea or EEZ, they can only be high seas.

I. Territorial Sea

The territory of a State with a sea coast includes the land within its boundaries and waters within or adjacent to its land boundaries, i.e. internal waters and territorial sea.[241] Territorial sovereignty extends over all land territory, the territorial sea, and the seabed and subsoil of the territorial sea.[242] In other words, territorial seas, have the same legal status as territorial land, and as such constitute part of State's territory. The coastal State has absolute sovereignty over these seas; though the coastal State's rights are restricted in some aspects, such as innocent passage for foreign ships. Thus, a territorial sea is described as those areas where "the sovereignty of a coastal State extends, beyond its land territory and internal waters ..., to an adjacent belt of sea".[243]

The legal status of the territorial sea is mirrored by the status of the air space above a State's territorial sea and the subsoil beneath it. Even though there were distinct schools of thought on the legal status of the superjacent air space after the Franco-Prussian war in 1870–1871, one of which demanded that the air be treated as high seas and the other asserting that the air is under the State's sovereignty, sovereignty over the superjacent air has been a firm principle of international law since 1919, when an agreement was reached at the Paris Conference on a Convention for the Regulation of Aerial Navigation.[244] The question of the status of the territorial sea bed was ignored until the last century, because of a lack of interest in its use and a lack of State practice. Until the 1930 Hague Conference, the territorial seabed and subsoil belonged to a coastal State's territory, an approach which was reiterated in the Territorial Sea Convention and UNCLOS.[245] The process of recognition for the

---

from the later treaty or is otherwise established that the parties intended that the matter should be governed by that treaty".

240 See Lagoni, Rainer, "Antarctica's Mineral Resources in International law", 39 ZaöRV 1(1979), footnote 60 at 23: surrounding waters do not include "subglacial basins", since they are under sea level parts of continent of the Antarctic pressed by the tremendous ice.

241 Oppenheim (note 142), at 415.

242 Brownlie (note 144), at 106.

243 Article 2 (1) UNCLOS regulates territorial sea as: "The sovereignty of a coastal State extends, beyond its land territory and internal waters and, in the case of an archipelagic State, its archipelagic waters, to an adjacent belt of sea, described as the territorial sea."

244 Churchill, Robin Rolf / Lowe, Alan Vaughan, *The law of the sea*, (2nd ed., Manchester University Press, 1988), at 63–64.

245 Article 2 (2) UNCOLS, "This sovereignty of a coastal State extends to the air space over the territorial sea as well as to its bed and subsoil."

legal status of air space and seabed of territorial sea is also a history of the general recognition of sovereignty over the territorial sea itself.[246]

Based on this unquestioned definition, it is clear that Antarctica does not have territorial sea. Article IV of the Antarctic Treaty prohibits any new territorial claims over Antarctica, and given that territorial seas are a form of territory, any argument suggesting that the Antarctic has territorial sea will violate the Antarctic Treaty. Some opposition to this argument is to be expected, presumably based on the notion that as territorial seas are automatically appurtenant to a coastal State, rather than requiring States to claim such sovereignty separately, any claim over territory in Antarctica automatically extends to territorial seas does not constitute a new territorial claim, and cannot therefore be a violation of the agreement.[247] Although this objection sounds reasonable, thorough analysis reveals that it is not tenable, or at least not suitable for Antarctica.

Sovereignty over coastal waters was not a natural right of a coastal State, as it was not recognized in the beginning phase of the law of the sea. Since the establishment of independent sovereign States in 16th century, coastal States have gradually gained the right to regulate activities in adjacent sea in their own interests. As rights over the sea could be divided into a right of property and rights of jurisdiction or control, Grotius only permitted coastal States jurisdiction in adjacent waters, which could be effectively controlled from the land, however, property rights over coastal waters was refused.[248] Although freedom of the high seas and sovereignty of the coastal States over coastal waters was established on the basis of Grotius' theory by the end of 17th century, the question of the juridical nature of coastal State's right over the territorial sea remained unresolved until 20th century. Some writers argued that coastal States had a proprietary right, or sovereignty, or plenary jurisdiction over their territorial sea, which was supported by many States' practices, i.e. British and American, however, others claimed that States owned only a "bundle of servitude" (*faisceau de servitudes*) over a specific area, an approach supported by France and Spain. With the passing of time, support for the principle of coastal State sovereignty over territorial seas grew, and it became finally established in UNCLOS in 1982.

Based on the process of development of sovereignty over territorial seas, it is clear that the main consideration of sovereignty theory was the interest relationship between home land and the territorial sea of coastal States. The coastal State had sovereignty over territorial seas in order to protect land interests, thus it could adopt laws for navigation, protection of cables and pipelines, fisheries, pollution, scientific research, customs, fiscal, immigration and sanitation.[249] The enumeration of the coastal State's legislative competence illustrates that a costal State could

246 Churchill and Lowe (note 244), at 64.
247 See Triggs (note 236), at 90.
248 Churchill and Lowe (note 244), at 59.
249 Article 21 UNCOLS.

maintain an advantage through the creation and enforcement of legislation over these interests. As there is no government or inhabitants in Antarctica, the water surrounding it cannot be construed according to the true interpretation of territorial sea, instead as an expression of a claimant States' objective to broaden territory their into Antarctica. Therefore, Article IV of the Antarctic Treaty does not recognize the existence of a territorial sea in Antarctica.

II. Exclusive Economic Zone (EEZ)

The EEZ is a zone extending up to 200 nautical miles from the outer limit of territorial sea, in which the coastal State has sovereign rights over natural resources and jurisdiction rights, and third States enjoy freedom of navigation and overflight as well as the right to lay submarine cables and pipelines. The EEZ was initially proposed by developing States, striving for economic development and control of the economic resources off their coasts, particularly fish stock, which were usually exploited by the distant-water fleets of developed States. The rights of coastal States are listed in Article 56 of UNCLOS, and can be categorized into six broad headings: non-living resources, living resources, other economics resources, construction of artificial islands and installations, marine scientific research and pollution control.

The EEZ raises a problem similar that of territorial seas: Whether rights in an EEZ are inherent aspects of sovereignty which do not need to be asserted. Firstly, the EEZ is a recent concept, proposed in the 20th century, which indicates that rights in this zone differ from the notion of "sovereignty". Secondly, developing States generally advocated the establishment of an EEZ to preserve natural resources and prevent exploitation by third States, especially developed States. Thus, the EEZ in this special *sui generis* character is something between a territorial sea and high seas.[250] Thirdly, the declaration of an EEZ has been traditionally a State practice.[251] Since about 1976, the idea of the EEZ had been accepted by most participants of the Third United Nations Conference on the Law of the Sea, and a large number of States – some 74 as of 1987– have unilaterally claimed an EEZ before UNCOLS entered into force.[252] Therefore, it is clear that the EEZ is not a natural right of coastal States, but a right which has to be claimed by coastal States.

As the EEZ only arose as a concept after the adoption of the Antarctic Treaty, and as an EEZ must be created by an express declaration by the coastal state, application for an EEZ in the Antarctic Treaty area could be treated as a new claim under Article IV of the Antarctic Treaty, although it is hard to exclude the EEZ concept, which has become a part of customary international law, from the express "law of the sea" implied by Article VI of the Antarctic Treaty.[253] However, some authors

---

250 Churchill and Lowe (note 244), at 137.
251 Triggs (note 236), at 90.
252 Churchill and Lowe (note 244), at 134.
253 Vigni, Patrizia, "The Interaction between the Antarctic Treaty System and the Other Relevant Conventions Applicable to the Antarctic Area: A Practical Approach versus

who attempt to support the legitimacy of the delimitation of an EEZ in Antarctica, have emphasized that coastal states do not have sovereign rights in EEZs but rather exclusive rights over such areas.[254] This "exclusive right" argument is far from reasonable, firstly because of the non-recognition of territorial claims in Antarctica and secondly because under the Antarctic Treaty, all parties, especially consultative parties, enjoy the same rights and are subject to the same duties.[255]

However, there are some islands and island groups in the Antarctic Treaty area[256] where sovereignty is not in question[257] including Heard and McDonald Islands of Australia, Kerguelen and Crozet Islands of France, Bouvetoya Island of Norway, and Prince Edward Island of South Africa. The governments of France, Australia and South Africa have declared and delimited EEZs around these islands.[258] However, it is arguable whether these uninhabited islands have the same status as the continent. Article 121 UNCLOS recognizes that naturally formed islands have same legal status as a continent, which does permit an EEZ.[259] However, the records for the debate surrounding Article 121 reveal a strong requirement that delimitation of the EEZ should consider the size and population of islands and their contiguity to the metropolitan state.[260] Finally, paragraph 3 of this Article excludes rocks which cannot sustain human habitation or economic life from being the basis of an EEZ or continental shelf. Therefore, it is hard to grant maritime zones, especially 200 mile EEZs to these islands within the Antarctic Treaty area, which are not only far away from territorial possessions but also lack an indigenous population.[261]

Theoretical Doctrines", in Frowein, Jochen A./ Wolfrum, Rüdiger (eds.), *Max Planck Yearbook of United Nations Law*, Vol. 4, 481, (The Hague, London, Boston, Kluwer Law International, 2000), at 495.

254 Triggs (note 236), at 88 et seq.

255 Vigni (note 253), footnote 38 at 495.

256 Article VI of the Antarctic Treaty regulates that the Treaty applies to the area south of the 60° south latitude.

257 Watts, Arthur D., *International Law and the Antarctic Treaty System* – (Hersch Lauterpacht Memorial Lecture Series; Vol. 11), (Cambridge: Grotius Publications Limited, 1992), at 152–153.

258 See Joyner, Christopher C., / Lipperman, Peter J., "Conflicting Jurisdictions in the Southern Ocean: the Case of an Antarctic Minerals Regime", 27 Va. J. Int'l L. 1 (Fall 1986), footnote 32 at 10,; Auburn (note 161), at 219; and Bush, *Antarctica and international law* (note 96), at 63, Chile, at 447, France at 584.

259 Article 121 UNCLOS: "1) An island is a naturally formed area of land, surrounded by water, which is above water at high tide; 2) Except as provided for in paragraph 3, the territorial sea, the contiguous zone, the exclusive economic zone and the continental shelf of an island are determined in accordance with the provisions of this Convention applicable to other land territory."

260 Dyke, Jon M. Van / Brooks, Robert A., "Uninhabited Islands: their impact on the Ownership of the Oceans' Resources", 12 Ocean Dev.and Int'l L. 265 (1983), p. 286.

261 Joyner (note 68), at 88.

Moreover, a declaration of an EEZ for these undisputed islands is not a legitimate claim from an international law perspective, as all claimants, under *pacta sunt servanda*, have to observe the Antarctic Treaty in good faith. Since the Treaty has been signed by the states who have sovereignty over these islands, the Treaty therefore applies in the same way as it applies to the areas south of the 60° south latitude. Article VI of the Antarctic Treaty requires that the rights of the high seas be respected within its area. Therefore, the Antarctic Treaty area cannot be affected by national delimitation of maritime zones. In summary, delimiting EEZ in waters offshore from Antarctica is neither legitimate nor practicable.

Under UNCLOS, beyond the EEZ and territorial sea are the high seas, the fact that neither territorial sea nor EEZ can be applied to the waters surrounding Antarctica, leads to the probable conclusion that all the waters around the Antarctic continent are classified as the high seas. Article VI of the Antarctic Treaty seems also to support this opinion.

III. High Seas and Article VI of the Antarctic Treaty

The development of the freedom of the seas principle could be treated as the guideline of the history of the law of the sea.[262] In order to contest Portuguese monopoly of the Eastern spice trade and thereby obtain a share of the huge profits, Grotius developed a maritime custom for Asian Countries and elaborated them in the form of a doctrine supported by logical arguments, Christian theology, and the International Seabed Authority (ISA) of venerable Roman Law.[263] The keen logic of proving that the seas are free for everyone relied on two propositions, first being "that which cannot be occupied, or which never has been occupied, cannot be the property of any one, because all property has arisen from occupation"; second "all that which has been so constituted by nature that although serving some one person it still suffices for the common use of all other persons, is today and ought in perpetuity to remain in the same condition as when it was first created by nature".[264] Since "the seas are not susceptible of occupation", and "its common use is destined for all man",[265] Grotius concluded that the all man are free to navigate and fish in seas which can in no way become the private property of any one person. However, as the Dutch defeated the Portuguese and obtained a monopoly over trade in the East,

---

262 Anand, Ram P., "Freedom of the Seas: Past, Present and Future", in Gutiérrez-Girardot, Rafael / Ridder, helmut / Sarin, Manohar Lal / Schiller, Theo (eds.), *New Directions in International Law Essays in Honour of Wolfgang Abendroth: Festschrift zu seinem 75. Geburtstag*, (Frankfurt/New York: Campus Verlag, 1982), 215 at 215. It is generally assumed that it was Hugo Grotius who propounded the freedom of seas principle for the first time in his famous book the *Free Seas (Mare Liberum).*

263 Ibidem, at 219.

264 Grotius (note 157), at 27.

265 Grotius (note 157), at 28.

neither Grotius nor Holland persisted with the freedom of the seas as a principle.[266] It was a British Scholar John Selden whose *Mare Clausum, sen de Domino Maris Libri Duo* (*The Closed Sea or Two books concerning the Rule over the Sea*) provided the basis for European countries controlling as much ocean as their power would permit in the next 200 years.[267] However, as Great Britain and other powerful European States established a large number of colonies in Asia and Africa in the eighteenth and nineteenth centuries, it became clear that open and free seas would benefit further exploitation of unexplored areas of the world. Therefore it was more attractive for these States to revive and approve the freedom of the seas, thereby eclipsing Selden's *Mare Clausum*. Neither the *Mare Liberum* or *Mare Clausum* have been accepted as holy writ by any means. Effectively, a few dominant maritime powers' interests influenced their practices, and their practices formed the basis of the rules of modern maritime law. After the Second World War, what had become all the more evident was that coastal States, especially in the Third World, were struggling to protect their offshore resources – both living resources and non-living resources. As the term "freedom of the seas" changed to "freedom of the high seas" but remained a principle of maritime law, the meaning also came under challenge. Moreover, since the argument about the inexhaustibility of fishery and other resources of ocean has been proved false, "the purely negative conception of the freedom of the high seas ceased to be sacrosanct".[268] However, it was a confusing situation until UNCLOS brought some order into the divergent standpoints in the field, formalising the definition of territorial seas, fishery jurisdictions, continental shelves, and other issues.

Article VI of the Antarctic Treaty is the closest provision to the law of the sea system,[269] and regulates the scope of application to the area south of 60° south latitude, including the whole landmass, extensive sea adjacent to the continent, as well as all ice shelves. The ice shelf covering the landmass is tantamount to land and subjected to a regime of land law.[270] For the landmass including all ice shelves, the application of the Treaty is beyond question. However, offshore jurisdiction is not yet so clear. As Article VI provides that the Treaty will not infringe on any state's rights of the high seas, some authors, in order to avoid the delimitation of the marine zones in Antarctica, attempt to affirm that all waters around Antarctica should be

266 Anand (note 262), at 220.

267 Ibidem.

268 United Nations' "Memorandum on the Regime of the High Seas" (reputed to have been prepared by the French jurist Gidel), UN General Assembly Doc. No. A/CN. 4/38, 14 July 1950, at 13 and 16. (hereafter UN Memorandum)

269 Article VI of the Antarctic Treaty reads: "The provisions of the present Treaty shall apply to the area south of 60° south latitude, including all ice shelves, nothing in the present Treaty shall prejudice or in any way affect the rights, or the exercise of the rights, of any State under international law with regard to the high seas within that area" (emphases added).

270 Pharand, A.D., *The law of the sea of the Arctic: with special reference to Canada*, (Ottawa: University of Ottawa Press, 1973), at 181.

treated as high seas.[271] As discussed above, the inability to delimit maritime zones in the Southern Ocean around Antarctica does seem to support the argument that the whole offshore area should be viewed as high seas.

A High Seas Convention[272] had been adopted in 1958, one year before the Antarctic Treaty. In this Convention, the high seas were defined as all parts of the sea not included in a State's territorial sea or internal waters.[273] The development of marine technology revealed a new world within the seas, with deposits of natural resources and minerals in quantities beyond what had previously been imagined. In order to claim possession of these vast treasures, many States sought wider area of sea under their control. Therefore, the high seas have been redefined in Article 86 of UNCLOS as: "... all parts of the sea that are not included in the exclusive economic zone, in the territorial sea or in the internal waters of a State, or in the archipelagic waters of an archipelagic State."

Although the scope of the high seas has been reduced, the rights of freedom have been extended in UNCLOS. In the 1958 High Seas Convention, the freedom of the high seas was divided into four aspects: navigation, fishing, laying submarine cables and pipelines, and flying over the high seas. After the adoption of UNCLOS, two more kinds of freedom on the high seas were added, the freedom to construct artificial islands and other installations[274] and freedom of scientific research[275].

Although UNCLOS provides that the high seas shall be reserved for peaceful purposes, there is no specific prohibition of military activities,[276] such as naval manoeuvres and weapons testing, being conducted on the high seas, although these have been explicitly forbidden by the Antarctic Treaty[277]. This reveals an inherent contradiction in the Treaty: Article VI of Antarctic Treaty provides that nothing in the Treaty shall prejudice or in any way affect rights of the States with regard to the high seas. Logically, if nothing in the Treaty including Article I interferes with the rights of freedom of the high seas, subject to UNCLOS, States could exercise military activities for peaceful purposes. However, this conclusion is obviously irrational and unreasonable in the Antarctic situation. Firstly, as Professor Gidel claimed in 1950: "the expression 'freedom of the high seas' is in reality purely negative...; it has no

271 Brinie, Patricia, "Effect of Article VI of the Antarctic Treaty on Scientific Research", in Wolfrum, Rüdiger (ed.), *Antarctic Challenge III, Conflicting Interests, Cooperation, Environmental Protection, Economic Development : Proceedings of An Interdisciplinary Symposium*, (Berlin: Duncker & Humblot, 1988), at 105–120.

272 Convention on the High Seas, Geneva, 29 April, 1958.

273 Article 1 Convention on the High Seas.

274 Article 87 (1) (d) UNCLOS.

275 Article 87 (1) (f) UNCLOS.

276 Article 88 UNCLOS.

277 Article 1 (1) of the Antarctic Treaty reads: Antarctica shall be used for peaceful purposes only. There shall be prohibited, inter alia, any measure of a military nature, such as the establishment of military bases and fortifications, the carrying out of military manoeuvres, as well as the testing of any type of weapon.

meaning for us, except as the antithesis of another, a positive concept, which has long since disappeared. The idea of the freedom of the high seas is paradoxically, a survival of the idea – long since dead – that the high seas are subject to dominion and sovereignty, just like any territorial dominion."[278] Therefore, the essential idea of the high seas is to ensure that a ship in peace time will not be interfered with by any nation except its flag State.[279]

Secondly, if we read the provisions of the Treaty carefully, the real meaning of Article VI is to protect a state's "rights with regard to the high seas", which are those specifically defined in both the Convention of the High Seas and UNCLOS and cannot be extrapolated. Thus, in 1950s, the rights of high seas provided freedom of navigation, fishing, laying cables and overflight, the freedom to establish artificial buildings and the freedom to conduct scientific research were added after 1982. Subject to international law, these rights are limited by the requirement of "due regard", which is itself an amendment to safeguard not only the interests of other States exercising the freedom of the high seas, but also rights with respect to activities in the area.[280] Moreover, coastal States may regulate submarine cable and pipeline laying activities and the establishment of non-economic installations on the sea bed on their continental shelf.[281] Freedom of research is limited by the rights of coastal States over their continental shelf, and by the International Seabed Authority (hereafter ISA). Freedom of fishing is subject to a general obligation to adopt national measures[282] and to seek to protect living resources[283] in the high seas. It seems that the 1958 High Seas Convention and UNCLOS only list non-exhaustive activities under the freedom of the high seas, adding some limits and leaving rooms for negotiation.[284]

Analysis of Article VI of the Antarctic Treaty is not to judge but to reiterate the high seas status of the Southern Sea, which reads: "The provisions of the present Treaty shall apply to the area south of 60° South Latitude, including all ice shelves, but nothing in the present Treaty shall *prejudice* or in any way *affect* the rights, or the exercise of the rights, of any State under international law with regard to the high seas within that area." (Emphasis added) "Prejudice", which means "affect adversely or unfavourably; in jure or impair the validity of (a right, claim, etc.)[285]; and "affect", which means "move, touch, (in mind or feelings); influence; make a material impression on"[286] are normally used where activities have the potential to change

278 See UN Memorandum (note 268), at 2–3.
279 See UN Memorandum (note 268), at3–4.
280 Article 87 (2) UNCLOS.
281 Article 80 and 60 UNCLOS.
282 Article 117 UNCLOS.
283 Article 118 UNCLOS.
284 Churchill and Lowe (note 244), at 167–168.
285 Brown, Lesley ed., *The New Shorter Oxford English Dictionary on Historical Principles*, (Oxford: Clarendon Press, 1993), at 2333.
286 Brown (note 285), at 35.

the original situation, i.e. something must already exist before it can be "prejudiced" or "affected" by something else. In the Antarctic case, the notion of the high seas already existed before the Antarctic Treaty was adopted, and the latter can neither "prejudice" nor "affect" the high seas rights of any state. Therefore, Article VI implies that the seas within 60° south latitude were already being treated as high seas before the Treaty was adopted, although it only refers to "rights" of the high seas. Moreover, it is clear that the issue of territorial claims in Antarctica cannot affect the legal status of its surrounding waters. It is worth noting that if one day the Antarctic Treaty expires, several states, not only the seven existing claimants but other States who have some basis for a claim to territorial sovereignty, would claim territories in Antarctica. Even though they could, theoretically, provide historical or legal bases for their claims on land territories, any claim of sovereign right over the adjacent sea would be absurd, due to the high seas status of the waters around Antarctica.

We have reached therefore the following conclusions. Firstly, it is explicitly stated within the Antarctic Treaty area that the surrounding waters are free to all States under the freedom of the high seas. Secondly, the specific rights for states over the subjacent water of Antarctica are exactly the same as the rights defined in UNCLOS. Thirdly, except for the rights of the high seas, States cannot claim or exercise any further right in this area. Even though there are still risks if we define the status of superjacent water around Antarctica as the high seas, there is no doubt that there is not a more appropriate designation for this area than high seas. Therefore, the resources and opportunities are equally open to all States, whether claimants or non-claimants. The high seas status also indicates that this area could be freely used by all States of the world no matter whether parties of the Treaty or not, and no States could process any superior ranking.

However, the offshore area of Antarctica is not the traditional or normal high seas under the law of the sea. Article 30 (2) of the Vienna Convention of the Law of Treaties stipulates that "when a treaty specifies that it is subject to ... an earlier or later treaty, the provision of that other treaty prevails."[287] For the Antarctic, there are two systems relating to the legal status of the sea area south of 60° south latitude, the Antarctic Treaty and the Law of the Sea. Of these two legal systems, the Antarctic Treaty System should have priority. Moreover, the Antarctic Treaty and other components of the ATS have established a regional regulatory regime to deal with maritime jurisdictional issues, especially for the preservation of living resources and the protection of the environment. The ATCPs have proven their ability to manage Antarctic matters and control human activities within the Antarctic Treaty area. Thus, offshore Antarctica comes within the jurisdiction of the ATS, which is subject to the high seas regime under the law of the sea. It is more appropriate to discuss offshore Antarctica as "high seas in the Antarctic context".

287 United Nations, Vienna Convention on the Law of Treaties, done at Vienna on 23 May 1969, entered into force on 27 January 1980, *Treaty Series*, vol. 1155, p. 331.

IV. Antarctic Continental Shelf

The continental shelf is natural prolongation of a costal state's land territory, and the coastal state exercises sovereign right to explore and exploit mineral resources within this area. In Antarctica, although lacking official statements on possible Antarctic hydrocarbon resources, Antarctic claimant states have made either full or partial submissions to the CLCS to claim that they have these sovereign rights over continental shelves around their claimed sectors. According to these continental shelf claims, conflict opinions arise from claimants and non-claimants. On one hand, claimant states believe that they own continental shelf offshore their claimed Antarctic sectors, because a continental shelf is nature prolongation of a costal state's land territorial, and the coastal state exercises sovereign right to explore mineral resources within such area. On the other hand, it is not supervised that non-claimants hold that any claim over the Antarctic continental shelf violates Article IV of the Antarctic Treaty, which prohibits new assertions of territorial sovereignty in Antarctica. Moreover, no territorial claims have been recognized by the international community. Therefore, there are legal and practical impediments to identifying national jurisdictions over the continental margin surrounding Antarctica. Non-claimant States, including the US, Russia, Japan, the Netherlands, Germany, and India have explicitly rejected of any rights over the seabed and subsoil of the submarine areas beyond and adjacent to the continent of Antarctica.[288] The ambiguity of the sovereignty situation of Antarctic makes these claims of offshore zones arguable and complicated.

If there are no legitimate coastal sovereignties in Antarctica, its continental margins could be classified as either deep seabed or a *de facto* condominium. The deep seabed belongs to the common heritage of mankind under UNCLOS and is under the control of ISA. Were the common heritage principle applied to the Antarctic, no State would be allowed to stake a national claim over the deep seabed. Mineral resources from the Antarctic continental shelf and sea floor would be treated as common heritage, and could be exploited subject to UNCLOS norms. This would not require the establishment of any jurisdiction or oversight body, as the ISA exercises its powers on behalf of mankind as a whole. In other words, the Antarctic continental shelf would be governed by the international community through the ISA, which would have all resource rights in the area. Once mineral resources had been exploited from the Antarctic shelf, the benefits would have to be shared internationally. Under the deep sea mining regime, the Antarctic continental margin can only be used for peaceful purposes,[289] and any military activities are forbidden, as is any

288 For rejection of Australia, see http://www.un.org/Depts/los/clcs_new/submissions_files/submission_aus.htm; for New Zealand, see http://www.un.org/Depts/los/clcs_new/submissions_files/submission_nzl.htm.

289 Article 141 UNCLOS.

form of station or base. Scientific research[290] however is regulated in the same way as under the Antarctic Treaty: as being free and open to all States, unless there is a threat to the environment or ecology of the area. States should disseminate research results and analysis. However, the critical problem with this scenario is that there are two overlapping international legal regimes covering the Antarctic continental shelf. On the one hand, the ATCPs have obtained rights to establish regulations relevant to Antarctic issues, such as mineral resource activities. A series of measures and conventions centred around the Antarctic Treaty have been set up to govern the Antarctic Treaty area, and north to the "Antarctic Convergence" for matters of environmental preservation.[291] On the other hand, the ISA, as the representative of the whole of mankind, may exercise rights to organize and control activities in the Antarctic continental margin.

In order to avoid conflicts of these legal regimes, the latter form, condominium regime, is more appropriate to Antarctic situation. In fact, the Convention on the Regulation of Antarctic Mineral Resource Activities (CRAMRA) found that a regime to manage mineral resources under the ATS was feasible, although this has been shelved for the moment. CRAMRA's failure to enter into force does not eclipse its significant value in providing experience and the potential basis for a future mineral resources regime. The ability of the ATS to develop appropriate measures to solve specific issues through its negotiation mechanism has been proven by with solid position *vis-à-vis* UNCLOS in Antarctic Treaty area.

V. Conclusion

The issue concerning the applicability of the law of the sea to the Antarctic marine area, is still the basic question of whether there are territorial sovereignties in Antarctica. If the sovereignty claims are recognised, this will give rise to territorial sea, executive economic zone and continental shelf appurtenant considerations. Taking the opposite tack, if there are no legal coastal states in the Antarctic, all the Southern Sea should be treated as high seas. Therefore, both claimants and non-claimants could provide arguments to support their interests.

1. Arguments in Favour of Maritime Zone Delimitation

The seven claimant States are not unanimous with regard to a delimitation of maritime zones. Some States have claimed part of the continental shelf in the Antarctic, and Australia has even claimed territorial sea and an offshore EEZ.[292] There are a

290 Article 143 UNCLOS.

291 Articles 1 and 4 CCAMLR.

292 Under Australia's Seas and Submerged Land Act of 1973, sovereignty is expressly extended to territorial seas adjacent to all Australian claimed territories, implicitly including the Australian Antarctic Territory. In 1990, Australia's territorial sea was extended by proclamation to 12 nautical miles, and its claimed territory was

number of points to be considered in this debate. Firstly, even though sovereign claims in Antarctica have been frozen by the Antarctic Treaty, this does not mean these claims have been denied or ceased to exist. If a coastal State is found to exist legally, this automatically confers ownership over the continental shelf.

Second, the fact that claimant States possess rights over the continental shelf does not violate the Antarctic Treaty. The Antarctic Treaty provides that "no new claim...to territorial sovereignty in Antarctica shall be asserted"[293]. However, the legal status of the continental shelf is not as the same as the coastal sovereign territory.[294] According to 1958 Convention and UNCLOS coastal States may only exercise "sovereign rights" but not "sovereignty" on their continental shelf. Thus, when coastal States exercise rights over their continental shelf in the offshore Antarctic, it is difficult to say that this kind of right is a new claim in the Antarctic.

Thirdly, some scholars declare that an assertion of rights over an EEZ is not contrary to Article IV of Antarctic, as an EEZ exits as a result of territorial sovereignty, therefore exercising a sovereign right in an EEZ does not constitute a claim of territorial sovereignty.[295] Similarly, claims over the continental shelf adjacent to Antarctic do not violate the Antarctic Treaty, as UNCLOS states that rights over the continental shelf "do not depend on occupation, effective or notional, or on any express proclamation".[296]

Lastly, claimant states have variously applied the law of the sea to their "Antarctic territories". In 1940, Chile specified the limits of its Antarctic territory including "all lands, islands islets, reefs of rocks, glaciers, already known, or to be discovered, and their respective territorial waters."[297] Later, in 1947 Chile established 200 mile EEZ.[298] France restricted fishing rights for foreign vessels within a twelve mile band of its Antarctic territories in 1972.[299] Additionally, Australia has issued regulations for three nautical miles of territorial sea adjacent to the Australian Antarctic Territory, and has submitted claims over a wider territorial sea under UNCLOS.[300] By the

---

not explicitly excluded. Moreover, in 1994, proclamation of the Maritime Legislation Amendment Act 1994 announced that EEZ and contiguous zone applied to Australian territories, including that in the Antarctic. See Campbell, William "International Law Notes: Maritime Legislation Amendment Act", 5 Public Law Review 141 (1994), at 141–143.

293 Article IV paragraph 2 Antarctic Treaty.

294 Joyner (note 68), at 121.

295 Triggs (note 236), at 91.

296 Article 77 (3) UNCLOS.

297 Chile, Decree No. 1747 of 6 November 1940 declaring the limits of the Chilean Antarctic territory, reprinted in Bush, *Antarctica and International Law* (note 96), at 311.

298 Bush, *Antarctica and International Law* (ntoe 96), at 447.

299 France, Order No. 5 of 13 January 1972, reprinted in Bush, *Antarctica and International Law* (note 96), at 579.

300 Harry (note 91), at 730.

expiration of the submission made by coastal States to the CLCS seeking to extend the limits of the continental shelf beyond 200 nautical miles, all seven claimants had declared the continental shelf area appurtenant to Antarctica.[301] These declarations were explicitly rejected by other States such as Germany, India, Japan, the Netherland, Russian and United States of America.[302] Under UNCLOS, the outer limits of the continental shelf should be based on CLCS recommendations.[303] In advance of these recommendations however, one point that can be safely made is that giving the continental shelves around Antarctica the same status as continental shelve in other areas does not make sense.

2. Arguments against maritime zones

Non-claimant States perceive the legal status of the Antarctic offshore area differently. They deny that coastal States exist in the Antarctic, as territorial claims cannot be said to have been recognised by international law or the international community. Based on this premise, non-claimant States insist that no territorial seas, EEZs or continental shelf areas may be legally declared offshore from Antarctica. There a number of elements to this claim.

Firstly, maritime zones offshore from Antarctica may support a claim to territorial sovereignty in Antarctica, which definitely violates the Antarctic Treaty, raising a recurrent paradox. The Antarctic Treaty, as a compromise between the claimants and the non-claimants, created a "constructive ambiguity", which founded the platform for cooperation in the Antarctic. Therefore, any step that interferes with this ambiguity is not permissible as long as the Treaty is valid. According to non-claimant State reasoning, acceptance of maritime zones under UNCLOS in Antarctica would "thaw out" the seven territorial claims frozen under the Antarctic Treaty, as specific legal consequences derive from those claims. This would clearly compromise the current equilibrium. Additionally, new concepts, such as the EEZ, lack a legal foundation, as the advent of the EEZ as part of customary law came

301 More detailed information please see Part 1 A "Sovereignty claims of seven States".

302 According to Australia's submission to the CLCS, notification No. CLCS.03.2004. LOS of 15 November 2004, the USA rejected in "United States Mission to the United Nations", POL 10–04, December 3, 2004; Russia rejected in "Permanent Mission of the Russian Federation to the United Nations", No. 739/n, December 9, 2004; Japan rejected in "Notification regarding the submission made by Australia to the Commission on the Limits of the Continental shelf", Reference: CLCS.30.2004. LOS/JPN, February 4 2005; the Netherland Statements expressed on March 31, 2005, No. NYV/2005/690; Statement of Germany was expressed on April 5, 2005, Note No. 88/2005; India rejected in "Notification regarding the submission made by Australia to the Commission on the Limits of the Continental shelf", Reference: CLCS.03.2004.LOS/IND, July 13, 2005. To the submission of the New Zealand to the CLCS, the Netherland rejected in DJZ-IR 178/2006, 19 December 2009; Japan rejected in SC/06/459, June 28, 2006.

303 Article 76 (8) UNCLOS.

long after the Antarctic Treaty. It is difficult to justify the creation of an EEZ under the scope of the Antarctic Treaty, as the EEZ is based upon the premise that a coastal State is granted rights of jurisdiction, including exclusive rights to natural resources.[304] This is inconsistent with the operational premise of the Antarctic treaty, which suspends all claims to sovereignty granting all States the same rights, and subjecting them to the same duties.

Secondly, the Antarctic Treaty System itself renders the delimitation of maritime zones in Antarctica unnecessary. The ATS is a special legal system, that creates an international area for "the interest of all mankind",[305] and which has effectively managed activities carried out in Antarctica since its introduction. Introduction of another international legal system into the Antarctic Treaty area would cause unnecessary conflicts between authorities and complications in the unclaimed areas in the Antarctic.

3. *No Maritime Zones in Antarctica*

As the settlement of disputes over the legal status of maritime zones adjacent to Antarctica is contingent on the thorny issue of sovereignty, resolution does not seem likely in the foreseeable future. The preservation of Antarctica as an international zone for peaceful uses in the face of increasing international interaction suggests that the UNCLOS norms on delimitations of maritime zones are inappropriate for the situation in Antarctica[306], for the following reasons:

Firstly, the unique geophysical factors of Antarctica, i.e. immense ice sheets, irregular coastlines and extremely deep continental shelf,[307] make UNCLOS difficult to apply as it is hard to determine the initial line of maritime zones.[308] Secondly, the institution of a dual system for claimant sectors and a non-claimant sector in the Antarctic would be problematic at best.[309] For example, applying UNCLOS to the non-claimant sector, would make seas adjacent to offshore waters high seas, in turn, classifying the continental shelf beneath as the common heritage of mankind, and therefore under the control of ISA – a seemingly fanciful arrangement. Thirdly, the establishment of maritime zones in the Antarctic would serve no real purpose, territorial sea acts as a buffer between territory and high seas to help protect a coastal State from outside dangers. The lack of real sovereignty in Antarctica deprives the territorial sea of this purpose. The EEZ provides the coastal state with the right to explore, exploit, conserve and manage natural

---

304 Churchill and Lowe (note 244), at 133–152.

305 The Preamble of the Antarctic Treaty.

306 Vigni (note 253), at 498.

307 The average depth of Antarctic continental shelf is about 500 meters, nearly eight times the world shelf depth average. See: Joyner (note 68), at 109.

308 Some scholars have provided measures to solve the baseline problem. See: Joyner (note 68), at 79–87.

309 Vigni (note 253), at 498.

resources however, relevant conventions for the conservation of living resources have already been established in the area south of 60° south latitude. According to Article VI of the Antarctic Treaty, the waters around the Antarctic continent could be treated as special high seas, as the rights of high seas are respected within the Antarctic Treaty although human activities are subject to the ATS.

D. Options for a Potential Minerals Regime

Antarctic mineral resource exploration is a controversial issue, which directly touches the question of territorial sovereignty. Claimants retain their territorial sovereignties and national jurisdictions for mineral resources activities which would be taken place in their sectors. Non-claimants reject any sovereign claims in Antarctica and any priority in potential mineral regimes. Non-aligned states attempt to treat the Antarctic as a common heritage for mankind. Therefore, there are different theoretical assumptions on a future Antarctic minerals regime.

I. National Regime

National solutions would grant absolute control to those individual claimant States in Antarctica over their territories. Thus, each claimant-State could build up their own legal regime for the utilization of the mineral resources within its sector, moreover, the legal regime would apply to not only to the sovereign territories of Antarctica but also to territorial seas, an EEZ and even a continental shelf. Under this regime, claimant-States could allocate resources as they will.

It is beyond doubt that the national regime is not feasible because: Firstly, if this regime is supported, that support will only come from claimant States; it has been uniformly opposed by all non-claimant States, which are the overwhelming majority. Secondly, this solution would abolish the Antarctic Treaty or at least disobey Article IV of the Treaty.[310] Thirdly, there is a sector in Antarctica which has been over-claimed by three States, and another sector claimed by none. Therefore, the use of national regimes for the exploitation of mineral resources in Antarctica is, in a word, impossible.

II. Open Use Regime

The second alternative for mineral regime is to open Antarctica to free appropriation of mineral resources. The USA, the most faithful advocate of this regime, argues that Antarctica, in the absence of valid territorial claims and no clear regulations governing utilization of resources, must be treated as a part of high seas and all

310 Wolfrum, Rüdiger, "The Use of Antarctica Non-Living Resources: The Search for A Trustee?", in Wolfrum, Rüdiger (ed.), *Antarctic Challenge: Conflicting Interests, Cooperation, Environmental Protection, Economic Development*, (Berlin: Duncker & Jumblot, 1984), 143 at 144.

States should have free access to the exploitation of Antarctic resources.[311] To apply uniform and non-preferential rules to all States for the development of resources is the cornerstone of this regime. This approach has been used for the oceans outside the jurisdiction of coastal States, and has worked well so long as there was no pressure on the resources or uses of the ocean.[312]

However, disastrous consequences could result from an open use regime. Firstly, developed States, such as the USA, possessing procurement technology and capital for exploitation will take the lead over developing States for the exploitation of Antarctic mineral resources. Formally imposing equal access conditions will cause *de facto* discrimination, as developing States lack currently the capability to exploit mining activities, by the time they are in a position to take advantage, non-renewable mineral sources may have been exploited entirely. Thus, freedom of access cannot be accepted by developing States, unless mechanisms for benefit-sharing are established.

Moreover, the pursuit of individual interests in common spaces often leads to common disaster, as demonstrated by illegal, unreported and unregulated fishing. In an open regime, each State will only consider their own interests, rarely demonstrating concern for environmental protection or preservation. Therefore, it is not difficult to imagine that, if this regime were to be established in Antarctic, accessing resources would take precedence over environmental protection.

As Antarctica is the common heritage of mankind, free appropriation of these resources is illicit even though such activities are under Consultative Party guidelines.[313] According to the common heritage principle, only fully representative machinery is capable of determining whether and how the resources in Antarctica could be exploited.

III. Condominium

International law recognizes the condominium, which is defined as "a piece of territory consisting of land or water is under the joint tenancy of two or more States, these several States exercising sovereignty conjointly over it, and over the individuals living thereon."[314] Some scholars try to apply this method to Antarctica, arguing that establishing a common organization, comprised of the 12 original signatories of the Antarctic Treaty, is the best possible way to manage the Antarctic continent.[315] According to the condominium approach, the original Consultative Parties would

311 Thaw (note 153), at 841.

312 Westermeyer, William E., "Alternative Regimes for Future Mineral Resource Development in Antarctica", 8 Ocean Management 197 (September 1983), at 205.

313 Thaw (note 153), at 842.

314 Brownlie (note 144), at 116.

315 Hambro, Edvard, "Some Notes on the Future of the Antarctic Treaty Collaboration", 68 A.J.I.L. 217 (April 1974), at 225.

declare that Antarctica is "their joint and exclusive property"[316], and they would exercise joint and equal "sovereign right" over the Antarctic continent and offshore area. Future, additional Consultative Parties would possess some right in administration, but no sovereign rights.[317] The legal system to govern the exploitation of Antarctic mineral resources and the proportion of benefit-sharing would be established by the original Consultative Parties alone,[318] while under condominium approach co-owners in common could give concessions for resources exploitation and share the royalties.[319] So far, there is no great difference between the national regime and condominium, the latter merely extends the number of States who have right to exercise sovereignty in Antarctica and treats Antarctica as private property.[320]

The condominium plan could enhance the authority of the ATCPs in Antarctica and would resolve the difficult problem of overlapping claims held by Argentina, Chile and the United Kingdom, as these three Parties could gain access to the whole of Antarctica. Moreover, "an undeclared condominium" such as the cooperation between New Zealand and the USA in Ross Dependency[321] has been raised a number of times as an alternative to individual claims. However, the exercise of an exclusive condominium scheme in Antarctica would be impossible. Firstly, the other claimant States rejected the idea outright when the USA proposed an Antarctic condominium in 1948, as such a regime would diminish their individual interests.[322] Secondly, the international community feared that some States, primarily developed States, were seeking to carve up the Antarctic resources for their own use,[323] therefore a condominium system in Antarctica would raise international concerns, which would derogate the role of the ATCPs in managing Antarctic resources activities, and would probably require UN intervention. Thirdly, there is a demonstrated lack of success with condominium in practice,[324] with it seemingly leading to a "piling up of casuistic constructions and fallacious arguments, often mutually exclusive

316 Thaw (note 153), at 842.

317 Mitchell, Barbara, *Frozen Stakes: The Future of Antarctic Mineral*, (London: International Institution for Environment and Development, 1983), at 88.

318 Thaw (note 153), at 844.

319 Hambro (note 315), at 224.

320 Wolfrum (note 310), at 144; Thaw (note 153), at 842.

321 See Auburn, F.M., "The White Desert", 19 Int'l & Comp. L. Q. 229 (1970); Auburn, F.M., "The Ross Dependency: An Undeclared Condominium", 1 Auckland University law Review 89 (1970); Auburn, F.M., "A Sometime World of Men: Legal Rights in the Ross Dependency", 65 American Society of International Law 578 (July 1971).

322 Peterson, M. J., "Antarctica: The Last Great Land Rush on Earth", 34 International Organization 377 (Summer, 1980), at 396.

323 U.S. Antarctic Policy: U.S. Policy with respect to Mineral Exploration and Exploitation in the Antarctic: Hearing Before the Subcommission on Oceans and International Environment of the Senate Commission on Foreign Relations, 94th Congress, 1st Session 1975. p. 1, cited from Thaw (note 153), at 843.

324 Jessup /Taubenfeld (note 101), at 11–49.

and contradictory called upon to justify the predatory policy of imperialist states in the Antarctic".[325] In addition, taking the position that there is a joint Antarctic sovereignty would violate the Antarctica Treaty, especially Article IV, as such a declaration necessarily implies that all these original Consultative Parties assert common sovereignty over the whole Antarctica, including the continental shelf of Antarctica. As such, a declaration of "common sovereignty" goes beyond prior sector claims and prior claimants; it must be treated as a new territorial claim. These considerations make it difficult, if not impossible to simply adopt the condominium approach to Antarctica.

However, the condominium approach would resolve the issue of the lack of a national regime and settle the sovereignty question, and lead Antarctica into a limited international cooperation between Consultative Parties. In order to maintain the effective management of Antarctic Treaty System in Antarctic mining activities, and to avoid "being upstaged by the international community in the Antarctic area",[326] the condominium approach needed to be adapted to the special characteristics of Antarctica. Firstly, to prevent initial objections to a condominium *per se*, the model for adoption needed to be a *de facto* condominium which excluded "sovereignty", only granting ATCPs the right to exercise jurisdiction over potential mining activities in Antarctica. Secondly, membership of this group needed to extend beyond the 12 original members to include all ATCPs. However, it was unlikely that claimant States will approve this approach without some special considerations to their benefit e.g regarding the proportion of claimant States in institutions and decision-making procedures. Thirdly, in order to get approval from international community outside of the Antarctic Treaty System, equal access to Antarctic mineral resources activities for every State needed to be set up, while, the ATCPs retained the right to elaborate detailed provisions on access and revenue distribution.[327] Approved by the followed Antarctic mining system, i.e. the Convention on the Regulation of Antarctic Mineral Resource Activities (CRAMRA), the *de facto* condominium was adopted as the appropriate solution for the mining questions.

IV. Moratorium

Some commentators, especially environmentalists, have called for a moratorium on all commercial mining in Antarctica and for the preservation of Antarctica as an "international park".[328] Because of the unique ecological environment in Antarctica and marine ecosystems in Southern Ocean, proponents of this theory argue

325 Molodtsov, S.V., *Sovremennoe Mezhdunarodno-Pravovoe Polozhenie Antarktiki*, (Moscow: Gosyurizdat, 1954), at 43. Cited from Toma (note 124), at 622; *cf.* also Mikhailov, S., "Imperialisticheskaya Bor'ba za Antarctiku", 10 Voprosy Ekonomiky 66 (1949), at 66–73.

326 Joyner (note 235), at 437.

327 Wolfrum (note 310), at 144; Thaw (note 153), at 844.

328 Thaw (note 153), at 839; Wolfrum (note 310), at 146; Parriott (note 152), at 100.

that the entire area should only be used for scientific purposes.[329] The advantage of moratorium is that it protects the Antarctic continent and surrounding waters from irreversible environmental harm.

The international community first became enthusiastic about discussion on the possibility of exploitation and potential regimes on mineral resources in Antarctica in the 1970's, however after the Protocol on Environmental Protection to the Antarctic Treaty (the Antarctic Environmental Protocol) was adopted in 1991, this pressure was reduced.[330] The conversion stemmed from the Article 7 of the Antarctic Environmental Protocol, which stipulates that "any activity relating to mineral resources, other than scientific research, shall be prohibited". This Protocol entered into force in 1998 and will be reviewed in 2048.[331] It would appear that the problem of exploitation in Antarctica has been solved, or at least a temporary moratorium has been constituted. However, the reality is not so simple; on the contrary, rapidly improving technology and the increasing visibility of the energy crisis means it is not impossible that States will try to build another treaty to replace the Protocol in order to exploit Antarctica in the future.[332] According to the Protocol, which will be analysed in detail in the following chapter, it can be said the moratorium cannot become a permanent solution, because third party States are not likely to support the idea of locking up valuable resources, moreover, from an international law perspective, they lack any obligation to obey the Protocol.

329 Kindt, John warren / Parriott, Todd Jay, "Ice-Covered Areas: The Competing Interests of Conservation and Resource Exploitation", 21 San Diego L. Rev. 941 (1983–1984), at 982–983.

330 This trend is reflected by the amount of relevant articles, which peaked from late of 1970s to late of 1980s. See: Wolfrum Wolfrum (note 310); Lagoni (note 240); Kindt / Parriott (note 329); Alexander, Frank C. (Jr.), "Legal Aspects: Exploitation of Antarctic Resources – A Recommended Approach to the Antarctic Resource Problem," 33 Miami Law Review 371 (1978–1979); Joyner, Christopher C, "The Evolving Minerals Regime for Antarctica", 19 Ocean Dev. & Int'l L.73 (1988); Heim, Barbara Ellen, "Exploring the Last Frontiers for Mineral Resources: A Comparison of International Law Regarding the Deep Seabed, Outer Space, and Antarctica", 23 Vand. J. Transnat'l L. 819 (1990–1991); Behrendt, John C, "Are There Petroleum Resources in Antarctica?", 68 AAPG Bulletin 453 (1984); Rowley, Peter D. / Williams, Paul L. / Pride, Douglas E., "Mineral Occurrences of Antarctica", in Behrendt, John C (ed.), *Petroleum and Mineral Resources of Antarctica*, (USA Geol Surv, 1983) 25–49; and etc.

331 Article 25 (2) Antarctic Environment Protocolsays that after the expiration of 50 years from the date of entry into force of this Protocol, any Consultative Parties may requests to review the operation of this Protocol.

332 Article 25 (5)(a) Antarctic Environment Protocol expresses that the prohibition on Antarctic mineral resource activities could be changed until a binding legal regime on Antarctic mineral resource activities has been established.

V. Common Heritage Model

Although there is no single definition of the common heritage principle, several considerations have been generalized by most international lawyers when the notion is applied to common space.[333] Firstly, the world as a community is the owner of the space.[334] According to Part XI of UNCLOS, the Area and its resources are regarded as the common heritage of mankind and represented by the ISA as its trustee.[335] Thus, the appropriation of an area and its resources as well as a claim and exercise of sovereignty or of sovereign rights over any part thereof are prohibited.[336] This has led to it also being known as the principle of "non-appropriation" by some scholars.[337] In short, theoretically, States cannot claim territorial sovereignty in specific areas. In addition, it is not only a confirmation of existing international law, but also a State obligation to exclude any use of the Area except in accordance with Part XI.[338] According to the original common heritage concept, Professor Wolfrum maintains that coastal States are not allowed, to establish exclusive economic rights or to enlarge areas of continental shelf.[339]

Secondly, the management of a common space should be shared by all people, i.e. States do not administer it but participate as representatives. For example, UNCLOS requires that all rights in the Area are vested in "mankind as a whole" instead of all States.[340] There are at least two reasons to use term "mankind" here, which contains broader sense than "State". This is because a) "Mankind" indicates that the interests of future generations have to be considered in making use of the Area; furthermore, the use of the Area should avoid undue waste of resources and protect the environment[341]; and b) to represent those groups of people who have not attained full independence or other self-governing status. Therefore, expressing the true goal of the notion is more adequately captured by the term "mankind". However, as it is impracticable to have all of "mankind" acting in concert, a coordination agency such as the ISA is needed to act on behalf of mankind. However, the ISA is the organization "through which States Parties shall... organize and control activities

333 Heim (note 330), at 819–827.
334 Wolfrum, "The Principle of the Common Heritage of Mankind", 43 ZaöRV 312 (1983), at 316–319.
335 Article 137 (2) UNCLOS.
336 Article 137 (1) UNCLOS.
337 Frakes, "The Common Heritage of Mankind Principle and the Deep Seabed, Outer Space, and Antarctica: Will Developed and Developing Nations Reach a Compromise?", 21 Wis Int'l L. J. 409 (2003), at 411); Joyner, Christopher C., "Legal Implications of the Concept of the Common Heritage of Mankind", 35 Int'l & Comp. L. Q. 190 (1986), at 191.
338 Article 137 (3) UNCLOS.
339 Wolfrum (note 334), at 317.
340 Article 137 (2) UNCLOS.
341 Wolfrum (note 334), at 318–319.

in the Area"[342], pushing the humanitarian ideals somewhat to the sidelines, as the real participants with respect to the utilization of the Area are actually States, as only they and international organizations have the capability to act directly in international legal society.[343]

Logically, there is no contradiction between Article 137 and 157. UNCLOS names the beneficiary as "mankind" is to express the notion of non-discrimination, however, while ideals are one thing, reality is another. In the process of achieving this lofty ideal in the real world, States play the critical role in implementing the principle, and the ISA is a valid solution.

The third element of cultural heritage is *de facto* equity. Many scholars refer to this element as "benefits sharing", which means that any State should share benefits gained from exploitation of the common resources.[344] However, Professor Wolfrum points out that developed and developing States have two very different schools of thought, on preferential treatment and compensation.[345] Developed States assert that royalties levied on activities in a common heritage area are part of development aid and they support a loose framework to facilitate access; however, developing States, on the contrary, insist on a *de facto* equality. According to developing States, financial and technical restrictions prevent them from exploiting common area resources. If a formal equality regime is established, meaning that every State could enter into it freely, developing States will not be capable of participating, and therefore *de facto* inequality will be the end result. Although both approaches seem to result in the same amount of money to be paid, the difference in reasoning will determine what kind of regime should be established. *De facto* equity derives from the common heritage concept of "treating all States on the same footing and accordingly benefiting all States".[346]

Moreover, *de facto* equity also implies equality across generations. Principle 3 of Rio Declaration on Environment and Development 1992 proclaims that "The right to development must be fulfilled so as to equitably meet developmental and environmental needs of present and future generations." This provision requires that resources are conserved and the environment is protected, whereas improvement of technology provides huge impetus to further exploit resources in common spaces. However, once these resources have been used, they cannot be renewed or repurchased, a consideration that must be taken seriously in light of the heritage of mankind, and the consideration of *de facto* equity for future generations.

Fourthly, common spaces must be limited to peaceful purposes exclusively, because any military use would only serve national interests, not the benefit of all

342 Article 157 (1) UNCLOS.

343 Wolfrum (note 334), at 319.

344 Heim (note 330), at 827; Frakes (note 337), at 412–413; Joyner (note 337), at 192; White, Mary Victoria, "The common heritage of mankind", 14 Case W. Res. J. Int'l L. 509 (1984), at 537.

345 Wolfrum (note 334), at 321–323.

346 Wolfrum (note 334), at 323.

mankind, thus violating the requirements of the principle.[347] Some general aspects have been achieved: no military bases or installations;[348] no weapon tests and no manoeuvres;[349] no weapons or weapons systems;[350] and no military conflicts are allowed.[351]

Fifthly, there should be international machinery to control the common spaces. The discussion on building international machinery for common spaces has been going on since 1968, and the Moon Treaty arranged a future resources regime; and UNCLOS has built up the ISA.[352] The demand for international machinery stems from the possibility of exploitation, the potential prevention of free access to resources in common spaces for individual States, moreover, international machinery would treat all States at the same level, which contributes to achieving *de facto* equality. In other words, the *de facto* equality among all States is protected by "institutional matters such as equal rights for all States with respect to the decision-making process".[353]

This application of common heritage principle to Antarctica has been supported by some authors and approved by many non-aligned states, not party to the Antarctic Treaty. Therefore the question of whether the common heritage principle is appropriate for the legal status of the Antarctic deserves some analysis.

At the end of IGY, the United Nations affirmed "the common interest of mankind in outer space" and "the common aim that outer space should be used for peaceful purposes only".[354] In 1960s, technological improvements resulted in political conflicts between developing and developed Sates. On September 21, 1967, Ambassador Arbid Pardo of Malta announced that the seabed and the ocean floor beyond national jurisdiction were the "common heritage of mankind",[355] and that "the exploitation of its resources shall be carried out for the benefit of mankind as a whole, irrespective of the geographical location of States, whether landlocked or coastal, and taking into particular consideration the interests and needs of the developing countries."[356] The Outer Space Treaty of 1967[357], does not mention the new common heritage principle, but specifically enunciates "recognizing the common interest of

347 Wolfrum (note 334), at 319.
348 Frakes (note 337), at 413; Joyner (note 337), at 192.
349 Joyner (note 337), at 192.
350 Frakes (note 337), at 413; Joyner (note 337), at 192.
351 Frakes (note 337), at 413; Wolfrum (note 334), at 319–320.
352 White (note 344), at 536.
353 Wolfrum (note 334), at 324.
354 UNGA Res. 1348 (1958) GAOR 13th Session Supp. (No. 18) at 5, UN Doc. A/4090.
355 White (note 344), at 514–516.
356 Herber, "The Common Heritage Principle: Antarctica and the Developing Nations", 50 American Journal of Economics and Sociology 391 (1991), at 393.
357 Treaty on Principles Governing the Activities of States in the Exploration and Use of Outer Space, Including the Moon and Other Celestial Bodies, Jan. 27, 1967, 6 ILM 386.

all mankind"[358] and "for the benefit and in the interests of all countries".[359] The treaty, using *res communis*, maintains global property rights to outer space, and stipulates the interests of developing States.[360] The UN General Assembly (UNGA) formally proclaimed the common heritage principle in a Declaration of Principles in 1970, in which the Assembly declared that "the seabed and ocean floor, and the subsoil thereof, beyond the limits of national jurisdiction... as well as the resources of the area, are the common heritage of mankind".[361]

Seeking a more equitable distribution of resources and income, and to close the gap between developing and developed States, the UN adopted a Declaration on establishment of a New International Economic Order[362] in May 1974, which provided developing States with an increasingly important role in international decision-making. In particular, the Group of 77, an alliance of developing States, endeavoured to improve the common heritage principle to make it the cornerstone of the law of the sea.[363] After tough contest between different beneficiary groups, the principle, although not yet established in the forum of the law of the sea, did attain legal status in December 1979, by adoption in the Moon Treaty,[364] declaring that: "The Moon and its natural resources are the common heritage of mankind".[365] By the time UNCLOS was successfully adopted in 1982, the common heritage principle had achieved tremendous success. The Part XI of the Convention, not only provided that the seabed and ocean floor and soil beyond the limits of national jurisdiction and its resources are the common heritage of mankind, but also provided elaborate and complex machinery to implement the principle.

This question has been answered in a variety of opposing ways by various international lawyers. Some authors advocate that the common heritage principle should be applied in Antarctica, because Antarctica, while not clearly mentioned in UNCLOS, has been considered one of the primary targets for the application of the common heritage principle.[366]

Firstly, the Antarctic Treaty implements the common heritage principle in many ways although it fails to state it clearly.[367] The Treaty enunciates that Antarctica can only be used for peaceful purposes in keeping with the element of peaceful purpose. The consultative structure of the Treaty reflects the common governance of all of Antarctica, as the Consultative Meetings operate on a basis of consensus decisions and no special powers of States within claimed sectors. Moreover, the

---

358 Preamble The Outer Space Treaty.
359 Article I The Outer Space Treaty.
360 Herber (note 356), at 393.
361 UNGA Res. 2749 (XXV).
362 UNGA Res. 3201 (1974) GAOR 6th Special Session, Supp. (No. 1), U.N. Doc. A/9551.
363 White (note 344), at 516.
364 White (note 344), at 530.
365 The Moon Treaty, Article 11 (1).
366 Herber (note 356), at 391.
367 Frakes (note 337), at 428–429; Thaw (note 153), at 828–833.

Treaty sets no restriction on the rights of persons to access Antarctic regardless of nationality; indeed, scientific observations and results shall be exchanged and made freely available. Finally, the ATS established numerous environmental conservation measures to balance the interests between different generations. In sum, from a legal perspective, the principle could easily be applied in Antarctica.

Secondly, the common heritage principle could assist in protecting the future of Antarctica. The most serious issue, that of territory cannot be solved unless the principle is applied, not only because of the lack of acceptance and the level of overlap in these seven claims, but also because there are several other States who maintain that they possess historical right, which might induce more territorial claims in future. In the past fifty years, the Treaty has brought a peaceful resolution in Antarctica; and history has demonstrated that international cooperation is the most appropriate approach to develop Antarctica. The Treaty seeks to avoid a worst case scenario, with Antarctica becoming the scene or object of international discord. In addition, the principle would prove adequate to protect the Antarctic environment.[368] This argument was supported by most developing States, who were dissatisfied with the Antarctic Treaty in 1980s.[369]

Thirdly, the application of the principle in Antarctica could satisfy the Antarctic social function requirement.[370] According to a general principle of civilised national law, regarded as a source of international law,[371] territorial sovereignty in Antarctica should be treated the same as private property. For example, both the Italian and German Constitutions regulate that use of property should serve the public interest.[372] Presuming that this provision could be elevated into international society, the common heritage principle is the only way this could be applied in Antarctica to ensure its social function.

Fourthly, the common heritage principle applied to Antarctica conforms to the requirement of international community. Non-aligned developing States objecting to the Antarctic Treaty and the ATS did so for several reasons, including:[373] a belief that the Treaty did not sufficiently protect the Antarctic environment; non-recognition the seven territorial claims; opposition to the self-designated ISA; and a requirement to participate in the decision-making process. Use of the common heritage principle in the Antarctic would remove these criticisms, because Antarctica would be under the international community's common control and not under an exclusive club of ATCPs'.

368 Joyner, Christopher C., *Governing the Frozen Commons: The Antarctic Regime and Environmental Protection*, (University of South Carolina Press, 1998), at 235.
369 Joyner (note 368), at 235–237.
370 Conforti (note 181), at 258.
371 Article 38 (1)(c) The Statute of the International Court of Justices.
372 Conforti (note 181), at 258.
373 Frakes (note 337), at 429–432.

Although there are many elements and factors of the common heritage principle that do correlate with its application to Antarctica, there are some major factors which make this impossible.

Firstly, the fact that there are different territorial sovereignties claimed by several States, means that the area in question is not "beyond the limits of national jurisdiction" and not consistent with the definition of "the Area" in UNCLOS.[374] Despite the fact that seven claims of territorial sovereignty over certain sections of the continent of Antarctica have not been recognised by the majority of States, these claims still exist and cannot be simply negated. Additionally, Article IV of the Antarctic Treaty merely freezes these territorial claims; thus, applying the common heritage principle to Antarctica would prejudice the position of the claimant States.

Secondly, in Antarctica, contrary to the deep seabed, a detailed legal regime had been established before the common heritage principle was adopted in UNCLOS. Therefore, there are clear differences between the situation in Antarctica and that of "the Area" under UNCLOS.[375] Moreover, as shown above, the ATS, along with the specific treaties relating to the Antarctic matters, should be construed as prevailing over UNCLOS, especially for those states who are parties to both legal systems.[376]

Thirdly, the common heritage principle requires the exploration and exploitation of mineral resources be governed by an international organization the ISA, however, the ATCPs already existing in Antarctic legal system have on occasion expressed their responsibilities to the world community[377], throwing the need to establish another independent international bureaucracy into doubt. Adding another international organization to the mix would not only cause problems in coordinating operations and an assessing competencies but also endanger the status of the Antarctic Treaty as legal basis for non-militarization, scientific cooperation, and environmental protection in Antarctica.[378] These factors mean the common heritage principle cannot directly or entirely apply in Antarctica; however, it provides us a large number of precious experiences in governing specific international areas.

This special legal status of Antarctica determines that the norms of UNCLOS, in terms of the common heritage principle or delimitation of maritime zones and the continental shelf regime, are not appropriate for the Antarctic Treaty area. This due to a combination of factors, the seven territorial claims in Antarctica, of which three overlap, and the fact that approximately 20% of the Antarctic Continent, from 90° to 150° west latitude, has not been claimed by any State. In addition, Antarctica has unique geographical characteristics, which also impedes application of the UNCLOS norms concerning the delimitation of maritime zones. This complicated political and geographical situation means Antarctica can only be controlled under the Antarctic

374 Triggs (note 236), at 103.
375 Wolfrum (note 310), at 145; Lagoni (note 240), at 34.
376 Vigni (note 253), at 501.
377 Wolfrum (note 310), at 145–416.
378 Lagoni (note 240), at 34–35.

Treaty and the ATS. It is neither necessary nor prudent to delimitate maritime zones in Antarctica, at least at present. As the legal status of the Antarctic continental margin is not the common heritage of the mankind, mineral resources exploration at that area do not fall into the jurisdiction of the ISA. Therefore, ATCPs are authorized to establish the Antarctic minerals regime under the ATS, a legitimate jurisdiction proven by the adoption of the Convention on the Regulation of Antarctic Mineral Resource Activities (CRAMRA), despite its not yet entering into force. At the very least, a clear line of authority on Antarctic mineral resource activities has been drawn between the ISA under UNCLOS and the ATCPs under the ATS.

In conclusion, all these options could be summarized into two groups, which are under the control of either the UNCLOS or the ATS. If UNCLOS is the appropriate legal system, the logical consequence would be that the International Seabed Authority (ISA) will regulate the exploitation of the Antarctic continental shelf the same way as it does deep seabeds. However, conflicts between ISA and Antarctic Treaty Consultative Parties (ATCPs) eclipse this opportunity, as the latter will never give up its authority to govern the Antarctic issues. It is therefore, extremely unlikely that the ATCPs would accept the common heritage status for Antarctic continental shelf. Considering legal, political and economic aspects, a convention with regard to mineral resources activities in Antarctica concluded by the ATCPs and applying to the scope of the ATS might be an ideal solution. Although CRAMRA has failed to enter into force, and the Protocol prohibits mineral activities for fifty years,[379] there still is a huge possibility for mineral resource exploration in future. The only feasible method of regulating mining activities, which seem inevitable, is to establish a minerals regime specific to Antarctic mineral activities under the Antarctic Treaty, as done in CRAMRA.

E. The Antarctic Treaty and Antarctic Treaty System

I. The Antarctic Treaty

The improvement of weapon technologies during World War II led to advances in peacetime scientific research afterwards. A formal proposal for an international year of coordinated intensified investigations during the maximum of solar activity in 1957–1958 was made during the Joint Commission of the Ionosphere of the International Council of Scientific Unions in 1950. Several organizations, associations and commissions planned and provided finance for IGY. Ultimately, the IGY included 67 States and about 4000 stations.[380] Among these stations, 55 were operated by

379 Article 7 The Antarctic Environmental Protocol.

380 Berkman, Paul Arthur, *Science into Policy: Global Lessons from Antarctica*, (California, Academic Press, 2002), at 54.

twelve States[381] who are the original signatory States of the Antarctic Treaty.[382] The success of the research cooperation during IGY led to the establishment of Scientific Committee on Antarctic Research (SCAR) in 1958.[383] The purpose of the IGY was to encourage international cooperation in scientific research in the Antarctic region and to resolve political tensions.[384] A "gentleman's agreement" was entered into by all participants, stating that no actions during the International Geophysical Cooperation would be used as the basis of a territorial claim.[385]

The positive outcomes of the IGY prompted further negotiation on an international treaty on the Antarctic. The United States proposed on May 2, 1958 to other Sates actively interested in Antarctica that they should sign a treaty "designed to preserve the continent as an international laboratory for scientific research and ensure that it be used only for peaceful proposes".[386] The conference on Antarctica was held through October 15 to December 1, 1959, and ended by signing the Antarctic Treaty, which entered into force on June 23, 1961.

The Antarctic Treaty (the Treaty) was a unique success in alleviating political unrest in the Antarctic in the midst of the Cold War.[387] The Treaty was signed in Washington on 1 December 1959 and entered into force on 23 June 1961. Through it, concessions were made by twelve States who held opposing views and interests in Antarctic territory. Currently, the Treaty consists of 50 parties representing many more different positions on Antarctica, however, they all agree that Antarctica shall be used for peaceful purposes only[388] and be demilitarized through prohibiting nuclear explosions and the disposal of nuclear waste[389]. In order to promote international scientific investigation and cooperation,[390] information of scientific

381 They were Argentina, Australia, Belgium, Chile, France, Japan, New Zealand, Norway, South Africa, United Kindom, USA and USSR.

382 Bilder, Richard B., "Control of Criminal Conduct in Antarctica", 52 Va. L. Rev. 231 (1966), at 234.

383 Trolle-Anderson, Rolph, "The Antarctic Scene: Legal and Political Facts", in Triggs, Gillian D. (ed.), *The Antarctic Treaty Regime: Law, Environment and Resources*, (Cambridge: Cambridge University Press, 1987), 57–64 at 58.

384 Myhre (note 134), at 31.

385 Myhre (note 134), at 31; Westermeyer, William E., *The politics of mineral resource development in Antarctica*, (Boulder: Westview Press, 1984), at 15.

386 Cited in Bergin (note 156), at 21; Shusterich (note 106), at 805.

387 Shusterich (note 106), at 806; Trolle-Anderson (note 383), at 59; Auburn (note 161), at 85; Sollli, Finn, "The Development of the Antarctic Treaty: Trends and Issues", in Wolfrum, Rüdiger (ed.), *Antarctic Challenge II*, (Berlin: Duncker & Humblot, 1984), 17–37, at 25. Even though there are still some opposing opinions about the Antarctic Treaty. One of these negative comments is that the Treaty is a temporary solution but not the definite answer to the Antarctic question. See: Quigg (note 201), at 153; Parriott (note 152), at 90–93.

388 Preamble and Article I Antarctic Treaty.

389 Article V Antarctic Treaty.

390 Article II Antarctic Treaty.

programs, scientific personnel and scientific observations and results shall be exchanged to the greatest extent feasible and practicable by Contracting Parties.[391] Dr. Paul Berkman of the Scott Polar Research Institute commented: "The Antarctic Treaty is a very elegant document. It's only 14 articles long and yet those 14 articles have effectively managed ten percent of the Earth for peaceful purposes alone for the past 50 years."[392] The Treaty has established a series of regulations for conduct on Antarctica through The Antarctic Treaty Consultative Meeting (ATCM).

The Treaty achieved the four main aims sought by the US.[393] Firstly, the Treaty is mainly focussed on "recognizing that it is the interest of all mankind that Antarctica shall for ever to be used exclusively for peaceful purposes..."[394] In addition, Article I explicitly declares that "Antarctica shall be used for peaceful purpose only", which is achieved by the blanket ban on "military installations, weapons tests and military manoeuvers". However, in paragraph 2, military personnel or equipment are allowed to be used for scientific research or for other peaceful purposes. The historical background of this exception stems from the fact that many early expeditions of Antarctica were conducted by the military. The principle of peaceful use of Antarctica is enhanced by Article VII, relating to observers and inspection of "all stations, installations and equipment..., and all ships and aircraft" within all areas of Antarctica.[395] This article is a huge historical milestone for the Cold War era, since both the US and the USSR agreed that observers could inspect the installations of the other. However, such inspection would not truly have a substantive effect, as the severe climatic conditions in the Antarctic would mean that advance warning is necessary, which would leave a backdoor for violation of the terms of the Treaty.[396] In addition, Article V made Antarctica the first continent in the world without nuclear weapons which does not ban the presence of nuclear power in Antarctica, e.g. American McMurdo station had a fission reactor,[397] but nuclear explosions and disposal of radioactive waste material is totally forbidden.

Secondly, Antarctica should be open and free for scientific activities. The Preamble and Article II express agreement among signatories that the cooperation of scientific research began during the IGY should be extended. Article III goes on to state that the cooperation would be developed by "the greatest extent feasible and practicable" exchange of information, scientific personnel and scientific observations and results from Antarctica.

---

391 Article III Antarctic Treaty.
392 See the website of the British Antarctic Survey at: http://www.antarctica.ac.uk//about_antarctica/geopolitical/treaty/anniversary.php.
393 Plott, Barry Merrill, *The Development of United States Antarctic Policy*, (Fletcher School of Law and Diplomacy, 1969), p. 197.
394 Preamble Antarctic Treaty.
395 Wolfrum (note 181), at 63.
396 Myhre (note 134), at 37.
397 Wolfrum (note 181), at 68.

Thirdly, Antarctica should be free from political tension. Territorial questions have been key in conflicts between claimants, such as Argentina, Chile and UK and between non-claimants and claimants, however these questions have been frozen by the most important and controversial Article of the Treaty is Article IV.[398] It freezes territorial claims and puts off solving the claim issue, because participants' interests are in conflict in the negotiations preceding the ratification of the Treaty. In the non-claimant group, with the exception of the USA and USSR, who reserved their right to claim territorial sovereignty in the future, others resolutely opposed any claim. The range of views in the claimant group is revealed by examining the range of suggestions: New Zealand was prepared to relinquish her claim in favour of an international regime; however, Chile clearly rejected any form of internationalization; Argentina objected to the institution of regimes or the creation of any specific structure.[399] Since it was not easy to reconcile these varied interests at that time, and some form of agreement between these parties was necessary, the treaty sought to set the problem aside by freezing existing positions and prohibiting new claims until the treaty expires.

According to Article IV (1), the Treaty shall not be interpreted as affecting 'previously asserted rights of or claims to territorial sovereignty in Antarctica',[400] nor as a 'renunciation or diminution by any Contracting Party of any basis of claim to territorial sovereignty in Antarctica which it may have whether as a result of its activities or those of its nationals in Antarctica'[401]. The former subsection of this Article appears to protect the interests of seven claimant States, and the latter one is for potential claimants, i.e. the USA and the USSR. The Article then continues to announce that the Treaty shall not be interpreted to prejudice the position of any Contracting Party regarding "its recognition or non-recognition of any other States right of or claim or basis of claim to territorial sovereignty in Antarctica"[402]. This subsection tries to protect contracting parties' adherence to the Treaty by clearly stating that recognition of the Treaty does not constitute an official recognition of other States' claims.[403]

The second paragraph of Article IV denies that contracting parties' activities during the period the Treaty is in force "constitute a basis for asserting, supporting or denying a claim to territorial sovereignty in Antarctica or create any rights of sovereignty in Antarctica", and prohibits any "new claim, or enlargement of an existing claim, to territorial sovereignty in Antarctica" while the Treaty is in force.

---

398 Myhre (note 134), at 36; Trolle-Anderson (note 383), at 59; Fox, Hazel, "The Relevance of Antarctica to the Lawyer", in Triggs, Gillian Doreen (ed.), *The Antarctic Treaty regime: Law, Environment and Resources*, (Cambridge: Cambridge University Press, 1987), 77–87 at 81.

399 Myhre (note 134), at 36.

400 Article IV (1)(a) Antarctic Treaty.

401 Article IV (1)(b) Antarctic Treaty.

402 Article IV (1)(c) Antarctic Treaty.

403 Auburn, (note 161), at 106.

Article IV removed the most important obstacle to the Treaty by suspending sovereignty disputes.[404] It has served as the treaty's *modus Vivendi* to push the negotiations on related Antarctic issues to completion.[405] It is, however, doubtful that Article IV will be able to preserve the *status quo* of Parties prior to the Treaty.[406] Firstly, according to the principle *pacta tertiis nec nocent nec prosunt*, a treaty concerns contracting parties only, not binding on third States.[407] It has been a "general rule" that "a treaty does not create either obligations or rights for a third State without its consent."[408] The obligations under Article IV (2) cannot prevent States who are not parities of the Treaty from asserting a new claim or from enlarging an existing claim. Similar problems will happen once a contracting party has withdrawn from the Treaty. Under this situation, that party does not need to fulfil any obligation contained in the Treaty.[409] Secondly, the Parties can continue to use their Antarctic Bases as a foundation for a sovereignty claim in the event that the Treaty ends.[410]

Article IV, an agreement to disagree[411], does reconcile the clash between claimant States and non-claimant States; although it leaves the basic question unsolved. It is not possible that the freezing phrase will last forever. What will happen when sovereignty claims are released from the restraint of the Treaty, especially as resources in Antarctica are attracting more and more international interest? There appear to be three possibilities. Firstly, the legal situation returns to the *status quo ante*. This situation is the original intention which the draftsmen wanted to achieve, and is the most vulnerable. Secondly, as time goes by, claimants will eventually give up their claims. This is an ideal and unrealistic presentation of human nature, and international history does not paint a wholly altruistic picture. The final alternative involves a reasonable and popular resolution, such as the adoption of a CRAMRA or CRAMRA-like regime elaborated under the Antarctic Treaty System that would be acceptable to the international community.

Fourthly, a system of management and consultation among the interested States should be established. To achieve this aim, Article VI defines the area of application of the Treaty as the area south of 60°S, including ice shelves, and it also requires that rights of the high seas should be respected by all States. Article X binds signatories to ensure that no one violates principles and intents of the Treaty. In order to facilitate States' management, Article VIII demands that observers, scientific personnel and members of their staff are under the jurisdiction of the State of which they are nationals. Article XI requires that peaceful means are used for dispute resolution. In

---

404 Beck, Peter J., *The International Politics of Antarctica*, (New York: St. Martin's Press, 1986), at 116.
405 Joyner (note 68), at 64.
406 Triggs (note 236), at 89.
407 Oppenheim (note 142), at 831; Brownlie (note 144), at 622.
408 Article 34Vienna Convention on the law of Treaties.
409 Parriott (note 152), at 91.
410 Triggs (note 236), at 89.
411 Auburn (note 161), at 109; Joyner (note 68), at 63.

addition, according to Article XIII the Treaty is open to all UN members and other approved States, and all States signing or acceding to the Treaty are Contracting Parties. However, according to Article IX, Antarctic Treaty Consultative Parties (ATCPs), who are among the initial twelve signatories, and who can "demonstrate its interest in Antarctica by conducting substantial research activity there, such as the establishment of a scientific station or the despatch of a scientific expedition"[412] can make decision in the Antarctic Treaty Consultative Meeting (ATCM), which is the machinery to implement and adjust the Treaty. There are currently 29 consultative parties including the twelve original signatories.[413] For the ATCM, the rules are different for the two State groups of ATCPs, the twelve original signatories and all other interested States. For the first group, their consultative status is granted automatically by paragraph 1 of Article XI, they can only lose it by amendment to the Treaty; the second group States can be stripped of such status if they cease their scientific activities in Antarctica. Therefore, the twelve original signatories possess consultative status independent on their scientific research (for example Norway has not undertaken scientific research for a long time since the end of the IGY, but is still an ATCP[414]), while all other ATCPs must continue such activities to keep their position in the ATCM.[415] In addition, according to Article IX, all the members of the Antarctic Treaty could be divided into three groups: the original twelve signatories; ATCPs, who demonstrate their interest in Antarctica and are granted ATCPs status; and other States. Only the first two groups of States may cooperate in making decisions and performing inspections, while the last group of States are excluded from those activities.[416]

Except ATCPs, ATCM is also open to non-Consultative Parties; three Observers – Scientific Committee on Antarctic Research (SCAR), the Commission for the Conservation of Antarctic Marine Living Resources and the Council of Managers of National Antarctic Programs (COMNAP); and invited Experts[417]. All these participants contribute to the discussions, but only the ATCPs take part in decision-making.

---

412 Article IX (2) Antarctic Treaty.

413 Some States criticised restricting decision-making to the Consultative Parties, because they cannot afford the high entrance fee required for Consultative Parties. See: Triggs (note 236), at 95.

414 Hanevold, Truls, "Inspections in Antarctica", 6 Cooperation and Conflict 103 (1971), at 110.

415 Wolfrum (note 181), at 70–71; Lagoni (note 240), at 8; Myhre (note 134), at 40.

416 Wolfrum (note 181), at 71.

417 Experts are normally from: the Antarctic and Southern Ocean Coalition (ASOC), the International Association of Antarctica Tour Operators (IAATO), the International Hydrographic Organization (IHO), the Intergovernmental Oceanographic Commission (IOC), the International Maritime Organization (IMO), the United Nations Environment Program (UNEP), the International Union for Conservation of Nature (IUCN), the World Meteorological Organization (WMO), and the World Tourism Organization (WTO).

From 1961 to 1994 the ATCM generally met once every two years, but since 1994 the meetings have occurred annually. The ATCM was designed to increase communication of parties and ensure the smooth operation of the Treaty including: "a. use of Antarctica for peaceful purposes only; b. facilitation of scientific research in Antarctica; c. facilitation of international scientific cooperation in Antarctica; d. facilitation of the exercise of the rights of inspection provided for in Article VII of the Treaty; e. questions relating to the exercise of jurisdiction in Antarctica; f. preservation and conservation of living resources in Antarctica."[418]

In order to implement the principles of the Antarctic Treaty and the Environment Protocol and to provide regulations and guidelines for the management of the Antarctic Treaty area and the work of the ATCM, Measures, Decisions and Resolutions could be made at the ATCM by consensus.[419] Decisions address internal organizational matters of the ATCM and measures are legally binding on the Consultative Parties once they have been unanimously approved. Resolutions however are intended as hortatory texts and as such are not legally binding. All these agreements have a role to play in updating and supporting the Treaty and solving problems encountered in Antarctica. Before 1 September 2004, the Antarctic Treaty Secretariat[420] was established in Buenos Aires, Argentina, the ATCM had been the "sole forum for administering the Antarctica".[421]

II. Antarctic Treaty System

The primary purpose of the Antarctic Treaty is to ensure that "in the interests of all mankind that Antarctica shall continue forever to be used exclusively for peaceful purposes and shall not become the scene or object of international discord."[422] The fact that the Antarctic has been modified by human activities as technology and the economy have developed, compounded by the fact that the Treaty alone is not enough to coordinate complicated relations among States, has led to the creation of a number of agreements with respect to Antarctica. The ATCPs have been concerned

418 Article IX (1) Antarctic Treaty.

419 Article IX (4) Antarctic Treaty. Some may criticise that it is difficult to achieve an agreement based on consensus decision, and decisions would based on the lowest common denominator. However, the facts show the opposite. Once a decision is reached, it means that all Parties approve it and will implement it. See: Trolle-Anderson (note 383), at 61; Auburn (note 161), at 104.

420 The tasks of Secretariat were divided into the following: supporting the annual ATCM and the meeting of the Committee for Environmental Protection (CEP); facilitating the exchange of information between the Parties required in the Treaty and the Environment Protocol; collecting, storing, archiving and making available the documents of the ATCM; providing and disseminating information about the Antarctic Treaty system and Antarctic activities.

421 Myhre (note 134), at 42.

422 Preamble Antarctic Treaty.

with protecting native plants and animals from the beginning of the ATCM.[423] In 1964, Agreed Measures for the Conservation of Antarctic Fauna and Flora (the Agreed Measures) were adopted, to be followed by many other measures dealing with related issues in subsequent ATCMs. Then, ATCPs focused their attention on a specific aspect of conservation of marine living resources, the protection of seal species. In 1972, the Convention for the Conservation of Antarctic Seals (CCAS) was adopted in London and entered into force in 1978, which effectively put an end to commercial sealing. In order to preserve the Antarctic ecosystems, the Convention for the Conservation of Antarctic Marine Living Resources (CCAMLR) was initiated to in 1978, concluded in 1980, and entered into force in 1982. This aimed to prevent an increase in krill catches in the Southern Ocean, which could have a serious effect on populations of krill and damage the food chain for other marine species, such as birds, fish and seals that mainly depend on kills for food. So far, the Antarctic Treaty System had made considerable progress in this aspect of protecting living resources. Moreover, in 1980s, negotiations on the Antarctic minerals regime were convened and the Convention on the Regulation of Antarctic Mineral Resource Activities (CRAMRA[424], Wellington) was adopted in 1988. However, CRAMRA has not yet entered into force, and has been frozen by the Protocol on Environmental Protection to the Antarctic Treaty (The Protocol, Madrid, 1991). All agreements together, of which the Treaty is the core, compose the Antarctic Treaty System. Regarding the issue of sovereignty, other agreements in the Antarctic Treaty System, i.e. CCAS and CCAMLR, also insist on the *status quo* regime because Article IV of the Treaty provides a mechanism by which territorial claims are voided in subsequent conventions.[425]

The Agreed Measures for the Conservation of Antarctic Fauna and Flora apply to the same area as the Antarctic Treaty, namely south of 60° South Latitude, and respects the right of the high seas under international law.[426] Expect for special purposes for human living, study and education, the Agreed Measures prohibit disturbing, harvesting or destroying native species, and ban bringing in non-indigenous species.[427] Moreover, "Specially Protected Areas" (SPAs) and "Specially Protected Species" (SPSs) were designated and accorded special protection by the joint action of the parties as a whole.[428] The issues of the "special protected" area and species were the subject of lively discussion at the ATCM VI, which decided fifteen regions[429]

423 Recommendation I-8 "recognizes the urgent need for measures to protect living resources from uncontrolled destruction by man". The same concern was expressed by Recommendation II-2.
424 The dissertation uses the terms "CRAMRA", "minerals regime" and "minerals convention" synonymously.
425 Triggs (note 236), at 89.
426 Article I Agreed Measures.
427 Article VI, VII and IX Agreed Measures.
428 Article VI and VIII Agreed Measures.
429 Recommendation ATCM IV 1 to 15 (1966).

and two species[430]. According to Article IX of the Antarctic Treaty, the Agreed Measures would be effective when approved by all the ATCPs at that time. This agreement took a long time, particularly the recommendations regarding the SPAs and SPSs. Finally, almost twenty years later, in 1982 the Agreed Measures finally became effective. However, as the Agreed Measures are domestic regulations rather than convention under the ATS, they are not open but only apply to the contracting parties of the Antarctic Treaty. (More discussion can be found in Part 4.A.III.)

Antarctic seal populations were in seriously depleted by the 1820s, but recovered during World War II due to the suspension of economic exploitation, which had been the key driver of seal hunting.[431] In 1964, the Third ATCM suggested that parties should voluntarily regulate sealing activities at the national level.[432] In 1966, Recommendation ATCM IV-16 and IV-17 designated the fur seal and the Ross seal as Specially Protected Species. Since the Agreed Measures, like other provisions of the Antarctic Treaty, only apply to the land and floating ice-shelves but not to the sea around the continent, it was necessary to set up a regime for marine conservation. In 1972 the Convention for the Conservation of Antarctic Seals was singed in London. The CCAS applies to some stocks in the seas south of 60° South Latitude and all species of seals in Antarctic.[433] It recognizes that the stocks of Antarctic seals are an important living resource in the marine environment, thus, this resource cannot be over exploited.[434] Hunting has been banned for three species, and the hunting of other three abundant species is limited in strict catch quantities.[435] Moreover, it elaborates closed seasons, sealing zones and seal reserves, and requires that sealing methods should be quick, painless and efficient.[436] As knowledge and technology have developed, the CCAS provides opportunity to make further suitable regulations in the future, and many measures indeed have been adopted by ATCMs. (More discussion can be found in Part 4.A.IV.)

In 1970s, an upswing in krill fishing rang alarm bells for potential damage to the Southern ocean ecosystems, since the food chain in Antarctic is quite short and krill plays a fundamental role in it.[437] CCAMLR is concluded to safeguard the environment and conserve the integrity of the ecosystem in the seas surrounding Antarctic, but it does not prohibit harvesting in a rational manner. It is no doubt

430 Recommendation ATCM IV-16, Specially Protected Species: Fur Seals; and Recommendation IV-17, Specially Protected Species: Ross Seal, 1966.

431 Couratier, Josyane, "Regime for the Convention of Antarctica's Living Resources", in Vicuña, Francisco Orrego edited, *Antarctic resources policy: Scientific, legal and political issues*, (Cambridge: Cambridge University Press, 1983), 139–149 at 143.

432 Recommendation ATCM III-XI, 1964.

433 Article 1 CCAS.

434 The Preamble CCAS.

435 Paragraph 1 and 2 of Annex to CCAS.

436 Paragraph 7 (1) of Annex to CCAS.

437 Couratier (note 431), at 145.

that CCAMLR is a milestone in the development of the Antarctic Treaty System;[438] however, it is more a philosophical scientist's convention than a convention for fisheries managers, thus, its practical effectiveness still needs to be improved.[439] CCAMLR has three main characters. Firstly, it extends the area of application of the Antarctic Treaty. Secondly, States may access the CCAMLR by participating in its institutions. Thirdly, it applies the precautionary approach and ecosystem approach to the maintenance of living resources. (More detailed discussion about CCAMLR could be found in the following Part 4.A.V.)

In conclusion, the Antarctic Treaty is a huge triumph in keeping the peace in an international region in the cold-war era. It freezes the seven existing claims over territories in the Antarctic and prohibits any new claims. However, this compromise between the claimants and non-claimants cannot solve the long-term issues of Antarctic legal status. The unresolved sovereignty issue has been influencing and complicating other issues related to Antarctic, especially the issues of Antarctic mineral resource activities. Theoretically, after the Treaty was adopted and entered into force, all States in the world could be divided into three groups: States with territorial claims; States without territorial claims who are parties of the Antarctic Treaty; States who have neither claim nor are parties of the Antarctic Treaty. The different interests of these groups provide different standing points for the legal status of Antarctic. It is not easy to reconcile the various attitudes and to conclude agreements on substantial issues, which might involve or effect the Antarctic sovereignty claims. Therefore, Antarctic mineral resource exploration, as a sensitive and controversial issue, has been confronted with enormous difficulties and challenges as a result of the complicated situation which has been built up since the Treaty was born.

438 Shusterich (note 106), at 811–812.

439 Bonner, W.N., "Recent developments in Antarctic Conservation", in Triggs, Gillian Doreen (ed.), *The Antarctic Treaty regime: Law, Environment and Resources*, (Cambridge: Cambridge University Press, 1987), 143–149 at 145. See also: De la Mare, W.K., "CCAMLR's Management Task and the Definition of its Operational Objectives", in Kock, Karl-Hermann (ed.), *Understanding CCAMLR's Approach to Management* (2000) <http://www.ccamlr.org/pu/E/e_pubs/am/am-all.pdf>, at 29; Commission for the Conservation of Antarctic Marine Living Resources, *CCAMLR's Management of the Antarctic*, p. 20.

Part 2: CRAMRA

Despite being mentioned at the Antarctic Treaty Conference, the original Atlantic Treaty does not deal with mineral exploration and exploitation.[440] CRAMRA, as a component of the Antarctic Treaty System, fills a gap in the system. CRAMRA covers mineral resources, meaning all non-living natural, non-renewable resources, including fossil fuels, metallic and non-metallic minerals.[441] Although it failed to enter into force, it is still worth intensive study to draw out the lessons which might be useful for a future mineral resources regime. CRAMRA not only provides the general principles for mining activities in Antarctica, it also establishes institutions and regulates operational provisions.

A. Background

I. Preface to Mining Negotiations

The issue of Antarctic mineral resources was first considered by the VI Antarctic Treaty Consultative Meeting (hereafter ATCM) in 1970; however, due to a lack of the necessary technologies and economic support, it was only raised at an informal session.[442] That it was raised at all can be partially attributed to claims by several representatives that their governments had received inquiries regarding Antarctic mineral resources from private firms.[443] The fresh and sensitive nature of the subject

440 May 30- June 10, 1973, Nansen Foundation sponsored an expert meeting on the issue of whether possible mining activities in Antarctic would be compatible with the Antarctic Treaty. The Result was that such activities would violate the Antarctic Treaty. Nansen Foundation, Antarctic Resources, Report from the Meeting of Experts, reprinted in *U.S. Policy with Regard to Mineral Exploration and Exploitation in the Antarctic: Hearing before the Subcommittee on Oceans and International Environment of the Senate Committee on Foreign Relations*, 94th Congress, 1st Session 68 (1975). See: Wolfrum, R., *Die Internationalisierung staatsfreier Räume*, (Springer-Verlag, Berlin Heidelberg New York Tokyo, 1984), at 85; Auburn, F.M., *Antarctic law and politics*, (London: C. Hurst & Company, 1982), at 256. However, an opposite opinion was expressed in Francioni, Francesco, "Legal Aspect of Mineral Exploitation in Antarctica", 19 Cornell Int'l L. J. 163 (1986), at 165.

441 Article 1 paragraph 6 CRAMRA. Other living and nonliving resources have been covered by the Seals Convention, CCAMLR and the Agreed Measures for the Conservation of Antarctic Flora and Fauna.

442 At the VI ATCM in 1970, the first informal discussion considered possible interest in mineral development in Antarctica and the need to develop a regime that would prevent disorder in the Antarctic.

443 In the late 1960s, the Australian, New Zealand, U.K. and U.S. government were approached by commercial entities who were interested in prospecting in Antarctica.

may go some way to explaining why the Report and Recommendations of the VI ATCM do not address the mineral issue.[444]

The VII ATCM, first put the topic "Antarctic Resource – Effects of Mineral Exploration" on the meeting's agenda in 1972. This meeting also saw the adoption of Recommendation VII-6[445], at the urging of Chile, France and the UK who stated that it was necessary to consider relevant mineral exploration and exploitation questions. This Recommendation acknowledged that technological developments in polar mineral exploration had increased the possibility of Antarctic mineral resources activities, and raised a risk of a degeneration of the Antarctic environment.

Following the VII ATCM, four significant events took place that increased potential interest in Antarctic mineral resources exploration. Firstly, in 1972–1973, the *Glomar Challenger* drilled in the Ross Sea and provided the first evidence of hydrocarbons in Antarctica, not only from the continent but also on the Antarctic continental shelf.[446] Secondly, in 1973, the *Glomar Challenger* discovered traces of methane and ethane, normally associated with the presence of oil resources, in the Ross Sea continental shelf. The third driver was the massive hike in the price of oil that resulted from the Organization of Petroleum Exporting Countries (OPEC) oil embargo. Fourthly, on 17 December 1970, the UN General Assembly decided to convene a third conference on the law of the sea in 1973, and instructed the Committee on the Peaceful Uses of the Sea-Bed and the Ocean Floor beyond the Limits of National Jurisdiction to act as preparatory body for the conference.[447] There was

---

New Zealand government received an application to explore, prospect and develop an area of the Ross Sea. In the U.S. in 1970, Texaco inquired how to obtain a licence for oil exploration on the Antarctic continental shelf. Moreover, in late 1970s, three U.S. oil companies, Atlantic Richfield, Gulf and Esso, were represented on the U.S. Antarctic Advisory Committee. See: Mitchell, Barbara / Kimball, Lee, "Conflict over the Cold Continent", 35 Foreign Policy 124 (Summer, 1979), at p. 130; Colson, David A., "The Antarctic Treaty System: The Mineral Issue", 12 Law & Pol'y Int'l Bus. 841 (1980), at 884.

444 More detailed analysis please see Colson (note 443), at 884–885.

445 Recommendation VII-6, Antarctic Treaty Report of Seventh Consultative Meeting, Wellington, 3 October-10 November, 1972. In Bush, W.M., *Antarctica and International Law: a Collection of the Interstate and National Documents II, (New York: Oceana Publications, 1982)*, at 275–276. In this meeting, SCAR was invited to assess the possible impact of mineral activities on the environment; and a special preparatory meeting was provided which was convened in 1976.

446 See Spivak, J, "Now the Energy Crisis Spurs Idea of Seeking Oil at the South Pole", Wall Street Journal, 21 Febuary 1974, at 1. Also in Lucas, P.H.C., "Finding Ways and Means to Conserve Antarctica", in McNeely, Jeffrey A. / Miller, Kenton (ed.), *National Parks, Conservation, and Development: The Role of Protected Areas in Sustaining Society*, (Smithsonian Institution Press, 1984), 369–374 at 372.

447 UN Resolution 2750 C (XXV) at: http://legal.un.org/diplomaticconferences/lawofthesea-1982/docs/res/a_res_2750_xxv.pdf.

some trepidation that the new law of the sea and regime of the seabed would impair the capacity of the ATS to govern the Antarctic mineral issue. Thus, in 1975 the issue of mineral resources was reconsidered in the VIII ATCM, which decided to convene a special Preparatory Meeting before the IX ATCM. The issue of mining was the discussed at each subsequent ATCM.[448]

The VIII ATCM concluded Recommendation VIII-14, based on Recommendation VII-6 which developed new elements for consideration as part of the mineral issue,[449] and established a group of experts who submitted a report on mineral exploration and exploitation in the Antarctic.[450] Although there was lack of concrete proof that significant deposits existed south of latitude 60° South, the Group agreed that exploration for hydrocarbons on the continental margins around Antarctica was foreseeable, with a potential for longer-term commercial exploitation, although land exploration for mineral resources was not likely for the foreseeable future.[451] The Group could not achieve a consensus on how far technological developments might permit seabed mining in the Antarctic, stating that further detailed geological and geophysical studies were necessary to investigate environmental factors before any exploratory drilling was undertaken.[452] To safeguard the unique Antarctic environment and ecosystems, the Group suggested establishing not only international agreements on safety at sea, pollution from shipping, dumping at sea and other marine environmental matters but also the adoption of special rules specific to the demanding climatic conditions of the Antarctic. It was clear that Antarctic mining activities would require focussed attention to safety precautions, both in the design of the equipment and installations and in its operation, meaning that measures for the protection of the Antarctic environment should be worked out prior to any commercial exploration or exploitation of mineral resources. Mining activities in the Antarctic would require both the operations themselves and any consequential changes in the environment to be monitored. Additionally, a warning system would be required that could provide immediate warning of an accident leading to significant pollution. The system would also have to monitor the dispersion and effects of the pollutant released, along with the effectiveness of any measures for containment or recovery, although this would be extremely difficult under Antarctic

448 For an excellent assessment of these ATCMs, see Colson (note 443), at 841–902.

449 Recommendation VIII-14 states that the purposes and principles of the Antarctic Treaty should be considered; it notes the technological developments in polar mineral exploration and exploitation and the adverse effect on the environment and ecosystems of the Antarctic; it also provides governments some procedures to follow, such as fully studying the questions relating to the mineral issue, and a special preparatory meeting should be held on the IX ATCM. http://www.ats.aq/documents/ATCM8/fr/ATCM8_fr001_e.pdf, at 44–45.

450 Report of the IX ATCM, London, 19 September – 7 October, 1977, at 8. http://www.ats.aq/documents/ATCM9/fr/ATCM9_fr001_e.pdf

451 Report of the Ninth ATCM, p. 61.

452 Report of the Ninth ATCM, p. 58.

conditions.[453] As oil pollution might be caused by human error and no technology can eliminate this, the Group demanded that special attention should be given to thorough training of technical personnel and to the development and enforcement of strict codes of conduct governing drilling operations.[454]

In 1976, in accordance with the Recommendation VIII-14, the Special Preparatory Meeting was held in Paris to consider "Antarctic Resources – The Question of Mineral resource Exploration and Exploitation".[455] This meeting created two working groups, one working on technical and environmental questions, and the other working on legal and political aspects of mineral development. The latter group set out four basic principles[456] that were incorporated into Recommendation IX-1 at the IX ATCM in 1977 and subsequently carried over into Recommendation XI-1, to become the basis for negotiation on the minerals regime. Recommendation XI-1 respected the Antarctic Treaty system and emphasized the ATCPs' responsibility to protect the Antarctic environment. It also recommended governments of ATCPs keep studying the relationship between Antarctic environmental protection and mineral exploration, and pledged to convene a meeting to consider legal and political aspects of mineral resource issues at the X ATCM in 1979.

By the X ATCM, the ATCPs had gathered sufficient confidence to construct a regime to monitor and control Antarctic mineral resource exploitation under the Antarctic Treaty System.[457] Recommendation X-1 affirmed that resolving the mineral issue would strengthen the regime established by the Antarctic Treaty.[458] It specified that the mineral regime should involve a means of assessing the possible environmental impact of mineral resource activities, determining whether approval should be granted, and governing all aspects of those activities if approved, including the establishment of rules relating to the protection of the Antarctic environment, with which all mineral resource activities undertaken pursuant to the regime must

453 Report of the Ninth ATCM, p. 69.

454 Report of the Ninth ATCM, pp. 69–70.

455 Report of the Eighth ATCM, pp. 44–45.

456 Report of the Ninth ATCM, p. 12. These principles are: i) the Consultative Parties will continue to play an active and responsible role in dealing with the question of the mineral resources of Antarctica; ii) the Antarctic Treaty must be maintained in its entirety; iii) protection of the unique Antarctic environment and of its dependent ecosystems should be a basic consideration; iv) the Consultative Parties, in dealing with the question of mineral resources in Antarctic, should not prejudice the interests of all mankind in Antarctica.

457 Colsom (note 443), at 896.

458 Preamble, Report of the X ATCM, http://www.ats.aq/documents/ATCM10/fr/ATCM10_fr001_e.pdf.

comply.[459] Using these basic requirements as a foundation, additional specifications in the mineral regime were dealt with in Recommendation XI-1.[460]

During the early discussion, the potential environmental effects of Antarctic mineral resource activities were the focus of attention, with environmental reports being submitted by several specially formed groups of experts.[461] The potential for damage arising from unregulated mining activities on the Antarctic environment and ecosystems gave the development of a regime on Antarctic mineral resources a greater level of urgency. In 1981, the Eleventh ATCM adopted Recommendation XI-1, to convene a Special Antarctic Treaty Consultative Meeting (SATCM) to elaborate a regime for Antarctic mineral resources. The SATCM was tasked with determining the form of the regime, establishing a schedule for negotiation, taking any other necessary steps for concluding the regime. Recommendation XI-1 provided five core principles for the purposed regime, which stemmed from the 1976 Special Preparatory Meeting. These principles required that: a)the Consultative Parties should continue to play an active and responsible role in dealing with the question of Antarctic mineral resources; b) the Antarctic Treaty must be maintained in its entirety; c)protection of the unique Antarctic environment and of its dependent ecosystems should be a basic consideration; d) the Consultative Parties, in dealing with the question of mineral resources in Antarctica, should not prejudice the interests of all mankind in Antarctica; e) the provisions of Article IV of the Antarctic Treaty should not be affected by the regime. It should ensure that the principles embodied in Article IV are safeguarded in application to the area covered by the Antarctic Treaty.[462]

Recommendation XI also stipulated that "a regime on Antarctic mineral resources should be concluded as a matter of urgency", despite this potentially being misconstrued as a push to expedite commercial Antarctic mineral exploration.[463] It would

---

459 Report of the X ATCM, paragraph 4, at 12–13.

460 Kimball, Lee A., "Special report on: the Antarctic Mineral Convention", in Splettstoesser, J.F. / Dreschhoff, G.A.M., (eds.), *Mineral Rsources Potentional of Antarctica* (Washington, D.C.: American Geophysical Union, 1990), 273–310 at 283.

461 The VIII ATCM request SCAR to establish a Group of Specialists, which produced and submitted to IX ATCM, "A Preliminary Assessment of the Environment Impact of Mineral Exploration /Exploitation in Antarctic" (EAMREA). The Report was considered by a group of experts who met at IX ATCM, and completed the "Holdgate Report" on mineral exploration and Exploitation. In 1979, a preparatory meeting for X ATCM produced the "Report of the Group of Ecological, Technological and Other Related Experts on Mineral Exploration and Exploitation in Antarctica". Recommendation X-1 and X-7 requested additional information from SCAR, which constituted a new Group of Specialists on Antarctic Environmental Implications of Possible Mineral Exploration and Exploitation (AEIMEE).

462 Report of the Eleventh ATCM, at 20. http://www.ats.aq/documents/ATCM11/fr/ATCM11_fr001_e.pdf.

463 Beeby, C.D., "An Overview of the Problems which should be addressed in the preparation of a regime governing the mineral resources of Antarctica", in Vicuña,

appear that fears of sovereignty disputes and potential threats to the ATS were driving the urgent demand for a mining regime in the Antarctic[464], together with a need to fill the gap in the ATS which was responsible for ensuring the protection of living marine resources by managing the non-living resources in the Antarctic. Moreover, the adoption of Recommendation XI-1 launched the Fourth Special Antarctic Treaty Consultative Meeting, which comprised a series of twelve negotiation sessions on the Antarctic mineral regime from June 1982 through June 1988.[465]

Between 1972 and 1981, ATCPs as the administrative mechanism for formulating policy[466] took a decade to consider the mineral issue and did manage to hammer out some key principles. However, it was still a long way from concluding a successful mineral regime in the Antarctic, which would not only require detailed provisions to govern Antarctic mineral activities but which must also balance the interests of the different groups of States in the Antarctic. Furthermore, the mineral regime should ensure that potential mineral resources activities would not disrupt the ATS and other international law systems.

II. Negotiations of Antarctic Mineral Resource Regime

Since the adoption of Recommendation XI-1 established guidelines for the Fourth Special Antarctic Treaty Consultative Meeting, twelve formal negotiating sessions on Antarctic mineral issue were convened between 1982 and 1988. The first and second session were held in New Zealand, in June 1982 and January 1983. By end of the second session, Mr. Christopher Beeby, the Chairman of these negotiations, issued a draft text, referred to as "Beeby I".[467] This personal proposal reflected his views on what had been agreed by the ATCPs. A third session was held in Federal Republic of Germany in July 1983, with the fourth taking place in the U.S. in January 1984. Before the fifth session in Japan in May 1984, "Beeby II" was produced,[468] which revised the original text and served as the main basis for the negotiation throughout

Francisco Orrego edited, *Antarctic resources policy: Scientific, legal and political issues*, (Cambridge: Cambridge University Press, 1983), at 191–198.

464 Beeby (note 463), at 297.

465 The Fourth Special ATCM began its work at Wellington from June 14–25, 1982. Further Sessions were held in Wellington (January 17–28, 1983), Bonn (July 11–22, 1983), Washington D.C., (January 18–27 1984), Tokyo (May 23–31, 1984), Reo de Janeiro (February 26 – March 12, 1985), Paris (September 23 – October 6, 1985), Hobart (April 14 -25, 1986), Tokyo (October 27 – November 12, 1986), Montevideo (May 11–20, 1987), Wellington (January 19–29, 1988) and wellington (May 2 – June2, 1988).

466 Article IX the Antarctic Treaty.

467 "The Antarctic Minerals Regime: The Beeby Draft", reprinted in Greenpeace International, *The Future of the Antarctic: Background for a Second U.N. Debate (1983)*, Appendix 8.

468 "Beeby Draft II", reprinted in Greenpeace International, *The Future of the Antarctic: Background for a Second U.N. Debate (1984)*, Appendix 8.

1984–1985.[469] After the sixth session was held in Brazil in February and March 1985, Mr. Beeby issued a package deal ("Rio-package"), which covered major aspects of structure and decision making.[470] This document was not officially published, and was distributed only to the sixteen ATCPs at that time. The following two sessions were held in France in September and October 1985, with the subsequent meeting being held in Australia in April 1986. In the ninth and tenth sessions, "Beeby III"[471] and "Beeby IV"[472] were discussed separately in Japan in October 1986 and in Uruguay in May 1987. The last two rounds of negotiations were held in New Zealand in January 1988, and in May 1988. On June 2, 1988, the Convention on the Regulation of Antarctic Mineral Resource Activities (CRAMRA) was adopted. All 20 ATCPs plus 13 of 18 non-ATCPs at that time attended the final session.

The negotiations were guided by the series of "Beeby Drafts", which went through six revisions, and gradually affected the process of negotiations despite their lack of legal status. The Chairman collected thoughts and suggestions about the mineral regime, and identified potential points of consensus, and then attempted to transplant them into the text of the convention. The final text and institutional design in particular bear a great similarity to "Beeby III", with the exception of the liability and dispute settlement provisions. Given the influence of these documents, it is worth examining the basis of the Beeby drafts.

1. *Basis of Beeby Drafts*

Firstly, in 1980s, the Antarctic Treaty System could not provide the mineral negotiating sessions a fixed timetable and venue. The lack of a permanent secretariat of the Antarctic Treaty[473] meant the participants had to arrange the negotiations on an *ad hoc* basis, for which one scholarly voice coined the phrase "nomadic negotiations".[474] This frequent foreign travel reduces the effective negotiating time over the course of the negotiations, especially given that some sessions lasted only

469 Joyner, Christopher C., "The Evolving Antarctic Minerals Regime", 19 Ocean Dev. & Int'l L.73 (1988), at 75.

470 "Beeby's Package Deal", 33 ECO (23 September-4 October, 1985), http://www.asoc.org/storage/documents/ECOs/1985/xxxiii.1_amr.pdf.

471 Beeby, C.D., Antarctic Mineral Resources: Chairman's Informal Personal Report: MR/17 RevII (19 September 1986).

472 Antarctic Mineral Resources Chairman's Informal Personal Report: MR/17, Rev. III Annex to Chairman's Informal Personal Report (April 1987) (mimeo).

473 The Secretariat of the Antarctic Treaty was established by Measure 1 (2003), ATCM XXVI-CEP VI, and effective in 2009. http://www.ats.aq/devAS/info_measures_listitem.aspx?lang=e&id=294.

474 Watts, Arthur D., "Lessons from Resource Negotiations", in Wolfrum, Rüdiger (ed.), *Antarctic Challenge III, Conflicting Interests, Cooperation, Environmental Protection, Economic Development : Proceedings of An Interdisciplinary Symposium*, (Berlin: Duncker & Humblot, 1988), 319–332 at 320–322.

two weeks.[475] Under these constraints, the continuous Chairmanship contributed "cohesion" to the negotiations would otherwise have been impossible.[476]

Secondly, Antarctic mineral resource exploitation was a completely new issue to all participants, and it was a sensitive topic relating to States' interests in the Antarctic, making the reaching of consensus in time-limited formal meetings a tough challenge for the participants. To obtain a range of original thinking from different participants, Mr. Beeby convened various informal meetings among groups of representatives, bringing a faint spark of hope to the darkness of concluding a common text for a mineral regime in the Antarctic. By end of the second negotiating session, the first personal proposal for the mineral regime was issued, reflecting the agreements the ATCPs had achieved. From then on, Mr. Beeby and his drafts essentially influenced the whole building process of the Antarctic mineral convention.

2. *Competitions of National Interests*

Since its discovery, the Antarctic has been a target for various national interests, and these interests have influenced the process of mineral regime negotiations. There are four groups of State interests in the Antarctic. The first group comprises the seven claimant States, who demand sovereign rights in their Antarctic sectors. The second group contains two States, the U.S. and USSR, who have a basis for a sovereignty claim in Antarctica. The third group includes all other parties of the ATS, who do not claim sovereignty or recognize any claims in the Antarctic. The final group consists of the states who are not members of ATS. Over more than fifty years of development, the positions of these groups have become clearly entrenched[477], creating two key hurdles for the new mineral resources regime in Antarctica: the internal accommodation of consultative parties who claim sovereign rights in the Antarctic and the other consultative States, and the accommodation of external interests between the ATCPs and the rest of the international community.[478] States outside the ATS would like to treat the Antarctic and its mineral resources as common heritage, under the control of the ISA. However, as claims to the Antarctic have never been denied by the Antarctic Treaty, claimant States will not accept the common heritage concept in the Antarctic mineral regime, and vice versa, non-claimants will definitely refuse any regime which is based on sovereignty under Article IV of the Antarctic Treaty.

---

475 Watts (note 474), at 322.

476 Watts, (note 474), at 323.

477 More details about the four groups of States see: Brennan, Keith, "Criteria for Access to the Resources of Antarctica: Alternatives, procedure and experience applicable", in Vicuña, Francisco Orrego edited, *Antarctic Resources Policy: Scientific, Legal and Political Issues*, (Cambridge: Cambridge University Press, 1983), 217–230 at 217–219.

478 The two accommodations were provided in a personal report by Mr. Busby who was the Chairman of an earlier meeting on Antarctic minerals, in Washington in December 1980. See: Beeby (note 463), at 194; Brennan (note 477), at 219.

3. *History of mineral regime negotiations*

The first two sessions focused on defining potential issues of an Antarctic mineral regime. At this stage, the U.S. submitted its proposal and suggested the following machinery[479]: a general Commission including all participants entitled to take part in decision making; panels covering four regimes into which Antarctica would be divided; and a Scientific and Technical Committee. The U.S. proposal also provided the decision-making process: free prospecting for any qualified applicant; and open areas for exploration by the Commission. Generally speaking, it insisted that a mineral exploitation regime must provide non-discriminatory access to all areas of Antarctica, and the Antarctic would be open and free to exploitation.[480] The rationale behind this proposed regime is that those developed States with advanced economic and technological resources would be in a better position to exploit resources, as it would be impossible for less developed States to truly compete under "equal conditions". This was, of course, rejected by the developing States. Moreover, claimant States, under the illusion of enjoying special rights in their area of claim, also disagreed with the unrestricted regime. This revealed another division, with claimant States expecting the regime to provide them with some privileges; and non-claimant States firmly objecting to anything that tacitly recognized special status.[481] A further division arose between the ATCPs and developing States. The UN, on behalf of the Non-Alignment Movement, was asked to prepare a study on Antarctica and to put the "Question of Antarctica" on the agenda of the 38th Session of the General Assembly.[482] The developing States sought the highest level of participation in Antarctic matters for themselves, and proposed that the "common heritage of mankind" principle and therefore, the regime of deep seabed mining should be applied to the Antarctic.[483] The ATCPs however emphasized that any mineral regime should be a part of the ATS thereby giving them decision-making rights.[484] Although this proposal was not supported widely, CRAMRA still reflected its ideas in many ways.

---

479 Wolfrum, Rüdiger, *The Convention on the Regulation of Antarctic Mineral Resource Activities: An attempt to Break New Ground*, (Heidelberg: Springer-Verlag, 1991), at 6–7.

480 Closon, David A., "The United States Position on Antarctica", 19 Cornell Int'l L. J. 291 (1986), at 297; Joyner, Christopher C, "The Antarctic Minerals Negotiating Process", 81 A.J.I.L. 888 (1987), at 894.

481 Joyner (note 480), at 901–902.

482 See UN Doc. A/38/132 (1983) sec.III, paragraph 122–123. More discussion about the Antarctic and the UN, see: Hayasif, Moritaka, "The Antarctica Question in the United Nations", 19 Cornell Int'l L. J. 275 (1986), at 275–290.

483 Statement of the representative of Pakistan, UN Doc. A/C.1/38/PV.44.

484 Newman, E. Paul, "The Antarctica Mineral Resources Convention: Developments from the October 1986 Tokyo Meeting of the Antarctic Treaty Consultative Parties", 15 Denv. J. Int'l L. & Pol'y 421 (1986–1987), at 424.

Based on the Antarctic Treaty, the Recommendation XI-1, CCAMLR and delegates' proposals, Beeby I (i.e. M/R 17) was issued by end of the second session. This draft did not deal with central provisions, such as application area, responsibility and liability, functions of the Commission and the Regulatory Committees, or settlement of disputes.

After the following two sessions, the Chairman revised his draft to reflect the broader agreements the delegations had reached. This facilitated the general acceptance of Regulatory Committees by the fourth session however, the functions and compositions of this institution required further discussion, as there were still gaps in responsibility and liability, decision making and settlement of disputes in the reviewed draft.

In 1985, a significant improvement, referred to as the "Rio-package", was achieved in the sixth session. The negotiating parties did not adopt the "common heritage of mankind" principle in Antarctic mineral regime, but agreed to insert certain elements into the regime, the Special Meeting of States Parties and the Regulatory Committees. The Special Meeting of States Parties would comprise all parties to the mineral regime and to the Antarctic Treaty. This institution would decide whether to open any area to mineral activities by a simple majority and a consensus of the ATCPs.[485] The Regulatory Committees were intended to comprise 10 members: the US, USSR, the relevant claimant State, three other claimant States and four non-claimant States, however, it provided no guidance as to how participants and "interested non-claimants" would be chosen.[486] The functions of the Regulatory Committees were to set forth the requirements for mineral resource exploration and development in the area opened, and to consider applications for mineral resource exploration, development and management schemes.[487] According to this design, decisions in Regulatory Committees would be made by a "qualified" majority, which means that the claimant States and non-claimant States as a group had a veto, but single claimant State was not accorded a right to veto. This led to dissatisfaction and controversy in the following negotiations, which planted the seed of failure for CRAMRA.

The next four sessions focused on issues such as the authorization and functions of these new institutions, as well as other legal issues. In terms of mechanisms, the Beeby Draft IV improved institutional components of the mineral regime, including the Antarctic Minerals Resources Commission; Special Meeting of States Parties; individual regulatory committees; Scientific, Technical and Environmental Advisory Committee; and a Secretariat. All these bodies had been incorporated by the final convention, with the addition of a Financial Arrangements group. Any

485 "Beeby's Package Deal" (note 470).
486 "Beeby's Package Deal" (note 470).
487 Wolfrum (note 479), at 8.

crucial questions arising in "Rio-package" were discussed in informal consultations conducted by the Chairman.[488]

Although an article-by-article debate had been undertaken since the eleventh session, several substantive questions had still not been solved by the last session, including the application area, inspection, the drafting and approval of the management scheme, relationship between Commission and Regulatory Committees, entry into force of this regime, liability and dispute settlement. The long-term negotiations were completed after the conclusion of the relevant package which dealt with issues relating to the Commission and Regulatory Committee, the management scheme and dispute settlement.[489] On June 2, 1988, CRAMRA text together with a final Act was ultimately adopted.

There were two main issues in the whole negotiation process, those of environmental protection and territorial claims. As Antarctic mineral resources activities should be carried out under universal standards in order to protect the Antarctic environment and ecosystem as a part of the global environment, CRAMRA attempted to protect the Antarctic from the pollution or wastage associated with mineral exploitation.[490] To reconcile various national interests, accommodation of internal and external interests had to be made. Internally, several interest groups needed to be considered: the claimants and those opposing them; the developing and developed States; those most able to exercise mining activities in the Antarctic; and those who just wanted to participate.[491] A "freezing clause" similar to that of the Atlantic Treaty was put forward in the Beeby draft as a means of dealing with the question of sovereignty in Antarctica. This clause should neither reject nor confirm claims in the new regime, nor should potential rights for mining activities give a State the foundation for future territorial claims.[492] The most logical conclusion was to establish an institutional arrangement to administer mineral resources activities and to provide opportunities for continued negotiation on substantive details. CRAMRA was just kind of framework, which needed to be improved through further supplementation.[493] Externally, CRAMRA tried to broaden cooperation in the international community in fields like scientific research, environmental protection and demilitarization. It emphasized that "... the Antarctic shall continue forever to be used exclusively for peaceful purposes" and "... Antarctic mineral resources activities should take into account the interests of the international community as

488 Wolfrum (note 479), at 8.
489 Wolfrum (note 479), at 9.
490 Wolfrum (note 479), at 12; Newman (note 484), at 425; Joyner (note 480), at 895.
491 Wolfrum (note 479), at 13.
492 Article VII, Beeby Draft I, in Greenpeace International, The Future of the Antarctic: Background for a U.N. Debate, App. 8 (1983) In ECO, Special Consultative Meeting on Antarctic Mineral Resources, Bonn, West Germany, July 11–22, 1983, Vol XXIII, No. 1.
493 Wolfrum (note 479), at 14; Wolfrum, Rüdiger (ed.), *Antarctic Challenge: Conflicting Interests, cooperation, Environmental Protection, Economic Development: Proceedings of anInterdisciplinary Symposium* (Berlin: Kumcker & Humblot, 1984), at 220–221.

this may be affected if the membership of the ATCM different from that of the three Commissions provided by the ATS conventions.[502]

The preservation and maintenance of the Antarctic environment as a critical function of the Antarctic mineral regime[503] highlighted the special responsibility of ATCPs. Mineral resources activities in the Antarctic would not only impact on the local environment but also affect dependent and associated ecosystems across the globe. The principle of environmental protection is established through a set of norms. The first step requires an impact assessment prior to any activities taking place.[504] Secondly, a series of preconditions must be met before the Commission can grant approval: no significant adverse effects on air and water quality should be evident; there will be no significant changes to Antarctic environments; no significant changes in or further jeopardy to any ecosystem; no degradation of or substantial risk to the Antarctic; and no significant adverse effects on global or regional climate or weather patterns would be caused by the proposed Antarctic mineral resources activities.[505] In addition, activities will not be permitted until technology and processes are available to cope with accidents and risks; to monitor critical environmental parameters and ecosystem components so as to identify any adverse effects, and to enable an effective response to accidents.[506] These requirements are sufficiently strict to delay actual mining activity in the Antarctic until detailed rules on environment protection could be incorporated into Article 4. Second, any mineral resources activity would be forbidden unless in accordance with CRAMRA and all other rules set up by the institutions of the regime.[507] As unregulated mining activities might adversely affect the environment and ecosystem of the Antarctic, the ATCPs are called upon to refrain from all exploration and exploitation of Antarctic mineral resources for the time being.[508] The requirement to comply with all other institutions and components of the ATS also applies the environmental standards in the Seal Convention and CCAMLR to Antarctic mineral resources activities. In short, nothing may supersede the maintenance of the *status quo ante* of the Antarctic environment.[509]

---

502 Wolfrum (note 479), at 17.
503 Beeby (note 463), at 192.
504 Article 2 paragraph 1 (1), CRAMRA.
505 Article 4 paragraph 1, CRAMRA.
506 Article 4 paragraph 4, CRAMRA.
507 Article 3, CRAMRA.
508 The moratorium was applied on mining activities in Antarctic was applied during the negotiations and has been incorporated into CRAMRA. Had CRAMRA entered into force, the moratorium would be replaced by the Article 3 and 4 paragraph 1 and 2, which provide very strict preconditions for approval of potential mineral resource activities in the Antarctic. Wolfrum (note 479), at 26.
509 Wolfrum (note 479), at 28.

II. Area of Application

Negotiating the area of application for CRAMRA was an arduous process, further complicated by its complex relationship with UNCLOS, particularly with regard to the extent of maritime areas. According to UNCLOS an exercise of territorial claim in the Antarctic would also provide ownership over the adjacent territorial seas, exclusive economic zone and continental shelf; however, without territorial sovereignties, the seas around Antarctica should be treated as high seas, and continental shelf below as deep seabed. The legal disputes on the status of the Antarctic seas could not be speedily resolved, as they related directly to the sensitive issue of territorial sovereignty in the Antarctic. Nevertheless, ATCPs whether claimant states or not, have reached a level of fundamental agreement to prevent interference with the ISA in mining activities on the seabed or subsoil adjacent the Antarctic landmass. In the final text, CRAMRA takes the Antarctic Treaty area as a basic definition of its scope, namely the area south of 60° south latitude, and adds conditions appropriate to the purpose of mineral resources activities in the Antarctic in Article 5. The first paragraph of Article 5 declares the Convention should apply to "the Antarctic Treaty area", which is defined as Article VI of Antarctic Treaty.[510]

The basic area of application is modified by the following paragraphs. The Second paragraph regulates that the Convention "shall regulate Antarctic mineral resources activities which take place on the continent of Antarctica and all Antarctic islands, as well as all ice shelves, south of 60° south latitude and in the seabed and subsoil of offshore areas up to the deep seabed". This provision extends the area in which the mineral regime can be applied from the landmass to the continental shelf, even though not using those exact terms, and provides the inner limits of the continental shelf – adjacent land areas and the outer limits – of the deep seabed. During the negotiations, parties were mindful of the terms of UNCLOS, therefore, paragraph 3 explains "deep seabed" as "the seabed and subsoil beyond the geographic extent of the continental shelf as the term continental shelf is defined in accordance with international law", which should be read as a reference to all the criteria and rules embodied in Article 76 of UNCLOS[511]. These provisions provide evidence of a balance between the positions of claimant and non-claimant states. On one hand, it uses the "geographic extent" of the shelf, a factual geographic line, as the outer limit of the area; on the other hand, the later term "continental shelf", which has become a well-known concept in international law, might be positively construed as supporting the notion of "coastal states" by claimant states.[512]

---

510 Article 1 (3), CRAMRA.

511 The Final Act of CRAMRA.

512 The claimant states might argue that their coastal states' status by virtue of the Final Act of CRAMRA, which recognizes that the geographic extent of the continental shelf would be determined by reference to all the criteria and the rules embodied in paragraphs 1 to 7 of Article 76 of UNCLOS, which is established based on existence of a coastal state. Mr. Watts, who goes further on this point, states

Based on the above provisions, CRAMRA applies to mineral resource activities taking place on the Antarctic land area including the territory of Antarctica and islands south of 60° south latitude. Additionally, if the geographic extent of the continental shelf adjacent to offshore areas extends north of 60° south latitude, the Convention still applies to mineral activities conducted in those areas.

Conversely, there are some islands with clear sovereignty located close to the 60° south latitude, which subject to UNCLOS should own continental shelves extending down to the 60° parallel. In these cases, CRAMRA does not cover mineral activities in the southward-extending areas of shelf. Instead, it is necessary to establish a delimitation between the area of application and the national continental shelf zones according to the international principles of delimitations for maritime zones.[513]

CRAMRA does not apply to mineral resources activities in the deep seabed, which fall under ISA jurisdiction. This was a compromise to preserve the jurisdiction of the ISA under UNCLOS and the legitimacy of the ATS.[514] However, since the responsibilities of the ATCPs should not be prejudiced, Article 4 of CRAMRA, which provides principles of judgment for such activities, should be applied to mineral resources activities undertaken in the deep seabed south of 60° south latitude.[515]

Even though CRAMRA's area of application might fall in some part within the area of the Antarctic Treaty, it does not prejudice the responsibility of the ATCPs under the Antarctic Treaty and measures pursuant to it.[516]

Another exemption is regulated in paragraph 4, which stipulates that the Article cannot limit the application of other Articles of the Convention in so far as they relate to possible impacts outside the area referred to in paragraph 1 and 2 above. This may arise, for example, in the situation of protecting the Antarctic or dependent or associated ecosystems, which would not be effective if management were limited by an arbitrary line, such as the 60° south latitude or the geographic extent of the continental shelf. Thus, the regulation of Antarctic mineral resource activities may be wider in scope than the Convention's actual area of application.[517]

---

that the outer line is not apart from "the underlying legal concept involving the notion of a 'coastal state'". See: Watts (note 257), at 154.

513 Wolfrum (note 479), at 32.

514 Article 5 paragraph 2, "... activities...up to the deep seabed". For more details about the compromise between two legal system see Müller, Scharnborst, "The Impact of UNCLOS III on the Antarctic Regime", in Rüdiger Wolfrum (ed.), *Antarctic Challenge II*, (Berlin: Duncker & Humblot, 1984), 169–176 at 175; Infante, María Teresa, "The continental shelf of Antarctica: Legal Implications for a Regime on Mineral Resources," in Vicuña, Francisco Orrego (ed.), *Antarctic Resources Policy: Scientific, Legal and Political Issues*, (Cambridge: Cambridge University Press, 1983), 253–264 at 259.

515 Wolfrum (note 479), at 32.

516 Article 5 (2), CRAMRA.

517 Watts (note 257), at 155–156.

To summarize these complicated provisions, there are three different levels of area: the first and basic level is the Antarctic Treaty area which extends as far north as 60° south; the second level contains Antarctica and its continental shelf, including the small portion which extends above 60° south latitude; the third level is an indeterminate area which depends on the legal status of the adjacent offshore areas.[518]

III. International Participation

Article 6 of CRAMRA promotes cooperation between the ATCPs and other interested Parties, particularly developing States. This based on the fact that developing States, who generally have a less developed economy and technological capacity for the pursuit of mineral resources activities as industrial States, wanted to participate to the fullest extent in the activities which might be taken by developed States. In this, the developing states, as the majority of the international community, were attempting to bring the Antarctic issue into the UN General Assembly, where it might be approved though simple majority. In order to gain broad acceptance, the consideration of the interests of developing States might be a wise move. Therefore, in the overall structure of CRAMRA, some operational aspects should be considered as part of the search for international cooperation.[519]

IV. Compliance

Article 7 requires that operators carrying out mineral resource activities in the Antarctic have to ensure compliance with CRAMRA and any pursuant measures in effect. However, the question here is which parties have jurisdiction[520] over Antarctic mineral resource activities? One proposition, preferred by claimants, is jurisdiction in the Antarctic on the basis of territory, i.e. those claimants exercise jurisdiction over their claimed sectors of Antarctic, or, the territorial principle. Obviously, this would not be accepted by non-claimant States. They support national principle, which means that mineral resource activities should be regulated on the basis of operator's nationality. Article 7 is too ambiguous to grant jurisdiction powers, and leaves the question open.[521] Parties must resolve any jurisdictional dispute to achieve

---

518 Watts (note 257), at 156.

519 Watts, A.D., "The Convention on the Regulation of Antarctic Mineral Resource Activities 1988", 39 Int'l & Comp. L. Q.169 (1990), at 171.

520 Jurisdiction is explanted as "...power or ISA in general; administration, control." In Brown (note 285), at 1465. Similarly, in international law, "jurisdiction" refers to judicial, legislative, and administrative competence of states. See Brownlie, Ian, *Principles of Public International Law*, (4th ed., Oxford: Oxford University Press, 1990), at 298. To ensure compliance with CRAMRA, parties have to exercise jurisdiction on Operators of Antarctic mining activities.

521 Wolfrum (note 479), at 36–37; Watts (note 519), at 171.

a mutually acceptable solution through consultation,[522] and the opportunity exists for the parties and the Commission to supervise each other[523] Although parties are not compelled to notify the Commission, its supervision of all mining activities is mandatory, regardless if conducted by a party or not.[524]

V. Liability

The liabilities of operators and sponsoring States are laid out in Article 8. The first paragraph of this Article requires that an Operator should take all action necessary in responding in a timely way to actual or potential threats arising from mineral resource activities to the Antarctic environment or dependent or associated ecosystems. The response may take the form of prevention, containment, clean up and removal measures. Damage is defined in Article 1 (15) as "any impact on the living or non-living components of that environment or those ecosystems, including harm to atmospheric, marine or terrestrial life, beyond that which is negligible or which has been assessed and judged to be acceptable pursuant to the Convention" – although negligible impacts and acceptable impacts are excluded. It does lack mechanisms for determining whether an impact is negligible or not, however, some provisions specific to the determination of acceptable loss can be found in the Convention. Firstly, the sponsoring State should provide an assessment of the possible environmental and other impacts of proposed mineral resources activities to the relevant institutions before the commencement of prospecting, exploration or development.[525] Secondly, these impacts are assessed, and a mineral resources activity will only be permitted, if it would not cause: significant adverse effects on air and water quality; significant changes in atmospheric, terrestrial or marine environments and in the distribution, abundance or productivity of populations of species of fauna or flora; further jeopardy to endangered or threatened species or populations of such species; degradation of other uses of Antarctic; or significant adverse effects on global or regional climate or weather patterns.[526] There is an unconditional and precautionary liability to take appropriate action to protect the Antarctic environment from possible damage, and if this fails, to restore the *status quo ante*.[527]

522 Article 7 paragraph 3, CRAMRA. Article 47 (q) regulates that "applicable law to the extent necessary" should be covered in the Management Scheme.

523 Article 7 paragraph 6 and 7, CRAMRA.

524 Article 7 paragraph 8, CRAMRA.

525 Article 37 paragraph 7(d); Article 39 paragraph 2 (e); Article 44 paragraph 2 (b) (iii); and Article 53 paragraph 2 (b), CRAMRA.

526 Article 4 paragraph 2 and 3, CRAMRA.

527 Wolfrum (note 479), at 63.

Paragraph 2 of Article 8 places a strict liability, i.e. liable in the absence of fault, for pollution damage caused by oil spilled from the ship as a result of incident[528], on Operators to cover the costs of any damage, loss or impairment to the Antarctic environment or dependent or associated ecosystems, or for the loss of an established use of these, even if this damage is an inability to restore the Antarctic environment *status quo ante*, and for loss of or damage to property of a third party or loss of life or personal injury of a third party. CRAMRA stipulates that the Operator has to pay the cost of this damage, loss or impairment. Any other liability incurred for any loss of life, personal injury or loss or damage to property is determined by the relevant national jurisdiction.[529] Moreover, the Operator, whose mineral resources activities result or threaten to result in damage to the Antarctic environment or ecosystem, has to unconditionally reimburse all reasonable costs of necessary response actions. However, Article 8 (4) does provide an exemption from these liabilities in two situations : 1) the Operator can prove that damage has been caused directly by an event constituting a natural disaster of an exceptional and unforeseen character in the Antarctic context; and 2) the Operator proves the damage was directly caused by armed conflict or an act of terrorism directed against the activities of the Operator, and no reasonable precautionary measures could have been effective against such action. Moreover, if an Operator can prove that damage is caused totally or in part by an intentional or grossly negligent act or omission of another party the Operator may seek relief totally or in part from its liability to pay compensation.[530]

Sponsoring States, as additional debtors, are also liable for damages. If the respective sponsoring States failed to respond to an actual or potential threat, resulting in damage to the Antarctic environment and ecosystem, the State carries the liability. According to Article 37 (3), during the prospecting stage, the sponsoring State must ensure that Operators possess the necessary financial and technical capacity to respond appropriately to actual or possible damage. If the Operator fails to respond appropriately, responsibility falls to the sponsoring State to ensure such action is undertaken. Moreover, sponsoring States must ensure the Operator has sufficient financial capacity to cover the cost of the damage, loss or impairment, or incur a subsidiary liability to supply the portion not satisfied by the Operator. However, based on *culpa in vigilando*, the sponsoring State only carries this responsibility if they have failed to properly supervise the Operators according to CRAMRA.

During the exploration and development stages, the Sponsoring States must certify, in addition to possessing technical competence and financial capacity[531], that the Operator has the capacity to comply with the general requirements of the Convention, measures, associated conditions and general guidelines adopted

528 Article III, 1969 International Convention on Civil Liability for Oil Pollution Damage in connection with Article 4 of 1992 Civil Liability Convention.

529 Article 8 paragraph 5, CRAMRA

530 Article 8 paragraph 6, CRAMRA.

531 Article 44 paragraph 2 (d) and Article 53 paragraph 2 (c), CRAMRA.

by the Commission when identifying the area.[532] Sponsoring States are not only liable for certification, but also for ensuring their Operators to comply with the requirement of general compliance in Article 7 paragraph 1, although it is difficult to draw specific obligations from this rather general provision. This makes proving that a sponsoring State has either, carried out all appropriate measures, or, that it did not, by no means an easy task – it is a laborious undertaking to apply Article 8 paragraph 3(a) to this situation.

CRAMRA does not provide detailed rules and procedures on liability but leaves elaboration of a separate protocol under the provisions on liability as set out in Article 8 to the Commission. The protocol is limited to rules and provisions appropriate to the liability provisions in Article 8, thus, leaving no space for discussion of the entire system of prospecting, exploration and development during the establishment of the protocol.[533] The purpose of the protocol is to enhance the protection of the Antarctic environment and associated ecosystems, and it would enter into force under the same conditions as CRAMRA itself.

Article 8 paragraph 7 (c) lays out some rules and provisions that should be involved in the protocol, including limits on liability; means and mechanisms, such as a claims tribunal or other fora for dealing with claims pursuant to this Article against an Operator; and a means to support immediate response action, *inter alia*, that an Operator cannot satisfy its financial obligation, that it exceeds any relevant limits of liability, that there is a defense against liability or that the loss or damage is of undetermined origin. The protocol must contain provisions that detail how claims against Operators should be dealt with, reflecting the dissatisfaction of some ATCPs with the dispute settlement rules.[534] The means and mechanisms should be dealt with in accordance with Article 57, as is the case for the without prejudice clause. It would appear that the protocol will give rise to special and independent tribunals, which will be on the international level rather in national fora.[535] The protocol is expected develop the necessary provisions, and to set up a Fund or Funds, (and/or other possible options), to be financed by Operators or on an industry wide basis; to ensure permanent liquidity. This will be subject to mandatory supplementation in the event of insufficiency and reimbursement of the costs of mitigating actions. The proposed fund(s) are intended to assist with response action and to additionally satisfy liability, which is rather a tall order for one institution.[536]

The International Oil Pollution Compensation Funds (IOPC Funds) provide a good example for the proposed fund(s) in Antarctic mineral resource activities. The IOPC Funds are part of an international liability and compensation regime for oil

532 Article 44 paragraph 2 (c) and Article 53 paragraph 2 (d) in connection with Article 43 paragraph 3, CRAMRA.
533 Wolfrum (note 479), at 66.
534 Wolfrum (note 479), at 69.
535 For more detailed analyses, see Wolfrum (note 479), at 69.
536 Wolfrum (note 479), at 67.

a whole".[494] However, this was not enough to obtain support from the rest of the international community. For example, there was the potential for disagreement on the role of the ISA given the definition of the Antarctic, and the mineral regime needed to be open to other States.[495]

Although CRAMRA was concluded by consensus among the participants, the governments of Australia and France did not sign the Convention, resulting in its failure to enter into force and leading to the drafting of the Protocol on the Environment Protection to the Antarctic Treaty.[496]

B. General Provisions

I. Objectives and Principles

As an integral part of the Antarctic Treaty system,[497] CRAMRA reiterated and strengthened[498] the "peaceful use" principle enshrined in the Antarctic Treaty and the cornerstone of the ATS.[499] The purpose of the provision was not only to maintain the status quo of territorial claims in the Antarctic, but also to prevent the ATS from any infringement by the majority of the UN General Assembly. This idea has been further elaborated in Article 9, which declares that neither the convention itself, nor any activity according to the convention would affect any party's rights or position.

In addition, all principles whether in the Antarctic Treaty, or in the Convention for the Conservation of Antarctic Seals (London, 1 June 1972, hereafter the Seals Convention") or Convention on the Conservation of Antarctic Marine Living Resource (Canberra, 20 May 1980, hereafter CCAMLR) are equally applied to CRAMRA, in which the "frozen principle" of territorial sovereignty and the obligations to allow inspections[500] are detailed in Articles 9 and 11. ATCP recommendations also apply. This approach is repeated in Article 10 stating that Antarctic mineral resources activities should be carried out in a manner consistent with the instruments and measures in effect pursuant to components of the Antarctic Treaty system – the Antarctic Treaty, the Seals Convention and CCAMLR. Therefore, all recommendations issued either under these treaties or by the ATCPs should be incorporated into CRAMRA if they have entered into force and been reviewed as supplementary to the conventions.[501] Although this solution appears unproblematic,

---

494 Article 2 CRAMRA.
495 Wolfrum (note 479), at 15.
496 Wolfrum (note 479), at 10.
497 Article 2 CRAMRA.
498 CRAMRA adds "forever" word in the peaceful use clause.
499 The Preamble and Article 1 Antarctic Treaty.
500 Article IV and VII Antarctic Treaty.
501 The legal statues of recommendations see Wolfrum (note 440), at 72. In addition, the effect can only become true when the idea of creating a consistent Antarctic Treaty system is to be seriously pursued. See: Wolfrum (note 479), at 17.

pollution damage caused by oil spills from tankers. There are at present three IOPC Funds: the 1971 Fund (established in 1978), the 1992 Fund (established in 1996) and the Supplementary Fund (established in 2005).[537]

Under this regime, the owner of a tanker has strict liability. Although the 1992 Civil Liability Convention does not afford full compensation for victims of oil pollution damage in all cases, as the owner is liable to pay compensation up to a certain amount for oil pollution damage following an escape of persistent oil from the ship, it is still an additional financial burden on ship-owners. However, the ship-owner is exempt from liability if he can prove that the damage: resulted from an act of war, or a serious natural disaster; or the damage was wholly caused by sabotage by a third party; or the damage was wholly caused by the negligence of any Government or other ISA responsible for the maintenance of lights or other navigational aids.[538] The system was introduced to provide compensation to victims and relieve ship-owners from additional financial burdens. The 1992 Fund will pay the compensation to any person suffering pollution damage if such person has been unable to obtain full and adequate compensation for the damage under the 1992 Civil Liability Convention, provided one of following requirements is met: a) the ship-owner is exempt from liability under the Convention; b) the ship-owner is financially incapable of meeting his obligations in full and has insufficient insurance to satisfy pollution damage claims; or c) the damage exceeds the ship-owner's liability under the Convention.[539]

The Supplementary Fund, as the name implies, supplements the compensation available under the 1992 Civil Liability Convention and 1992 Fund Convention. Any member of the 1992 Fund may join the Supplementary Fund.

The IOPC Funds are financed by contributions by any entity which receives, in a calendar year, total quantities exceeding 150 000 tonnes of crude oil or heavy fuel oil in ports or terminal installations in a member State, after sea transport.[540] Normally,

---

537 The framework for the regime was originally contained in the 1969 International Convention on Civil Liability for Oil Pollution Damage (entered into force in 1975) and the 1971 International Convention on the Establishment of an International Fund for Compensation for Oil Pollution Damage (entered into force in 1978). As membership of the 1971 Fund Convention decreased, it ceased to be in force on 24 May 2002. Therefore, the 1971 Fund no longer has any member States. In 1992, a new regime was established by two Protocols: the 1992 Civil Liability Convention and the 1992 Fund Convention. The former governs the liability of ship-owners for oil pollution damage with the principle of strict liability. However, the liability is limited to an amount linked to the tonnage of the ship. The latter, acts as a supplement to establish a regime for compensating victim when the compensation under the applicable Civil Liability Convention is inadequate. The three Funds have different maximum amounts of compensation and have different Member States.

538 Article 4 of 1992 Civil Liability Convention.

539 Article 4 paragraph 1, 1992 Fund Convention.

540 Article 10 paragraph 1, 1992 Fund Convention.

it is the first physical receiver of the oil, i.e. a company, in a member State who is liable to pay contributions. However, member States may also choose to pay the contributions on behalf of the individual receivers.

In drafting the relevant provisions of the Protocol relating to the purposed Fund(s) under CRAMRA, valuable experience could be gained from the IOPC Funds. The importance of the protocol can be seen in paragraph 9 of Article 8, which declares that any application for exploration or development shall not be permitted unless the Protocol is in force.

VI. Inspection and Protected Areas

Besides the inspection system designed in the Antarctic Treaty, CRAMRA also lists a more appropriate system of inspection and monitoring for Antarctic mineral resources activities, to promote and ensure the observation of the provisions and measures in effect pursuant to it.

Firstly, all stations, installations and equipment related to mineral resource activities in the Antarctic, as well as the ships and aircraft supporting these activities at points of loading or unloading of cargoes or personnel, shall be open at all times to inspection by observers as designated under Article VII of the Antarctic Treaty.[541] This inspection duty applies to Antarctic mineral resources activities without supplementing or amending the system,[542] but only in the Antarctic Treaty area.

Secondly, where CRAMRA covers Antarctic mineral resource activities, all stations, installations and equipment relating to these activities shall be open at all times to inspection.[543] Since CRAMRA does not cover the deep seabed south of 60° south latitude or the continental shelf area appertaining to sub-Antarctic islands north of 60 ° south latitude, this inspection does not cover the activities undertaken in these areas.[544] For those areas not covered by CRAMRA, a two-tier system of inspection has been established, which allows the designation of observers by any single member of the Commission, the Commission itself or any relevant Regulatory Commission. Inspection, as an important factor for controlling Antarctic mineral resources activities, reinitiated tension between claimant States and non-claimant States. The claimant States attempted to gain inspection rights in their claimant sectors, although this was firmly rejected by the non-claimant States. Inspection rights were therefore attributed to all members of the Commission, while the relevant Regulatory Committees are also authorized to inspect activities in specific areas. The reports issued by these two institutions have the same weight and enjoy equal status.[545]

---

541 Article 11, CRAMRA.
542 Wolfrum (note 479), at 57.
543 Article 12, CRAMRA.
544 Wolfrum (note 479), at 57.
545 Wolfrum (note 479), at 57.

Professor Wolfrum points out that monitoring under management schemes constitutes a third type of inspection, which may become the most important.[546] Regulatory Committees, the actual functional and executive organ, are authorized to develop Management Schemes, provide the detailed regulations for mineral resources activities in identified areas; to monitor the compliance of Operators with Management Schemes; and to monitor and assess the effects on the Antarctic environment and on dependent and on associated ecosystems of Antarctic mineral resources activities.[547]

The system of inspections under CRAMRA implies that Antarctic mineral resource activities are not only subject to monitoring by the sponsoring State, but also international inspection. In other words, the international community, represented by commissions, may affect the national jurisdiction of their operators' mining activities in the Antarctic.[548] Especially noteworthy in CRAMRA is that the provision for inspection effectively becomes a precondition for exploration or development in designated areas.[549] It is possible that this constitutes an attempt by the ATCPs to control Antarctic mineral resource activities under the ATS and to dilute national jurisdiction.

Mineral resource activities are not allowed in a Specially Protected Area, a Site of Special Scientific Interest or other designated protected areas established according to the Antarctic Treaty or other components of the ATS. The Commission is also authorized to prohibit mineral resource activities in specific protected areas for historic, ecological, environmental, scientific or other reasons, if such areas have already been established.[550] Even for those non-established protected areas, the Commission may prohibit Antarctic mineral resource activities, if those activities will interfere with scientific investigation, conservation of Antarctic marine living resources, tourism, historic monuments and navigation and aviation.[551] Therefore, mineral resource activities must be carried out prudently, and prohibited where conflicts with other uses of Antarctica occur.

C. Institutions

CRAMRA provides five main institutions, a Commission, an Advisory Committee, a Special Meeting of Parties, Regulatory Committees and a Secretary. They follow a common pattern adopted by other international organizations, but also represent features designed for internal accommodation of conflicts among interested States.[552] Although the Seals Convention and CCAMLR have provided institutions, building

---

546 Wolfrum (note 479), at 57–58.
547 Article 52, CRAMRA.
548 Watts (note 519), at 172.
549 Article 12, paragraph 8, CRAMRA.
550 Article 13, paragraph 2, CRAMRA.
551 Article 15, CRAMRA.
552 Wolfrum (note 479), at 38.

new institutions is still a sensitive issue in the Antarctic, as claimants might perceive international organizations in the Antarctic as eroding their claims and potential privileges.[553] However, using the membership of institutions of the regime to deal with the complicated internal decision making processes of those institutions is a wise move.[554] To that end, in order to protect both claimant and non-claimant interests in Antarctic mineral resources activities, designers of the regime preferred an alternative approach which would leave some elements open, to be resolved by a process of negotiation on a case by case basis, and built up the necessary institutions to facilitate future negotiations.

I. The Antarctic Mineral Resources Commission

As for other international organizations, the Antarctic Mineral Resources Commission is the functional body authorized to make decisions on fundamental matters. The Commission has the authority to determine; whether or not to designate an area for exploration; whether to prohibit mineral resource activities; to adopt measures for the environmental protection; to adopt detailed standards related to prospecting; the deposition of revenues and whether to perform other functions as listed in Article 21. The Commission exercises its functions across the Antarctic as a whole, including protection of the Antarctic environment and dependent and associated ecosystems. Therefore, the Commission is authorized to adopt measures with respect to some special issues.[555]

Membership of the Commission reflects a feature of the Antarctic regime, expressed by other components of ATS. Only those States who can demonstrate an interest in the Antarctic may participate in the decision-making procedures.[556] Besides those ATCPs who possessed this status when CRAMRA was opened for signing,[557] any other Party who is not an ATCP but engages in relevant activities, such as substantial scientific, technical or environmental research in the Antarctic, could also

553 Wolfrum (note 479), at 39.

554 Beeby (note 463), at 195.

555 Article 21 paragraph 3, CRAMRA.

556 Article IX paragraph 2, The Antarctic Treaty, requires that "[e]ach Contracting Party... shall be entitled to appoint representative in the meetings ..., during such times as that Contracting Party demonstrates its interests in Antarctic by conducting substantial research activity there, such as the establishment of a scientific station or the despatch of a scientific expedition". Similarly, Article VII paragraph 2 (b), CCAMLR, reads: "each States Party... shall be entitled to be a member of the Commission during such time as that acceding Party is engaged in research or harvesting activities in relation to the marine living resources...."

557 CRAMRA was opened for signature on 25 November 1988, following States were ATCPs at that time: Argentina, Australia, Belgium, Brazil, Chile, China, France, German Democratic Republic, Federal Republic of Germany, India, Italy, Japan, New Zealand, Norway, Poland, South Africa, Spain, Sweden, USSR, the UK and the USA.

be a member of the Commission.[558] Moreover, the Commission includes those States whose operators undertake relevant mineral resource activities in Antarctica.[559] The role of Observer is also open to any Party to CRAMRA or party to the Antarctic Treaty. Thus, the membership of Commission is not limited to ATCPs. That reflects the trend in Antarctic mineral resource activities, moving from being a regional issue in the circle of present ATCPs to a more international focus drawing more broadly from the international community. An Antarctic mineral regime without recognition outside of the ATCP group would have great difficulty in functioning effectively, thus, an effort has been made to enlarge the scope of the regime and to accommodate external parties.[560]

The rules of decision-making are separated into substantial and procedural matters. Substantial matters require a three quarters majority of the members present and voting, however, consensus has to be achieved for some special matters, such as budget, notification fees, payable levies, disposition of revenues, elaboration of the non-discrimination principle and identification of areas for exploration and development.[561] Consensus is also deemed to exist in the absence of formal objection[562]. The consensus decision-making measure gives every member of the Commission a right of veto on key decisions. This is a double edged sword, while it will advance environmental protection and prudent mineral exploration in the Antarctic; any member may effectively hold the operation of CRAMRA hostage to its interests or preferences. It is therefore necessary for every member to endeavour to ensure the mineral regime functions without extraneous distributions.[563] Moreover, the three-quarter majority rule is motivated by a desire to make matters difficult to approve for either group.[564] In other words, this is an attempt to reconcile internal conflicts, as consultations among parties and multiple acceptable measures are required to solve substantial issues. Procedural matters are decided by a simple majority.

II. The Scientific, Technical and Environmental Advisory Committee

The Advisory Committee is a substantial body that exists to provide advice on any decision requiring scientific, technical, or environmental expertise. It provides advice to the Commission and to the Regulatory Committees throughout the implementation of principles related to decisions on Antarctic mineral resources activities.[565]

558 Article 18 paragraph 2(a) and (b), CRAMRA.
559 Article 18 paragraph 2(c) , CRAMRA
560 Wolfrum (note 479), at 40.
561 Article 22 paragraph 2, CRAMRA.
562 Article 22 paragraph 5, CRAMRA.
563 Watts (note 519), at 173.
564 Wolfrum (note 479), at 40.
565 Artilce 26 in connection with Article 4, CRAMRA.

The Advisory Committee must supply advice when the Commission seeks to: assess environmental impacts of Antarctic mineral resources activities; designate protected areas; adopt measures for protection of environment and ecosystem; identify an area for exploration and development; adopt measures relating to prospecting; consider prospecting proposals; consider monitoring reports; review the conduct of Antarctic mineral resource activities; and dispose of surplus from mineral activities.[566]

The Advisory Committee has to provide any Regulatory Committee with advice, when a Regulatory Committee: adopts or revises guidelines for general requirements for exploration and development in its area of competence; examines applications for exploration; prescribes, suspends, modifies or cancels, a Management Scheme; and monitors the compliance of Operators with Management Scheme.[567] Moreover, it is also required to provide a comprehensive environmental and technical assessment on decisions identifying an exploration area, examination of applications and the issue of development permits made by the Commission or the Regulatory Committee.

To promote international participation in Antarctic mineral resource activities, the Advisory Committee has the duty of providing advice to interested developing States who are Parties, concerning the availability of training programmes and opportunities for cooperation. Generally speaking, the Advisory Committee consults on all critical issues and provides its advice based on the requirements of the regime. In addition, the Advisory Committee also serves as a forum for consultation and cooperation among Parties.

The Advisory Committee comprises representatives of all States Parties, who possess suitable scientific, technical or environmental competence. Observer status is open to Parties of both the Antarctic Treaty and the CCAMLR.[568] With the exception of the first meeting, which is explicitly required, any meetings thereafter shall be convened only if necessary.[569] As the Advisory Committee is merely an advisory organ without decision-making power except for internal procedural matters, it must only present a report on each of its meetings covering all matters considered and reflecting any conclusions and dissenting views.[570] Dissenting views are included in the report to provide an opportunity for potential political pressure to be bought to bear depending on the inherent merits of the dissent.[571]

---

566 Article 26 in connection with Article 21 (a) to (f), (u) and (x), and Article 35 paragraph 7, CRAMRA.

567 Article 26 in connection with Article 43 paragraph 3 and 5, Article 45, 47, 51 and 52, CRAMRA.

568 Article 23, CRAMRA.

569 Article 24, CRAMRA.

570 Article 27, CRAMRA.

571 Watts (note 519), at 174.

III. The Special Meeting of Parties

The Special Meeting of Parties is also an advisory body, which considers whether the identification of an area for possible exploration and development satisfies relevant requirements in the Convention and reports its conclusions to the Commission.[572] As the decision to identify and area for potential mining activities is crucial to the implementation of these activities, approval can only be granted by consensus in the Commission. As the Commission is not a plenary body, but rather must represent the views of all its members[573], all Commission members participate in the Special Meetings, and observer status is open to all parties to the Antarctic Treaty.[574] Where possibly, the Commission will divide the application into two aspects requiring advice, scientific, technical and environmental aspects and political aspects, Scientific, technical and environmental advice is provided by the Advisory Committee, with political advice being provided by the Special Meeting of the Parties. Moreover, the Commission has to take full account of all views presented and give special weight to the conclusions of the Special Meeting.[575]

The Advisory Committee and the Special Meeting constitute a deliberate attempt by CRAMRA to accommodate the internal interests of other non-consultative parties, to compensate for domination of the ATCPs in the functionary bodies of the mineral regime.[576]

IV. The Regulatory Committees

The Regulatory Committees are a unique executive body found only in CRAMRA in the scope of the ATS. The relationship between the Regulatory Committees and the Commission is similar to that of the Security Council and the General Assembly of the UN.[577] The Commission's mandate for detailed regulation of mineral activities across the whole of Antarctica requires the establishment of a series of Regulatory Committees to deal with each area identified for exploration and development.[578] Once the Commission decided to open an area to mineral exploration and development, a smaller group with limited membership would be convened to inspect and regulate the potential activities. A limited membership was considered more effective than the whole Commission for making specific decisions, and better suited to concluding detailed regulations for monitoring activities in the identified area.[579]

---

572 Article 28 paragraph 1, CRAMRA.
573 Watts (note 519), at 174.
574 Article 28 paragraph 2 and 3, CRAMRA.
575 Article 41 paragraph 1, CRAMRA.
576 Wolfrum (note 479), at 41.
577 Wolfrum (note 479), at 41.
578 Watts (note 519), at 175.
579 Scullz, Tucker / Kimball Lee A., "Antarctica: Is There Life After Minerals? The Minerals Treaty and Beyond", 13 Marine Policy 87 (1989), at 92.

1. *Membership of Regulatory Committees*

Membership of the Regulatory Committees was a subject of fierce debate between claimant and non-claimant States, with the claimant States persisting with their argument that they should have the authority to regulate mineral resource activities in their claimant sectors and non-claimant States adamantly rejecting that option. There were two major considerations during the negotiations on the composition of the Regulatory Committee. The first one was the overlapping claims in the Antarctic between Argentina, Chile and the UK. An identified area were to be located in one of these areas, would result in at least two claimant States and probably three, with the same situation arising from an identified area straddling the boundaries of two areas. The second consideration addressed the unclaimed area of Antarctica. To accommodate the interests of all States it was decided that each Regulatory Committee should have a total of ten members, which would include four claimants, four non-claimants, the US and USSR, who assert a basic claim to the Antarctic and were granted automatic membership to any potential mineral area[580], and three developing States.

In the claimant group, some members of the Regulatory Committee would be decided by CRAMRA or by the decision of the claimants. Article 29 paragraph 2(a) states that any claimant States whose claim was directly affected was granted automatic membership, both for unclaimed sectors and areas claimed by more than one State. The claimant(s) would then nominate other claimants as members of the Regulatory Committee to the Chairman of the Commission.[581] This composition was intended result in equality of representation for claimants and non-claimants, however, as non-claimants outnumbered claimants among the ATCPs by a significant proportion, this design actually benefited the claimants[582], in acknowledgement of their concession on the establishment of an international legal regime for Antarctic mineral activities, effectively creating a composition that accommodates the various States' interests.[583]

The remaining four positions were granted to non-claimant States. Non-claimants as a group could recommend four of their number to become members of a Regulatory Committee. Their recommendations were however subordinate to those of claimant States, as Article 29 paragraph 4(b) only requires the Chairman of the Commission "to give full weight to the views which may be presented on behalf of those members of the Commission". Based on the requirement that three out of ten members of each Regulatory Committee must be developing States, there were three potential situations: a) either Chile or Argentina would be a member of any given Regulatory Committee, and two other developing States would be members as

580 Article 29 paragraph 2(b), CRAMRA. The U.S. and USSR play an intermediary role to balance claimants and non-claimants. See Wolfrum (note 479), at 41.
581 Article 29 paragraph 4, (a), CRAMRA.
582 Wolfrum (note 479), at 41.
583 Wolfrum (note 479), at 41.

a non-claimant group; b) both Chile and Argentina were members of a given Regulatory Committee, with only one position left for a developing State as a member at the Regulatory Committee; c) neither Chile nor Argentina were members, and three developing States would sit in the non-claimant group of Regulatory Committee with only one position left for developed State. CRAMRA does not divide States between developed States and developing States, rather between claimants and non-claimants. For the claimant group in particular, there is no distinction between developed States and developing States. This distinction which is unusual in international law, reflects the fact that the crucial conflict on the issue of Antarctic mineral resource activities is not between the developed States and developing States but between the claimants and non-claimants.

According to the Convention, it is the Chairman of the Commission who nominated the members of the Regulatory Committees, and the Commission would determine the final list. Article 29 paragraph 3(a) further complicated the compositions of the Regulatory Committees and enhanced the status of the Chairman, by requiring that the Chairman's recommendation include members of the Commission, who through prospecting, scientific research or otherwise, have contributed substantial scientific, technical or environmental information. Article 29 paragraph 5 regulates that the Chairman's recommendation is deemed to have been approved by the Commission only if it does not decide otherwise in the same meeting as the Recommendation is submitted. Therefore, as opposed to the Chairman's right to nominate, the Commission has substantial power over the composition of Regulatory Committees rather than just approving or objecting to the Chairman's recommendation.

Article 29 paragraph 6 allows for an extended Regulatory Committee membership in two specific situations. Firstly, when a member of the Commission who is not a member of the relevant Regulatory Committee has sponsored prospecting in the identified area and submitted the notification, it should be a member of the relevant Regulatory Committee until an application for an exploration is lodged. Secondly, if a Party is lodging an application for an exploration permit, it should be a member of the Regulatory Committee. This arrangement could be seen as a form of external accommodation.[584]

2. *Functions of the Regulatory Committees*

The Regulatory Committees was intended to be active throughout the whole mineral activity procedure. Firstly, in the preparatory period, each Regulatory Committee should divide its area of competence into blocks, determine which Parties could submit applications for exploration and development; establish fees, periods,

584 Wolfrum (note 479), at 44.

procedures of applications; and adopt and revise guidelines to identify the general requirements for exploration and development in the identified area.[585]

Secondly, Regulatory Committees should consider applications for exploration and development permits from potential operators. The Regulatory Committee should "seek and take full account of the views of the Advisory Committee",[586] however it could reject the application for exploration at any time of consideration.[587] At the same time, the Regulatory Committee should develop a Management Scheme, which describes specific terms and conditions for mineral activities in the identified area. When considering an application for a development permit, the Regulatory Committee should consider whether such activity would cause previously unforeseen impacts on the Antarctic environment or ecosystems, and make necessary modifications to the Management Scheme.[588]

Thirdly, during mineral activity operations, the Regulatory Committee should monitor and inspect all activities,[589] and should, according to specific situations of an Operator, modify, suspend, or cancel the Management Scheme or impose a monetary penalty where necessary[590].

3. *Decision Making Procedure*

Generally speaking, the Regulatory Committees decide substantial issues by a two-thirds majority, and procedural issues by a simple majority. The Final Act of the Fourth Special Antarctic Treaty consultative Meeting on Antarctic Mineral Resources agreed that it was desirable that the two-thirds majority should include at least one developing State. Some crucial issues have specific voting requirements, decisions on examination of applications and the issue of development permits and decisions about approval and modification of the Management Scheme need a two-thirds majority, including a majority in both the group of four claimant States and the group of six non-claimant States.[591] Therefore, those issues could only needed to be rejected by two claimant members or three non-claimant members of the Regulatory Committee to fail. To ensure no substantial decision could be made without majority support of both groups, decisions on the adoption and revision of guidelines identifying the general requirements for exploration and development of specific areas required a two-thirds majority including at least half of both the claimant and non-claimant groups.[592] The decision making procedure reflects the

585 Article 43, CRAMRA.
586 Article 45 paragraph 3, and Article 54, paragraph 6, CRAMRA.
587 Article 45 paragraph 2, CRAMRA.
588 Article 54 paragraph 3(b) and paragraph 4 CRAMRA.
589 Article 31 paragraph 1 (d) and (f), CRAMRA.
590 Article 51, CRAMRA.
591 Article 32 paragraph 1, CRAMRA.
592 Article 32 paragraph 2, CRAMRA.

internal structure of the mineral regime, any important decision required approval from both groups; making impossible for only one group to make such a decision.

V. The Secretariat

The Secretariat is an auxiliary body designed to serve the Commission, Regulatory Committees, the Advisory Committee, the Special Meeting of Parties and any subsidiary bodies established. The Commission must decide whether or not a Secretariat should be established, as there is little sense in committing resources to establishing an international bureaucracy which may or may not be required. Once founded, the Secretariat performs functions assigned to it by the Commission, and other institutions of CRAMRA, subject to a budget approved by the Commission.[593]

VI. Financial provisions

The funds for these institutions under CRAMRA come predominantly from the fees and levies paid by Operators, and other "payments in the nature of and similar to taxes, royalties or payments in kind", which may be paid to institutes.[594] Budget shortfalls must be supplemented by the members of the Commission.[595] It is also possible that the Operator sourced revenues may result in a surplus. If so, the Commission is authorized to determine the disposition of the surplus.

In determining the disposition of any surplus revenues, the Commission should promote scientific research in Antarctica, and should ensure that the interests of the most involved members of the Regulatory Committees are considered.[596] This entitles a claimant State to an extra share of revenues, should this be approved by consensus in the Commission.[597]

VII. Conclusion

Through analysing the institutions envisaged by CRAMRA, two characteristics may be concluded. Firstly, the mineral regime attempts to accommodate both the internal interests between claimant States and non-claimant States and external interests between the ATCPs and the other non-parties. The regime defines, on one hand, the central role of the ATCPs in regulating mineral resource activities by granting them membership of the Commission and the decision-making procedure of the Commission. To ensure balanced representation of interests, the Advisory Committee and Special Meeting may provide advice on scientific, technical, environmental

593 Article 33, CRAMRA.

594 Article 35 paragraph 4, CRAMRA

595 Article 35 paragraph 5, CRAMRA.

596 Article 35 paragraph 7, CRAMRA. The provision primarily protects the interests of the relevant claimant States. See Watts (note 519), at 177.

597 Article 22 paragraph 2(a), CRAMRA.

and political aspects, although these do not have a binding effect, they may bring some political pressures to bear on disputed issues. The membership design of the Regulatory Committees clearly expresses the CRAMRA effort to balance conflicts between claimant and non-claimant States. Secondly, the real purpose of the regime seems not to be the resolution of issues arising from Antarctic mineral resource activities, but rather to put some of them aside, i.e. to postpone the possibility of mining activities until all political, technical and economical conditions can be satisfied. Both the Commission and the Regulatory Committee set very high standards for decision making, requiring either consensus or two-thirds majority respectively. Issues that are not broadly supported are therefore best set aside, until some or all interested parties are willing to make the necessary concessions to ensure its success. There are two consultative methods for parties discussing substantial issues. One way is to insist on fulfilment of their own -self-interests, a stance almost always rejected by others, leaving the issue unresolved. The other way is to give up some benefits in exchange for a similar concession from the other parties, and finally achieve consensus or majority opinion. Therefore, the main role of these institutions is to provide different level forums for consultation on issues of Antarctic mineral resource activities.

D. Operational Provisions

In Recommendation XI-1, the ATCPs agreed that that no exploration or exploitation activities would be permitted before finalizing a regime for Antarctic mineral resource Activities. This is due to the lack of knowledge concerning the environmental effects of such activities, and the potential negative affects of unregulated activities on the unique environment and ecosystems in the Antarctic.[598] During the negotiation and establishment of the regime, a moratorium was declared on Antarctic mineral resource activities, including exploration, exploitation and prospecting at the last session of the Fourth Special ATCM, to coincide with the date of entry into force of CRAMRA.[599] Therefore, CRAMRA's entry into force is the first requirement for the initiation of Antarctic mineral resource activities.[600] Even after CRAMRA had entered into force, Antarctic mining activities are basically prohibited according to the convention unless stringent standards can be satisfied.[601]

Antarctic mineral resources refer to all "non-living natural non-renewable resources", excluding ice.[602] The whole structure of operational provisions are divided into three stages, namely "prospecting", aimed at identifying potential mining activities

598 Recommendation IX-1, 1977, London, effective on September 8 1983.
599 Final Act of the Fourth Special Antarctic Treaty Consultative Meeting on Antarctic Mineral Resources, adopted at Wellington, on June 2 1988.
600 Wolfrum (note 479), at 45.
601 Article 4, CRAMRA.
602 Article 1 paragraph 6, CRAMRA.

areas; "exploration", which is more substantial than prospecting; and "development" aimed at exploitation of specific mineral resources. During the process of mining activities, "Operator" refers to a State party, an agency of a State Party, a legal person or a joint venture.[603] To ensure an Operator implements the convention, especially in the case of a legal person or a joint venture, they must be effectively monitored by a "Sponsoring State" which has a substantial and genuine link with the operator.[604]

I. Prospecting

Prospecting is a surveying activity aimed at identifying areas of mineral resource potential for possible exploration and development, but excluding dredging and excavations or drilling exceeding 25 meters.[605] It does not confer upon the Operator any right to Antarctic mineral resources.[606]

Prospecting may be undertaken at all times without authorization by the institutions; however, such activities should comply with the convention and with measures in effect pursuant to it.[607] There is an explicit requirement that Operators should maintain the financial and technical ability to take response action necessary to preserve the Antarctic environment, and maintain the financial capacity to take responsibility for environmental damage arising from their prospecting activities.[608] The Sponsoring State must evaluate whether the Operator can satisfied these requirements. Additionally, the Sponsoring State has to notify the Commission at least nine months in advance of the proposed area where prospecting activities will take place; the mineral resource(s) that are to be the subject of the prospecting; the methods to be used and the general program of work to be undertaken and its expected duration; an assessment of environmental impacts, and measures to be adopted to avoid harmful environmental consequences; genuine link with the Operator; and other required information.[609] The sponsoring State must also notify if there are any changes to the above information, or if the prospecting ceases, and provide an annual report on prospecting activities.

While these requirements do not seem onerous in themselves, some standards have been provided in the Convention. Firstly, prospecting activities must comply with the interests of protection of the Antarctic environment. Secondly, the Commission could be convened by any other State, as a member of the Commission, even if not the sponsoring State, who considers that prospecting is inconsistent with CRAMRA or measures in effect pursuant to it.[610] Thirdly, prospecting is prohibited in

603 Article 1 paragraph 11, CRAMRA.
604 Article 1 paragraph 12, CRAMRA.
605 Article 1 paragraph 8, CRAMRA.
606 Article 37, paragraph 1, CRAMRA.
607 Article 37 paragraph 2, CRAMRA.
608 Article 37 paragraph 3, CRAMRA.
609 Article 37 paragraph 7, CRAMRA.
610 Article 38, CRAMRA.

Special Protected Areas or Sites of Special Interest[611], or any other area designated as a protected area for historic, ecological, environmental, scientific or other reasons.[612] Fourthly, the Commission may adopt measures to determine maximum drilling depth for specific cases and adopt measures to restrict or prohibit prospecting for specific purposes, such as environmental and cultural protection.[613]

One point must be noted here, the date and commercial value of the activities are not included in the notifications.[614] As Antarctic mining activities are commercial activities, they are not classified in the same way as deep seabed mining activities, as it is clear that Operators pursue benefits from such activities. Thus, data and information on commercial value may be kept confidential. However, the unique character of Antarctic mining activities permits the Commission to adopt measures relating to the release of commercial data and information.[615]

II. Exploration

Exploration in CRAMRA refers to activities "aimed at identifying and evaluating specific mineral resource occurrences or deposits"[616], and is a preparatory stage before exploitation. Generally, drilling, dredging and other surface excavations are allowed but pilot project or commercial production activities are excluded. As exploration is more complicated than prospecting, authorization is required.

Firstly, the Commission must determine whether an area is suitable for potential exploration and development activities. By submitting a notification to the Executive Secretary, any Party may require the Commission to identify an area of potential exploration. The Party submitting the notification must pay the relevant fees established by the Commission and submit relevant information including a precise delineation, specific resource(s), physical and environmental characteristics of the area; likely scale and methods of exploration; and a detailed assessment of environmental and other impacts.[617] After receiving notification, the Commission must first be convened[618]; followed by the Advisory Committee, who should meet and provide advice; and then, a Special Meeting of Parties to consider whether identification of a potential area is consistent with CRAMRA.[619]

After the Special Meeting of Parties has submitted its report, the Commission determines whether it is necessary to identify a potential exploration area. Several

611 Article IX(1), the Antarctic Treaty.
612 Article 13 paragraph 2 and 3 company with Article 21 paragraph 1 (a) and (e) (ii), CRAMRA. Wolfrum (note 479), at 46.
613 Article 21 paragraph 1 (e), CRAMRA.
614 Article 37 paragraph 9, CRAMRA.
615 Article 37 paragraph 13, CRAMRA.
616 Article 1 paragraph 9, CRAMRA.
617 Article 39, CRAMRA.
618 Article 19 paragraph 2(a), CRAMRA.
619 Article 40, CRAMRA.

matters have to be considered by the Commission before designating an area including: whether the area is part of the areas protected for a special purpose;[620] the potential opportunities for joint ventures or different forms of participation by interested ATCPs, in particular developing countries;[621] and the establishment of procedures for the settlement of disputes[622]. Following consideration of these matters, the Commission members must reach consensus.[623] It should be noted here that the designation of any area is only with respect to special mineral resource(s). If a Party requests the designation of an area, which overlaps in whole or in part with an area designated for different mineral resources, the notification has to be treated as a totally new request. However, the Commission has to consider it as "a coherent unit for the purpose of resource management", and to consider boundaries of the area which may be assigned by an established Regulatory Committee.[624]

Once an identified area has been established, the respective Regulatory Committee has to be set up as soon as possible by the Commission. Prescriptive and administrative functions are passed on to the Regulatory Committee,[625] which is authorized to adopt and revise guidelines and general requirements for exploration and development in its area of competence. Guidelines should be made under the provisions of CRAMRA and the resultant legal framework adopted by the Commission.[626] A Regulatory Committee may divide its area into blocks, in which Operators may apply to undertake exploration and development. It may establish fees to be paid with applications; periods in which applications may be lodged; procedures for dealing with applications; and a method of resolving competing applications.

In the designated area, any Party may apply for an exploration permit. An application should be accompanied by the necessary fees and information, which contains: an explanation of the proposed activities; Sponsoring State's certification of the Operator's capacity to comply with legal requirements; and Sponsoring State certifications of technical competence and financial capacity.[627] The relevant Regulatory Committee decides whether to reject an application based on its consistency with CRAMRA, related measures in effect and the guidelines made by the Regulatory Committee, or to prepare a Management Scheme, which contains specific terms and conditions for exploration and development activities within this block.[628]

620 Article 41 paragraph 1(b), CRAMRA.
621 Article 41 paragraph 1(d), in connection with Article 6, CRAMRA. It is hard to enforce the international cooperation, since the Final Act declares that the promotion and encouragement of international participation do not prejudice the right of Operators to freely choose partners.
622 Article 41 paragraph 1(f), in connection with Article 59, CRAMRA.
623 Article 41, CRAMRA.
624 Article 42(1) in connection with Article 41 paragraph 1(a), CRAMRA.
625 Wolfrum (note 479), at 52.
626 Article 43, (3), CRAMRA.
627 Article 44, CRAMRA.
628 Article 45(2) and 46, CRAMRA.

A Management Scheme covers almost every aspect of the mining activities, from the duration of mining activities, environmental protection, technical and safety standards, financial obligations, monitoring and inspection, to removal of installations and equipment.[629] As the Management Scheme regulates details about specific exploitation activities in the identified area, those responsible for establishing the Management Scheme also govern mining activities. Moreover, there is no explicit requirement for the terms and conditions in a Management Scheme to be consistent with CRAMRA, the general measures adopted by the Commission. Thus, the respective Regulatory Committee has an independent authority over a Management Scheme[630], giving further rise to conflicts between claimant and non-claimant States. Claimant States, with their guaranteed membership of the Regulatory Committee, want to control the terms and conditions on mineral activities in its claimant sector, a move clearly rejected by non-claimant States. As a compromise, the Regulatory Committee has recourse to Sponsoring States, relevant claimant States, and any others as necessary.[631] The rationale behind this design is that if an agreement can be achieved among these States, especially the Sponsoring States and claimant State(s), it means that the terms and conditions would be acceptable to most interested parties, and therefore to other States in the Regulatory Committee.[632]

The Management Scheme should be approved by a two-third majority including a simple majority of four claimant States and six non-claimant States. It is therefore impossible for any individual claimant or non-claimant State to block the approval of the Management Scheme. Once the Management Scheme has been approved, an exploration permit is issued. The Operator has exclusive rights to explore and develop mineral resource(s) under the terms of the Management Scheme in the area assigned.[633]

Summarily, the Commission may, but is not obliged to, adopt general regulations which would restrict any of the terms or conditions implemented by Regulatory Committees. It leaves Regulatory Committees some leeway in interpreting Convention provisions and standards needed for an application specific Management Scheme, to promulgate specific terms and conditions relevant to mineral type, equipment and methods used, etc. Moreover, to obtain Regulatory Committee approval, the claimant State and sponsoring state may negotiate additional agreements with respect to the financial obligations of operators or specific assignment of enforcement and dispute settlement responsibilities.[634]

On request by at least six members of the Commission or only one member of the Regulatory Committee, the Commission will review a decision to approve a

---

629 Article 47, CRAMRA.
630 Wolfrum (note 479), at 52–53.
631 Article 46, CRAMRA.
632 Wattes (note 519), at 179.
633 Article 48, CRAMRA.
634 Kimball (note 460), at 24.

Management Scheme or the issue of a development permit.[635] However, the Commission's supervision is quite limited, and cannot restrict the functions of the Regulatory Committee during the supervision period, and should the Commission decide that the Regulatory Committee's action is inconsistent with CRAMRA, it may only ask the latter to reconsider its decision.[636] This limited influence makes it difficult for the Commission to affect the Regulatory Committees' operation.

Generally, it is not possible to suspend, modify or cancel an approved Management Scheme and authorized permits for exploration and development without the consent of the relevant Sponsoring States. However, exceptions will be made when mineral activities result or might result in impacts on the Antarctic environment or dependent or associated ecosystems, or when an Operator has failed to comply with CRAMRA, measures in effect pursuant to it, or a Management Scheme.[637] Violations of the Management Scheme may incur financial penalties as regulated by the relevant Regulatory Committee.

It is worth noting that two or more qualified applicants in the same area must first resolve their differences before making a submission, or for irreconcilable differences, seek resolution from the Regulatory Committee in accordance with predetermined methods.[638] This provision reflects CRAMRA's preference for solving problem by consensus rather than by explicit rules.

III. Development

Development is carried out following exploration and seeks to exploit specific mineral resource deposits.[639] After a Management Scheme has been approved and an operator has been permitted to explore, the Sponsoring State could, on behalf of the Operator, submit a development application to the Regulatory Committee. The application should be accompanied by the required fees and relevant information including a development plan, an environmental assessment, and technical, financial and capacity certifications.[640] The Regulatory Committee may request the sponsoring State to submit further information, then must consider whether the application meets requirements, and determine whether or not to issue a development permit to the Operator. The procedure for handling a development application is quite similar to the method of application for an exploration permit. The Regulatory Committee may modify an existing Management Scheme rather than establishing a new one.[641]

---

635 Article 49 in connection with Article 19 paragraph 2(b) and (c), CRAMRA.

636 Article 49 paragraph 3, CRAMRA.

637 Article 51, CRAMRA. If the Operator ceases to have a genuine link with its Sponsoring States, the Management Scheme and the exploration and development should be cancelled by the respective Regulatory Committee.

638 Article 45 paragraph 4, CRAMRA.

639 Article 1 paragraph 10, CRAMRA.

640 Article 53, CRAMRA.

641 Article 54 paragraph 4, CRAMRA.

To summarize the relevant fees that apply to the whole process of mining Activities, three categories of fees are regulated in CRAMRA. The first one, the application fee, covers the costs of administration[642] and applies to all three stages prospecting, exploration and development, when an Operator submits a notification or applies for an exploration and development permit. The second category is levied on Operators, to cover the costs of the CRAMRA institutions themselves.[643] The third category deals with payments "in the nature of or similar to taxes, royalties or payments in kind", which also must be paid by the institutes.[644] All fees are established by the Regulatory Committees with the exception of notification fees and the levies on institutions. To prevent the anticipated, and in some cases actual blocking of Antarctic mining activities, and even the establishment of the mineral regime itself by claimant states, much consideration was given to the issue of assigning revenues from mining activities. Although not resolved by CRAMRA, the spirit of protecting the claimant States' interests was manifested through sharing revenues surplus to requirements for financing the budget and by elaborating the terms and conditions of a management Scheme.[645]

The Commission's key functions involve the initiation of mining activities, by identifying areas for possible exploration and development, and modifying the boundaries of identified areas[646]. Apart from these tasks, its competence is restricted, and the role of the Commission decreases. In contrast, the independence of the Regulatory Committees *vis-à-vis* the Commission has been underlined by the Management Scheme, the Commission's toothless review power of Regulatory Committee actions and Regulatory Committee's ability to impose "tax-like" fees on Operators.[647]

E. Dispute Settlement

Two categories of dispute are defined by CRAMRA. The first category deals with disputes concerning the interpretation or application of CRAMRA, which can be further subdivided into disputes between two or more parties concerning the interpretation or application of the Convention; and disputes concerning interpretation or application related to a measures in effect pursuant to this Convention or a Management Scheme. The second category deals with disputes concerning Regulatory Committee's decisions that allegedly violate the Convention.

Dispute resolution for the first category is initiated by consultation among parties to the dispute, to agree to a peaceful means of resolution, such as negotiation, enquiry,

642 Article 21 paragraph 1(p) in connection with Article 35 paragraph 4(a), Article 37, 39, 44 and 53, CRAMRA.
643 Article 21 paragraph 1(q) in connection with Article 47(k)(i) and Article 35 paragraph 4(b), CRAMRA.
644 Article 35 paragraph 4(c) in connection with Article 47(k)(ii), CRAMRA.
645 Article 37 paragraph 7 and 11, CRAMRA. Wolfrum (note 479), at 56.
646 Article 42 paragraph 2, CRAMRA.
647 See Wolfrum (note 479), at 52–56.

mediation, conciliation, arbitration, or judicial settlement. Disputes concerning the interpretation or application of the Convention will be passed on to the International Court of Justice (ICJ) or the Arbitral Tribunal if the parties do not find a means for resolving the dispute within twelve months.[648] Parties to CRAMRA may choose either the ICJ or the Arbitral Tribunal as the means for the settlement of disputes by written declaration, in the absence of a specific statement, or where not all parties unanimously decide on the ICJ, the matter will be passed to the Arbitral Tribunal.[649] The procedure is slightly more complicated for the other subcategory of disputes, concerning the interpretation or application of a measure in effect pursuant to this Convention or a Management Scheme. It is also initiated by consultation, but if no agreement has been reached after six months the disputed measure or Management Scheme is discussed in the institutions that adopted it. If relevant parties cannot resolve the dispute within twelve months, the dispute is submitted to the Arbitral Tribunal.[650]

There are some limitations to the competence of the ICJ and the Arbitral Tribunal: they do not have the legal competence to decide issues related to legal positions under the Antarctic Treaty,[651] or to make decisions regarding the exercise by an institution of its discretionary powers or to substitute their discretion for that an institution.[652] A third limitation allows any party to make reservation on the procedure for dispute settlement[653], with the exception of the seven specially listed disputes concerning the interpretation or application of provisions or measures related to the protection of Antarctic environment and ecosystem; compliance with CRAMRA and measures in effect pursuant to it within each party's competence; response actions and liability; inspection; non-discrimination; other uses of the Antarctic; and prospecting.[654] In these potential areas of dispute, the procedure for dispute settlement provided by CRAMRA must be applied, i.e. either through a specific mechanism as agreed between the relevant parties or through the international mechanisms provided.[655] In this situation, there is no chance for any party to avoid the settlement even by applying sovereign immunity.

As the Arbitral Tribunal was specially founded for dealing with the disputes arising from Antarctic mining activities, its competence is much wider than the ICJ's in the Antarctic mineral regime. For any party without a written declaration, the Arbitral Tribunal is the default-option.[656] Disputes concerning the interpretation or application of CRAMRA which relate to a measure in effect pursuant to the Convention or a Management Scheme, can only be resolved by the Arbitral Tribunal.

---

648 Article 57 paragraph 2, CRAMRA.
649 Article 56 paragraph 4 and 5, CRAMRA.
650 Article 57 paragraph 3, CRAMRA.
651 Article 57 paragraph 4 in connection with Article 9, CRAMRA.
652 Article 57 paragraph 5, CRAMRA.
653 Article 58 in connection with Article 63 paragraph 2, CRAMRA.
654 Article 58 paragraph 1, CRAMRA.
655 Wolfrum (note 479), at 76.
656 Wolfrum (note 479), at 76.

Arbitrators are appointed by the parties. Each party may appoint three Arbitrators and should maintain at least one Arbitrator at all times.[657] The Arbitral Tribunal is comprised of three Arbitrators, two of which are appointed separately by each disputing party, and the third, i.e. the Chairman of the Arbitral Tribunal, is appointed by both parties by agreement. If Arbitrators cannot be appointed within the regulated period, the President of the ICJ is authorized to appoint them.[658]

The Arbitral Tribunal resolves disputes on the basis of this Convention, other laws not incompatible with it, including other international law, measures or guidelines issued by the Commission or Regulatory Committees, national law, and *ex aequo et bono.*[659] Only measures or guidelines compatible with the Convention may be deemed a basis for decision, in effect, making the Arbitral Tribunal a balance for the legal competence of the Commission, Regulatory Committees and other institutions under the regime that can issue measures.[660]

At the request of any party to a dispute, the Arbitral Tribunal may prescribe provisional measures either to preserve the respective rights of the parties or to prevent serious harm to the Antarctic environment or dependent or associated ecosystem.[661] If the dispute relates to the seven specifically listed issues, any relevant party may request provisional measures at any time. The decision should be made within two months of the constitution of the Arbitral Tribunal and not later than 42 days following the initial request.[662]

Any party may intervene in the proceedings if it has a legal interest that may be affected by the award of an Arbitral Tribunal.[663] The award, which is made by a majority of the Arbitrators,[664] is final and binding on both the parties to the dispute and any party involved in the proceedings.[665] Moreover, it may make an award by default final under the jurisdiction of the Arbitral Tribunal,[666] which strengthens the compulsory feature of the mechanism.

The ICJ makes judgments on the basis of Article 38 of the Statute of the International Court of Justice, which regulates that the ICJ should apply international conventions, international custom, general principles or judicial decisions and the opinions of the most highly qualified publicists of the various nations, as subsidiary means for the determination of rules of law. Thus, the ICJ is not legally competent to decide whether measures or guidelines are compatible with this Convention or not.

657 Article 2 of the Annex for the Arbitral Tribunal, CRAMRA.
658 Article 3 of the Annex for the Arbitral Tribunal, CRAMRA.
659 Article 10 of the Annex for the Arbitral Tribunal, CRAMRA.
660 Wolfrum (note 479), at 74–75.
661 Article 6 paragraph 1 of the Annex for the Arbitral Tribunal, CRAMRA.
662 Article 6 paragraph 3 of the Annex for the Arbitral Tribunal, CRAMRA.
663 Article 7 of the Annex for the Arbitral Tribunal, CRAMRA.
664 Article 12 of the Annex for the Arbitral Tribunal, CRAMRA
665 Article 11 of the Annex for the Arbitral Tribunal, CRAMRA.
666 Article 9 of the Annex for the Arbitral Tribunal, CRAMRA.

For the second category of disputes, the procedure is initiated by an operator or a party which sponsored an operator against a Regulatory Committee. Disputes may arise from a decision to reject a Management Scheme; a decision to reject the issue of a development permit; or a decision to suspend, modify or cancel a Management Scheme or to impose monetary penalties. The Commission should elaborate "an additional procedure" to govern any potential dispute resolution issues before designating an area.[667] This is expected to pose a huge challenge for the authority of the Regulatory Committee, and claimant States considered this procedure as an infringement on their mechanisms of last resort to protect their interests.[668]

The CRAMRA mandatory dispute settlement procedure provides an opportunity to review those measures or Management Schemes formulated by institutions under the regime, allowing for a balancing of different internal interests to resolve conflicts. However, it is also true that the competence of the Arbitral Tribunal and the ICJ might erode some ATCPs' interests or rights. In particular, the introduction of additional procedure for resolving disputes between operators and Regulatory Committees further dilutes the privilege of the ATCPs. This formed a huge obstacle to having the compulsory dispute settlement procedure adopted, and may be one of the reasons behind CRAMRA failing to enter into force.

F. Interests Accommodated in the Negotiation Process

Generally, the negotiations for CRAMRA needed to accommodate three major concerns, involving internal interests, external interests and achieving a balance between exploitation, in the form of potential mineral resource activities and preservation, protecting the Antarctic environment and other uses of Antarctica from adverse effects of those activities.[669]

I. Internal Interests

Internal interests, between the claimant and non-claimant States, appeared intractable, and needed to be addressed. Claimants declared that they had the exclusive right to regulate mineral development in their Antarctic sector, because they had sovereignty. Sovereign States may promulgate law to regulate activities, such as controlling mineral activities, within their territorial extent, meaning that they are independent and free from foreign intervention in their internal affairs[670]. However, allowing the seven claimant States to treat the Antarctic as part of their territories, would have resulted in Antarctic mineral resources activities being regulated according to different national regimes. As this would never be accepted

---

667 Article 59 in connection with 41 paragraph 1(f), CRAMRA.

668 Wolfrum (note 479), at 75.

669 Kimball (note 460), at 5–8.

670 Jankovié, B.M., *Public International Law*, (New York: Transnational Publisher, 1984), at 114–119.

by non-claimant States, the ATCPs were forced to find a mutually acceptable solution, although satisfying the interests of claimant States and the flag state of the operator was not a simple matter.[671] The focal point of the struggle between these two groups concentrated on the role of the constitution established by the mineral convention, as shown in the debate on their composition, their powers and functions, the relationship between the institutions, and the voting procedures for decision making. The Regulatory Committees, of which some ATCPs were members, were the most disputed institutions, as these were designed to have executive power, with the right to elaborate specific rules of mineral resource activities in specific geographic areas of Antarctica.

After negotiations, these internal interests were accommodated by protecting the interests of a claimant State through obligatory membership of a claimant State on any Regulatory Committee established for an area over which a claim existed. Moreover, the US and the USSR also became obligatory members of all Regulatory Committees, as they assert a basic claim for territorial sovereignty in Antarctica, and were the largest political, economic and military players in the 1980's. Other States' interests were accommodated by providing for a Regulatory Committee to have four non-claimant states as members. In other words, a Regulatory Committee comprises four claimants, four non-claimants and the USSR and USA.[672] It would appear from this solution that the USSR and the USA are the primary parities, as they would be involved in all mineral activities across the whole Antarctic, a level of involvement that would have been difficult for any other State.

Secondly, when developing a Management Scheme and examining an application of exploration or application of development, the Regulatory Committee should consult with the Sponsoring State and relevant claimant states, to provide an opportunity for the claimant State and flag State to reach agreement.[673] This was based on the idea that Antarctic mineral activities differed vastly, and agreement would be easier to obtain between a smaller number of involved States applying similar approaches to the different subject matter, rather than across a large number of negotiating parties. Thus, both the respective claimant State and the sponsoring States were granted a special position in Antarctic mineral resource activities.

Thirdly, the convention also provided institutions, the Commission and the Advisory Committee to protect the general interests of all parties. Designation of an area for possible exploration and development of a particular resource or resources following a request from any party fell to the Commission.[674] Regulatory Committees would only be established for the specific area once this area for exploration and development activities had been identified. The Commission is charged with establishing guidelines for Regulatory Committee rule-making, and has the power to

671 Kimball (note 460), at 23.

672 Article 29 paragraph 2, CRAMRA.

673 Article 46, CRAMRA.

674 Article 39 and 41, CRAMRA.

review a decision to approve a Management Scheme or issues a development permit by that Regulatory Committee, either through the Commission itself[675] or through a dispute settlement procedure[676]. It should be noted however that the power to review is quite limited. Additionally, the regulations governing any application must be formulated and implemented before any application can be received. This has the benefit of providing fair and transparent conditions and requirements for all potential applicants.[677] The Advisory Committee, whose membership is open to all parties, is the organ that advises the Commission and Regulatory Committees at their request, or as required by the Convention. In adopting regulations and approving exploration and development activities, the Commission and Regulatory Committees should consider the Advisory Committee evaluation of relevant scientific, technical and environmental factors, although those advisory reports are not mandatory.

Fourthly, dispute settlement provisions also serve to protect the interests of all parties, by allowing any member of the Commission to inspect mineral resource activities, and initiate a binding dispute settlement procedure. Moreover, any member of a Regulatory Committee may request the review of a Commission decision to identify an area,[678] although subject to decision making provisions this requires a three quarters majority approval in the Commission.[679]

II. External Interests

The interests of the four State groups involved in Antarctic affairs can be roughly broken down as follows: the interests of developing States; the role of non-Commission "non-decision making" parties; observer participation; public availability of information; and most importantly, the relationship between the mineral regime and the ISA established by UNCLOS.[680]

The ATCM had not considered allocating a special position or privilege to developing States before the negotiations for the mineral regime. However, as the number of developing States involved in negotiations grew, Antarctic policymaking has had to give increasingly more consideration to their demands. Firstly, four developing States, Brazil, China, India and Uruguay, were granted the status of ATCP, meaning that the voice of developing States could be heard in the Antarctic forum. There is a general principle that the special situation of developing States should be taken into account when considering an Antarctic mineral regime.[681] Developing States participation is promoted and encouraged in Antarctic mineral resource activities,

675 Article 49 and 19.
676 Article 59, CRAMRA.
677 Article 43, CRARMA.
678 Article 49, CRAMRA.
679 Article 22, CRAMRA.
680 Kimball (note 460), at 24.
681 The Preamble of CRAMRA.

and the convention provides opportunities for joint ventures or different forms of participation, including a procedure for offering such participation.[682] When a Regulatory Committee determines a method of resolving competing applications, it should *inter alia* favour the application with the broadest participation of interested parties, in particular, developing States.[683] To promote international participation, the Advisory Committee shall provide interested developing States with advice concerning the availability of relevant information, training program and opportunities for cooperation with other parties.[684] Moreover, there are at least three developing States who are members of the Commission in each Regulatory Committee,[685] and Regulatory Committee decisions must include at least one developing State in the required majority.[686] Furthermore, the interests of developing States are also considered by the Commission in determining the distribution of any revenues.[687]

Non-Commission members are not granted decision-making rights in the Commission, however, they are allowed to undertake mineral resource activities and become members of the Advisory Committee and the Special Meeting of States Parties, if they are parties to the Convention. Moreover, it is also possible for them to attend the Commission and Regulatory Committees as observers.

Observers may participate in the activities of the institutions of the regime and obtain information from these institutions. Moreover, Observer status at a Special Meeting of Parties is open to international organizations, which have been accorded observer status in the Commission or the Advisory Committee. Thus, the role of Observer also provides a linkage between the mineral regime and international organizations. The interests of the international community are represented by the Advisory Committee, as it may, in preparing its advice, seek information and advice from other scientists and experts or scientific organizations as required on an ad hoc basis.[688] In addition, the Commission and the Advisory Committee, as appropriate, shall cooperate with the United Nations, its relevant Specialized Agencies, the International Union for the Conservation of Nature and Natural Resources, and other relevant international organizations including non-government organizations, with a scientific, technical or environmental interest in Antarctic.[689]

How authorities are to be allocated is disputed under the Antarctic mineral regime and that of the ISA under UNCLOS. Generally, CRAMRA restricts its

682 Article 6 and 41 paragraph 1(d), CRAMRA.
683 Article 43 paragraph 2 (e), CRAMRA.
684 Article 26 paragraph 6, CRAMRA.
685 Article 29 paragraph 3(b), CRAMRA. It is worth noting that Chile and Argentina, as developing States, are the members of the Commissions since their claimant state statue.
686 Final Act of CRAMRA.
687 Article 35 paragraph 7(a), CRAMRA.
688 Article 26 paragraph 5, CRAMRA.
689 Article 34, CRAMRA.

application to within the seabed and subsoil of adjacent offshore areas up to the deep seabed, which conflicts with the criteria identified in UNCLOS.[690]

In summary, CRAMRA attempts to balance different interests but does not challenge the substantial two-tier decision-making structure employed elsewhere in the ATS. The mineral regime provides observer status for non-Commission parties and international organizations during deliberations of all of the institutions under the regime. It also provides a substantial public record of the institutions, which will be discussed later in this analysis. Additionally, there are some provisions encouraging developing States to participate in Antarctic mineral resource activities; although a lack of concrete measures makes it difficult to implement.

III. Protection of Environment and Mineral Resource Activities

Protecting the Antarctic environment from adverse effects of mineral exploration has been the key driver since mineral resource activities were first considered at the VII ATCM[691], when it was first agreed among the ATCPs that the environment and ecosystems of Antarctic should take precedence over potential mineral resource activities.[692] Thus, the final Convention stipulates that all proposed mining activities should satisfy the fundamental standards and principles set forth in the Convention.

First of all, should mineral resource activities occur in the Antarctic, the respective parties should bear special responsibility for the protection of the environment and the dependent and associated ecosystems; and to respect Antarctica's unique role for the global environment.[693] This provides a general principle which stipulates that potential operators and states are under obligation to protect Antarctic environmental. More detailed principles are promulgated in Article 4, which requires that decisions on Antarctic mineral resource activities be based on sufficient information to enable informed judgments to be made about possible impacts, and which prohibit any mineral resource activity unless information is available for decisions relevant to those activities. Moreover, a mineral activity would be prohibited until it is judged as not causing significant adverse effects on any aspect of the environment and ecosystems,[694] and key environmental parameters and ecosystem components

690 Article 5 and the Final Act, CRAMRA.

691 Recommendation VII-6.

692 Wolfrum (note 479), at 23.

693 Article 2 paragraph 3(a) and (b), CRAMRA.

694 Article 4 paragraph 2, CRAMRA. The proposed activity would not cause: significant adverse effects on air and water quality; significant changes in atmospheric, terrestrial or marine environments; significant changes in the distribution, abundance or productivities of populations of species of fauna or flora; further jeopardy to endangered or threatened species or populations of such species; or degradation of , or substantial risk to, areas of special biological, scientific, historic, aesthetic or wilderness significance.

can be monitored with the capacity for a swift and effective response, to prevent or mitigate environmental effects.

Secondly, these standards and principles are to be given effect through regulations governing operations and institutions. Any prospecting notification, application for exploration or development, or notification for the identification of an area must be accompanied by a detailed assessment of the environmental and other potential impacts. The Advisory Committee should undertake a comprehensive environmental assessment when identifying an area, including consideration of the preparatory work by Regulatory Committees, and an examination of the application for an exploration and a development permit.[695] Although it is not required that the Advisory Committee should evaluate a notification of prospecting, Article 38 provides that a minimum six states[696] may request the Commission to convene and consider adopting regulations applicable to prospecting, in which the Commission should seek and take full account of the views held by the Advisory Committee.[697]

Moreover, according to Article 13, the Commission may prohibit all mineral resource activities in Sites of Special Scientific Interest (SSSIs) under Article IX (1) of the Antarctic Treaty as well as Specially Protected Areas (SPAs) which have been designated by Commission for historic, ecological, environmental, scientific or other reasons. The Commission may establish an SPA at any time, and should consider whether additional prohibition or restriction of Antarctic mineral resource activities would be prudent in adjacent areas. When the Commission considers prohibiting Antarctic mineral resource activities or considers identifying an area, it must contemplate whether mineral activities in any area protected pursuant to provisions of other components of the authority should be restricted or prohibited.

Thirdly, the problem of a hierarchy of uses for the Antarctic, covering mineral resource activities, scientific research, wilderness values, or tourism, is not resolved by the Convention, however, it has been explicitly declared is that Antarctic mineral resource activities should be carried out in a manner consistent with the ATC, including the Antarctic Treaty, the Convention for the Conservation of Antarctic Seals (and the Convention on the Conservation of Antarctic Marine Living Resources (CCAMLR) and the measures in effect pursuant to those instruments. This further requires that the Commission and the Advisory Committee should consult and cooperate with ATCPs, contracting parties to the Antarctic Treaty and the Seal Convention, the Commission on Antarctic Marine Living Resources, and the Scientific Committee on Antarctic Research.[698]

695 Article 26 paragraph 4, CRAMRA.
696 Article 19 paragraph 2, CRAMRA.
697 Article 21 paragraph 2 in connection with Article 26, CRAMRA
698 Article 10, 15, and 34, CRAMRA.

G. Responses to CRAMRA

I. The UK's Antarctic Mineral Act

The UK adopted the Antarctic Mineral Act (the Act) in 1989, which aims to regulate matters of exploration and exploitation of mineral resources in the "British Antarctic Territory".[699] The Act prohibits, in accordance with CRAMRA, the carrying out Antarctic mineral activities for, or for purposes connected with, the exploration or development of mineral resources.[700] However, "prospecting activity", which is defined in a strikingly similar manner to the definition of "prospecting" in Article 1 (8) of CRAMRA may be conducted if authorised by the government of the UK or other States party to CRAMRA.[701] The prohibition is effective for UK nationals, Scottish firms and bodies incorporated under the law of any part of the UK; and that class may be extended by Order in Council to bodies incorporated under the law of any of the Channel Islands, the Isle of Man or any colony.[702]

1. *Licensing of Prospecting Activities*

Any prospecting activity, according to the Act, should be initiated by a license for prospecting activities, which would be issued by the Secretary of State, authorizing any person to carry out such activities, if an application is accompanied by suitable fee and satisfies the international obligations of the UK arising not only from CRAMRA but also from measures in effect pursuant to it, as well as other international obligations, such as those under the Antarctic Treaty.[703]

With regard to CRAMRA, every Operator carrying mineral resources activities should have a sponsoring State, who shall take appropriate measures within its competence to ensure the Operator complies with CRAMRA.[704] Moreover, all Operators must take all actions necessary and in a timely manner, if its mining activity results in or threatens to result in damage to Antarctic environment or ecosystems;[705] and the sponsoring State is obliged to ensure that its Operators undertaking prospecting maintain the necessary financial and technological abilities to respond appropriately and to pay liabilities.[706] Reflecting the Convention, the Act provides that a license may be time limited and contain conditions, including: a) requiring the licensee to pay particular sums as may, with the consent of the Treasury, be determined by or under the license at such times as may be so determined; b) requiring the licensee

699 Antarctic Mineral Act 1989, can be found at http://www.legislation.gov.uk/ukpga/1989/21/introduction.
700 Section 1, Antarctic Mineral Act.
701 Section 2, Antarctic Mineral Act.
702 Section 1(3), Antarctic Mineral Act.
703 Section 3, Antarctic Mineral Act.
704 Article 7 paragraph 1, CRAMRA.
705 Article 8 paragraph 1, CRAMRA.
706 Article 37 paragraph 3 in connection with Article 8 paragraph 2, CRAMRA.

to take out insurance against any liability; and c) providing for the termination of the license following an event specified in the license.[707] However, the Act does not explicitly provide provisions for environmental assessments of proposed prospecting activities, although some clarification is given using the phrase "authorized activities will be consistent with the international obligation of the United Kingdom".

The prospecting license may be transferred or prescribed with the written consent of the Secretary of State, which ensures that a proposed transferee is a qualified British-sponsored Operator under CRAMRA and is able to comply with the Convention.[708] Moreover, the Secretary of State may revoke, vary or suspend a license with the consent of the licensee, or if the licensee does satisfy a condition of the license or regulatory provision, or if the revocation, variation or suspension is required to fulfill an international obligation of the UK.[709] On the one hand, the provisions grant the government of the UK authority to control and supervise its Operators' actions in compliance with CRAMRA, particularly regulating that a licensee must comply with the obligations of the license under the conditions of the license even when it has been suspended, revoked, terminated or expired[710], for instance, a licensee has to perform certain actions under Article 8 CRAMRA, such as clean up pollution, remove equipment, or pay for damage, loss or impairment. On the other hand, the ability to change policy imposes extra risk on Operators, if decisions are based on the vague notion of "international obligation of the UK", which is flexible to interpretation by the UK but difficult to judge and determine by others. Therefore, the licensing provisions provide the UK government with a greater benefit than its Operators in interpreting and applying CRAMRA and other international obligations.

Moreover, the Act requires that a licensee must indemnify the UK government against any claims brought for loss or damage arising out of authorized Antarctic prospecting activities.[711] Sponsoring States are liable to indemnify for any compensation gaps for damages not paid by their Operators according to Article 8 (3) (a) of CRAMRA; and Article (8) (5) states that the liability of an Operator for loss or damage is also governed by national law and procedures. Therefore, if the UK government supplements the non-satisfied liability for its licensee as obliged to by CRAMRA, according to the Act, it would get reimbursement from the licensee in other forms. However, the Act does not elaborate specific measures for recovery of damage, which would apply other relevant British national law. Moreover, provisions regarding the strict liability of an operator in CRAMRA have been incorporated into British domestic law by section 12 of the Act. According to CRAMRA[712],

707 Section 3 paragraph 5, Antarctic Mineral Act.
708 Hendry, I.D., "The Antarctic Minerals Act 1989", 39 Int'l & Comp. L. Q.183 (1990), at 187.
709 Section 4 (2), Antarctic Mineral Act.
710 Section 4 (3), Antarctic Mineral Act.
711 Section 5, Antarctic Mineral Act.
712 Article 8 paragraph 2, CRAMRA.

strict liability is imposed on an operator, meaning a licensee or the Crown,[713] for damage to the Antarctic environment and ecosystems; and for certain types of loss and damage. Once the Liability Protocol enters into force in the UK, claims of strict liability arising out of prospecting activities should be submitted to the British national courts.[714] For claims of damage to the Antarctic environment or ecosystems, section 17 of the Act grants the Commission the right to appear as a party in the UK's national court[715]. The operator could invoke exception clauses in CRAMRA to release them from their liabilities.[716]

2. *Other Controls*

The Act provides general rules for Antarctic mineral activities, section 6 (1) of the Act provides that the Secretary of State may make specific regulations involving a) prescribing anything required or authorized to be prescribed under the Act; b) securing compliance with the international obligations of the United Kingdom; c) securing the safety, health and welfare of persons engaged in prospecting activities; and d) generally for carrying the Act into effect. Under item a), the cover-all provision, the UK government is granted not only the right to legislate as permitted by the Act, but also to make rules on anything required. Moreover, power to create rules for the implementation of international obligations might, as discussed above, allow the UK to interpret and apply international law in its own way. These regulations should be applied to all licensees and applicants, as opposed to section 6 (2) which permits different provisions for different cases, in particular, the provision for the granting of exemptions and for anything failing to be determined. If an offence against section 6 has the sole effect of causing personal injury, including any disease, impairment of a person's physical or mental condition and fatal injury, the relevant civil liability will be applied to that offence.[717]

In emergency situations, the Secretary of State may direct the relevant person or persons to comply with the international obligations of the UK, with the conditions of the license or with the regulations. Urgent matters include activities which contravene the prohibition on Antarctic mineral activities, or which contravene a condition of the license or of a regulatory provision. The power to give directions is a useful precautionary measure which could be taken before making a decision to revoke, vary, or suspend a license in any field of mineral activities.[718]

The Secretary of State may appoint inspectors to assist in the execution of the Act and make special regulations for the detailed requirements of inspectors,

713 Section 12 (2). The term "the Crown" is defined in section 2(2) as the Crown in right of Her Majesty's Government in the United Kingdom.
714 Article 8 paragraph 10, CRAMRA.
715 Article 8 paragraph 10 in connection with Article 8 paragraph 2, CRAMRA.
716 Article 8 paragraph 4 and 6, CRAMRA.
717 Section 13, Antarctic Mineral Act.
718 Section 7, Antarctic Mineral Act.

such as powers to access to vessels, aircraft, installation or stations; powers to test equipment; power to require implementation of prescribed procedures; rights to require conveyance; duties to provide inspectors with reasonable accommodation and means; and powers to be exercisable by inspectors in case of immediate or apprehended danger. Inspectors are effectively managers in charge of the whole procedure of mining activities.[719] Moreover, British inspectors are granted extended right to inspect through Article 12 of CRAMRA, which allows observers designated by any member of the Commission to inspect all stations, installations and equipment, as well as vessels and aircraft used for Antarctic mineral resource activities in the area of application of the Convention.

Generally, data received pursuant to the Act, a condition of a licence, a provision of regulations, or information related to other persons cannot be disclosed. Therefore, the UK government is able to impose different obligations and conditions on different licensees confidentially, which is consistent with the "special regulations" for individual cases approach outlined above. However, Article 16 (1) and 37(12) provides that basic data and information generated by prospecting should be made available when not of commercial value. To implement the Convention, the Act sets some exemptions for non-disclosure, including, *inter alia*, that information should be disclosed in accordance with CRAMRA or regulations under the Act.[720]

Failure to comply with regulations, directions and disclosure is an offence according to Section 10 (d), (e) and (f), which would impose a fine on the offending individual. Moreover, the Act lists some offences, including carrying out prohibited activities; providing false information in a license application; and failure to comply with conditions of license. To enforce these punitive provisions, Orders in Council may set rules for the arrest of any person in any part of Antarctica, suspected of committing an offence under the Act; for conveyance in custody of any person so arrested; for the collection of evidence, and for attendance before any court.[721]

Section 11 provides that criminal law will be applied in two situations: where the criminal is a qualifying agent or employee of a licensee or the UK government, or inspector or observer who has been so designated by the UK under Article 12 of CRAMRA while they are in Antarctic for the purpose of exercising their functions; and the second, where the crime occurred in any part of the UK. There is no question about the jurisdiction for UK nationals; however, it is worth clarifying whether "any part of the UK" includes the "British Antarctic sector". Furthermore, Section 14 provides that Order in Council may confer criminal or civil jurisdiction on any British court regarding matters arising under the "law of the British Antarctic Territory". Thus, the UK is authorized to charge all Operations, irrespective of nationality, who carry out Antarctic mineral activities in the British sector. Although this provision

719 Section 8, Antarctic Mineral Act.

720 Section 9, Antarctic Mineral Act.

721 Section 15, Antarctic Mineral Act.

does cover the risk of no criminal law being applicable,[722] it is quite dangerous for the purpose of keeping the Antarctic *status quo*, and may even cause international disorder. It obviously reflects the desire of the claimant states to manage Antarctic mineral resource activities within their claimant territory, a desire to which the rest of the world may object.

3. Summary

The Antarctic Mineral Act, on the one hand, supports and implements CRAMRA by generally prohibiting mineral activities in Antarctic with the exception of prospecting activities; laying out the specific requirements for obtaining licence; and incorporating strict liabilities under the Convention into British law. To this extent, the Act does enable the UK to comply with the Convention; and it provides a framework for further legislation for detailed regulation of mineral activities in Antarctic. On the other hand, it envisages only a partial implementation of CRAMRA, as the Act grants the UK government the power to flexibly apply and interpret the Convention, and also provides that other international obligations should be respected the same way, which might cause conflict between the application of the Convention and of other international law.[723] Additionally, the "territorial approach" for liabilities allows the UK to control activities if they are carried out in its claimed section of the Antarctic, regardless of the nationalities of operators involved. This regulation is clearly aimed at declaring British territorial sovereignty in Antarctic, at the same time it constitutes a violation of the Antarctic Treaty and the Convention and impairs peaceful order in the Antarctic.

II. Attitude of the International Community toward Antarctica

1. "Question of Antarctica" inscribed on UN's agenda

It is obvious that the rich resources in Antarctica are the key reason for the international community's attention, although conclusive scientific evidence of these mineral resources is yet to be found, and the technological and economic requirements to exploit them are considerable. However, potential commercial benefits have led not only the States who had interests in Antarctica, but also those who had not yet established a relationship with the ice-covered land to enter the fray. In order to enable them to join into the economic exploitation of this area, both the Antarctic Treaty and its mechanism were criticized as inappropriate, using the reasoning that any regime governing the economic use of an area outside of existing territorial lines must ensure that all economic activities are for the benefit

722 Hendry (note 708), at 190.
723 Hendry (note 708), at 184–189.

of humankind. This therefore, required that all "humankind" be represented in the negotiation of this regime.[724]

Bowing to this pressure, discussions were eventually thrown open at a global level. During the negotiations for the Antarctic mineral regime, the question of Antarctica was submitted to United Nations debate in 1982 by the Prime Minister of Malaysia,[725] who considered that the Antarctic belonged to the international community and that the question should be entered onto the agenda of the 38th Session of the General Assembly. This led the Antarctic to be dragged out of the ATS context and promoted into wider international stage.

Meanwhile, debates on the question of Antarctica between the ATCPs and the non-aligned states were initiated. States, headed by Malaysia, stressed the need for the General Assembly to consider the question of Antarctica, agreed that the exploration of the Antarctic and its resources should be carried out for the benefit of all mankind, and sought that the United Nations should undertake a comprehensive study of Antarctica to satisfy the increasing international interest in the Antarctic.[726] The ATCPs and some other states, such as the German Democratic Republic and Italy,[727] firmly objected from both *de lege lata* and *de lege ferenda* perspective,[728] and maintain that the ATS is a valid mechanism for Antarctic questions.

The regime of Antarctic mineral resources was one of the critical issues discussed at the United Nations forum. Non-aligned states, such as Ghana, Malaysia, the Philippines, Tunisia, Thailand and Zambia, proposed that the Antarctic and its resources should be treated as common heritage of mankind, and suggested that the regime should fall under the UN's control. They strongly argued that Antarctic mineral regime should be negotiated with the full participation of all members of the international community. Finally, in 1986 and 1987, the UN required ATCPs to

---

724 Wolfrum (note 479), at 81. The author also analyses the validity of the UN to challenge the ATS and to intervene the negotiations on Antarctic mineral resource activities. He points out that according to Article 10 and 11 of the UN Charter the General Assembly could discuss any issue unless such discussion interferes with the functions of the Security Council. Thus, the UNGA has the power to enter the question of Antarctic on its agenda. Moreover, Article 14 of the UN Charter authorizes the General Assembly to "recommend measures for the peaceful adjustment of any situation, regardless of origin, which is deems likely to impair the general welfare or friendly relations among nations". However, it is true that the UN Charter should respect international agreements. Since the Antarctic Treaty has been playing an important role in maintaining peace and security of the Antarctic, it is not reasonable for the UN to qualify Antarctica or Antarctic mineral resources as a situation "likely to impair the general welfare or friendly relations among nations". pp. 81–82.

725 UN Doc. A/37/PV.10 (1982).

726 UN Doc. A/38/132 (1983), at section III, paragraph 122–123; and UN Doc. A/38/193(1983). More details see: Hayasif (note 482), at 275–290.

727 They were not Consultative Parties until 1988.

728 Hayasif (note 482), at 288.

"impose a moratorium on the negotiations to establish a mineral regime until such time as all members of the international community can participate fully in such negotiations"[729] and to invite the UN Secretary-General to meetings[730].

However, in 1988 the ATCPs ignored the UN resolution and adopted CRAMRA, leading to expressions of "utmost regret and deep concern"[731] from non-aligned states.[732] Meanwhile, the UN also deeply regretted that the ATCPs had proceeded with negotiations and adopted the new convention despite resolutions calling for the imposition of a moratorium on negotiations.[733] Similarly, in 1989, the General Assembly was still not invited to the XVth ATCM. The UN was convinced that any regime to be established for the protection and conservation of the Antarctic environment and its dependent and associated ecosystems must be negotiated with the full participation of all members of the international community, to ensure it is for the benefit of mankind as a whole and to gain universal acceptance.[734] Moreover, it urged all members of the international community to reject prospecting and mining in and around Antarctica and to ensure that all activities are carried out exclusively for the purpose of peaceful scientific investigations and environmental protection.[735]

Besides the rejection of the international community, there were also obstacles from inside the ATCPs. Two claimants, Australia and France, declared that they would neither sign nor ratify CRAMRA. According to Article 62 of the minerals convention, it would enter into force only if 16 of the 20 ATCPs who had participated in the final session of the Fourth Special Antarctic Treaty Consultative Meeting ratified or acceded to it; additionally the seven claimants as well as the US and the USSR must ratify or accede to it.[736] Thus, it is impossible for CRAMRA to enter into force unless Australia and France rescind their rejections.

Postponing the entry into force of CRAMRA might be the most appropriate result since "it could take a generation or more of dedicated research and technological development to ensure that minerals exploitation would not destroy the Antarctic's fragile ecosystem and its place in global environmental processes".[737] Although it

---

729 UN Doc. A/41/88 B, paragraph 2. The same expression was repeated in UN Doc. A/42/46. B, paragraph 3.

730 UN Doc. A/42/46 B, paragraph 4.

731 United Nations General Assembly Records 43re session, Letter from Permanent Malaysian Mission at the UN to the Secretary-General, 8 June 1988.

732 They are Antigua and Barbuda, Bangladesh, Brunei, Cameroon, Congo, Ghana, Indonesia, Kenya, Malaysia, Nepal, Nigeria, Oman, Pakistan, Rwanda, Sir Lanka, Sudan, Uganda, Zambia, and Zimbabwe.

733 UN Doc. A/Res.43/83 A, December 7, 1988.

734 UN Doc. A/Res. 44/124 B, December 15, 1989, paragraph 4.

735 UN Doc. A/Res. 44/124 B, December 15, 1989, paragraph 5.

736 The Final Act of the Fourth Special Antarctic Treaty Consultative Meeting on Antarctic Mineral Resources.

737 World Commission on Environment and Development, *Our Common Future*, (Oxford: Oxford University Press, 1987), p. 284.

cannot be denied that the Convention does herald a dawn of resolution for the issue of Antarctic minerals through negotiations, the reality proves again that reconciling the conflicts of interest between the various states will not be a straightforward process. Claimants wish to retain sovereignty in Antarctica, and developing states expect to have it replaced with the common heritage regime. CRAMRA neither clearly protects claimants expected privileges nor explicitly preserves developing states' interests; thus, neither side would be satisfied. Of course, it is not reasonable to imagine a regime which benefits only one group of states. The Antarctic will certainly never be national territory since the increasing international participation in Antarctic matters profoundly internationalizes the continent. Similarly, it is also impossible to treat the deep seabed as a pure common heritage for all human being, as the interests of claimants cannot be entirely ignored. It is not easy to envisage an exact regime for Antarctic mineral resource exploration; however, it must be a result that strikes a compromise between the two extremes.

2. "Question of Antarctica" retreats from the UN agenda

Since the "Question of Antarctic" was first placed on the UN's agenda in 1983, it had been an annual topic until it was modified to a biennial reference in 1994, and then to a triennial reference after 1996. Finally, in November 2005, the UN replaced the text requiring the "Question of Antarctica" be included in agenda of the next UN session, i.e. the 63th session, with a term stating that the UN should "remain seized of the matter",[738] which meant that Antarctica would no longer be discussed as a regular topic nor be a subject of regular reports made by the UN Secretary-General.[739]

The latest resolution adopted by UNGA in 2005[740] emphasizes the significance of Antarctica to the international community and declared that the management and use of Antarctica should be conducted in accordance with the purposes and principles of the UN Charter. Furthermore, it recognizes the Antarctic Treaty is in furtherance of the purpose and principles of the Charter. This resolution was adopted by the First Committee by consensus without a vote.

Over the 23 years of debates in UNGA, the non-aligned states represented by Malaysia have moderated their acrimonious critique on the ATS, although they still believe ATCPs could do better.[741] In 2002, Malaysia accepted the ATCPs' invitation to be an observer and admitted that "the circumstances currently prevailing concerning the question of Antarctic demand a modified approach from the international community".[742] This is a strong signal that the non-aligned camp has recognized the status and functions of the ATS and ATCPs. However, it is not safe to say that

738 UN Doc. A/Res/60/47.
739 Beck, Peter J., "The United Nations and Antarctica, 2005: The End of the 'Question of Antarctica'?", 42 Polar Record 217 (2006), at 217.
740 UN Doc. A/Res/60/47.
741 Beck (note 739), at 220.
742 UN Doc. A/C.1/60/PV23, November 1, 2005.

the international community has totally relinquished the matter of Antarctica. In the words of the Malaysian representative, "We will seek to ensure that the United Nations remains seized of the question of Antarctica, and we stand ready to discuss this subject in future sessions of the General Assembly, if necessary."[743] If the ATCPs cannot find an approach to obtain wider support from rest of the international community and if their actions encroach on interests of those non-aligned states, the "Question of Antarctica" may be raised again in the UN.

In order to prevent Antarctica returning to the UN's agenda, ATCPs should pay more attention to the following aspects, which are concluded in the resolution 60/47 and its predecessors.[744] Firstly, ATCPs should take account of the fact that they represent a much broader international community; thus, they should appropriately modify their responsibilities and actions. Secondly, protecting the Antarctic environment and preserving its dependent or associated ecosystems should be a fundamental premise which leaves rooms for discussing Antarctic matters beyond the extent of ATCPs. Thirdly, importance should be given to the creation of opportunities for cooperation between the ATCPs and the non-aligned states on Antarctic affairs or for the undertaking of scientific research therein. Fourth, the sharing of data and information resulting from research activities in Antarctica with the international community and enhancing access for the international scientific community is also of prime importance.

The removal of the "Question of Antarctica" from the UN agenda is a milestone for the ATS and ATCPs. This reflects the gradual acceptance from the international community of the ATS' role in maintaining peace and order in Antarctica, in protecting the environment and dependent and associated ecosystems, and in freely undertaking scientific research and exchanging scientific information. It is therefore necessary to affirm the achievements of the Antarctic Treaty. Meanwhile, it is worth recalling that the UN has not abandoned Antarctica; rather, it has modified the way it approaches the question – choosing to remain seized of the matter.

3. Non-government Organizations (NGOs) and Antarctica

It is not only the UN that is giving serious consideration to the significant impact Antarctica has for the global environment and ecosystems,[745] but also a good number of NGOs; for example the Antarctic and Southern Ocean Coalition (ASOC), Friends of the Earth, Greenpeace, the World Wildlife Fund (WWF), the Antarctic Project, the Australian Conservation Foundation, and the Cousteau Society have treated Antarctica as constituting a major global environment problem.[746]

---

743 UN Doc. A/C.1/60/PV23, November 1, 2005.

744 Beck (note 739), at 224.

745 UN Doc. A/Res/44/124 B.

746 Beck, Peter J., "The Future of the Antarctic Region: Is this Great White Continent Turning Green?", presented at the 33rd Annual Convention of the International Studies Association, Atlanta, Georgia, USA, (April 1992), at 3.

Although the spirit of the conservation of Antarctica has been embodied in the ATS since the inception of the Antarctic Treaty in 1959, the movement for the protection of the Antarctic and its ecosystems by the NGOs was stirred up in 1980s, prompted by negotiations for a minerals regime in Antarctica. The international voluntary conservation movement aimed to preserve the Antarctic as a global protected area.[747]

In the beginning of the 1980s, the identification of the "ozone hole" problem and as regards the climate change led environmentalists and politicians to highlight the significant role of Antarctica in the global environmental system.[748] The shipping accidents which occurred in 1989 in Antarctica and caused enduring damage to the adjacent environment attracted more international attention and increased focus on preserving Antarctica.[749] Thus, in the late 1980s environmental protection for Antarctica had become a hot topic came to rival the discussion on mineral resources exploration.

Environmentally focused NGOs recognized the intrinsic importance of Antarctic environmental protection; however, they have always treated the region as a "global spectacle", which often embodies exaggerated actions pursuing "a more symbolic purpose".[750] The ultimate goal and slogans of these actions was to establish a "World Park" in Antarctica. The concept of a "World Park" was proposed in 1972 at the Second World Conference on National Parks[751] and its development continued during the early 1980s. It aimed at establishing a comprehensive protection regime ensuring the conservation of the pristine Antarctic environment and independent or associated ecosystems by providing principles, policies, procedures and mechanisms to govern human activities for the benefits of human beings as a whole.[752] Environmental NGOs tried various tactics – publicizing questions, grabbing media attention, exploiting well-known personalities, and attempting to reach the man in the street[753] – with the aim of attracting the world's attention and imposing pressure on ATCPs in order to impede the entry into force of CRAMRA. As to the special legal

747 Davis, B.W., "Antarctica as a Global Protected Area: Perceptions and Reality", 23 Australian Geographer 39 (1992), at 41.

748 Beck (note 746), at 1.

749 On 28 January 1989, the Bahia Paraiso, the Argentine polar transporter, ran aground near Janus and Delaca Islands from Palmer Station. Although no passengers or crew members were injured as a result of the accident, diesel fuel and other petroleum products (estimated at 600,000 litres) began leaking from the hull. In February 1989, the British vessel H.M.S. Endurance hit an iceberg near Deception Island. And the BIC Humboldt ran aground in Fildes Bay, King George Island.

750 Beck (note 746), at 9.

751 May, John, *The Greenpeace Book of Antarctica: A New View of the Seventh Continent*, (Child & Associates, 1988), at 158.

752 Beck (note 746), at 12.

753 Beck (note 746), at 11.

status of the Antarctic, environmentalists championed Antarctica as a test case, to establish an Antarctic-precedent for environmental preservation in other areas in the world that could then be emulated in other regions.[754]

However, the adoption of CRAMRA escalated the dissatisfaction of NGOs, which highlighted the compromise nature of the Convention that mainly addressed political and economic rather than environmental concerns: the Convention does not provide a separate Antarctic Environmental Protection Agency; the Convention also does not include any provisions for NGO monitoring; the Convention emphasized regulation rather than conservation; and expressed an apparent leaning towards mining interests.[755] Furthermore, NGOs point out weaknesses in the Convention's means for ensuring environmental protection; definitions concerning environment assessment are neither explicit nor clear;[756] the role of the Scientific Advisory Committee is weak; the enforcement and compliance procedures are unclear; and the precautionary approach to environmentally safe mining is fundamentally wrong.[757]

On 22 May 1989 Australia refused to sign CRAMRA and announced: "dedicated to the comprehensive protection of the Antarctic environment and in that context our strong commitment is that no mining at all – including oil drilling – should take place in and around the continent".[758] Subsequently, a similar declaration was made by France. These two governments submitted a joint working paper "Franco-Australian Draft Working Paper on Possible Components for a Comprehensive Convention for the Preservation and Protection of Antarctic"[759] to the XVth ATCM, proposing the preservation of Antarctica as a "wilderness reserve". Although these governments decided not to sign CRAMRA, they were prudent to avoid the term "World Park", which is apt to be connected with the common heritage principle.

CRAMRA's failure to enter into force was seen as a tremendous achievement of the environmental NGOs' campaign on the preservation of Antarctica in its current state. However, whether the effects prove in reality to be as good as these groups' imagination is still an unanswered question that will be discussed later.

---

754 May (note 751), at 165.

755 Beck, Peter J., "Convention on the Regulation of Antarctic Mineral Resource Activities: A Major Addition to the Antarctic Treaty System", 25 Polar Record 19 (1989), at 26.

756 There are plenty of uncertainties in CRAMRA, such as the meaning of "acceptable" and "negligible" in Article 1 paragraph 15; "significant changes", "significant adverse effects" and "substantial risk" in Article 4 paragraph 2; the nature of "safe operations" in Article 4 paragraph 4; and the form of "appropriate efforts" in Article 7 paragraph 5. See Beck (note 755), at 26.

757 Beck (note 746), at 17.

758 Quoted in Redgwell, Catherine, "Antarctica", 39 Int'l & Comp. L. Q. 474 (1990), at 476.

759 "Working Paper submitted by Australia and France", XV ACTM/WP/3. http://www.ats.aq/documents/ATCM15/fr/ATCM15_fr004_e.pdf.

III. Period Following Refusal of Australia and France to Sign CRAMRA

Besides Australia and France, there were some other States who expressed their dissatisfaction with the environmental safeguards in CRAMRA and declared their intention not to sign it. Belgium, in June 1989, issued a statement by the foreign minister supporting the Australian proposal to establish a world nature reserve status in Antarctica and voicing doubt over the Belgian ratification of CRAMRA.[760] In September 1989 the Italian Parliament unanimously required the Italian government not to sign or ratify CRAMRA, and supported the proposal for the creation of a World Park in Antarctica.[761] Moreover, the non-signatory states also include the Federal Republic of Germany and India.

As a consequence of these international pressures, environmental considerations and political factors, the Preparatory Meeting of the XVth Antarctic Treaty Consultative Meeting and the following Consultative Meeting held in 1989 were the obvious focus of world attention. At the XVth ATCM, the item "Comprehensive Measures for the Protection of the Antarctic Environment and Dependent and Associated Ecosystem" was first considered in a general debate in plenary, with working papers were submitted not only by France and Australia, but also by Chile, New Zealand, the US and Sweden.

1. *French and Australian Proposal – a Comprehensive Convention*

France and Australia introduced the notion of a comprehensive environmental protection regime in Antarctica and suggested holding a Special Consultative Meeting to begin its elaboration. According to the French-Australian view, the "convention for the preservation and protection of Antarctica" should: 1) declare Antarctica as a "nature reserve, land of science"; 2) "regulate[...] or, where agreed as necessary, prohibit[...]" human activities on Antarctica having an impact on the environment; 3) take account of the requirement of scientific activity and principles established in Article II of Antarctic Treaty; 4) lay down bases for a strategy for the protection of the environment in Antarctica; 5) become an integral part of the ATS; and 6) cooperate with world scientific programs. The proposal also emphasized "certain zones requiring special protection", in which activities should be restricted and, "where necessary, prohibited". Moreover, it pointed out that the proposed environmental convention must contain rules for the protection of the Antarctic continent and standards for surveys of the environmental impact of all activities carried on in the Antarctica.

The most important part of the French-Australian proposal is that it envisaged a system of institutions responsible for devising practical measures and monitoring their application. It designated three institutions, namely an "environmental

760 Quoted in Redgwell (ntoe 758), at 478.
761 Quoted in Redgwell (ntoe 758), at 478.

commission", a "scientific and technical committee" and an "arbitration body" to be assisted by a secretariat and an inspection and monitoring corps. The "environmental commission" includes the ATCPs and have authority for the following: to decide on all measures required to implement principles concluded by the proposed convention; to define the zone(s) in which the different types of activity may be carried on (e.g. siting of research stations, areas for tourism); to ensure coordination of the environmental protection policy; and to prepare an annual report on the state of the environment in the Antarctic. The scientific and technical committee is obliged to deliver an opinion on the measures and on scientific programs. The inspection and monitoring corps should be completely free to investigate and monitor throughout the application area of the proposed convention, and the arbitration body was intended to settle disputes arising concerning the interpretation and application of the proposed convention.

The French-Australian proposal went beyond the extent of the existing ATS. The new system of institutions would render the functions of ATCM unfeasible, as all of mankind's activities might impact the Antarctic environment which would be under the control of the "environmental commission". Theoretically, the "environmental commission" could modify any components of the ATS not consistent with an environmental purpose. While such a design might be effective for environmental protection, it would, however, stir up conflicts between the two systems, namely the ATCM and that of the "environmental institutions".[762]

2. *Chilean Proposal*

In contrast to the French-Australian proposal, Chile indicated a preference for a new regime on environmental protection under the ATS. Thus, it, submitted a working paper titled "Comprehensive Measures for the Protection of the Antarctic Environment and its Dependent and Associated Ecosystems"[763] to the Preparatory Meeting of the XVth Consultative Meeting. These comprehensive measures highlighted the importance of applying the "Antarctic Treaty System's invariable policy of anticipating possible events or damages that can occur on the frozen continent". Chile based their submission on the belief that the 1964 Recommendation III-VIII designated the "Treaty area as a Special Conservation Area" and that the ATCPs' work on the subject has always been governed by this fundamental principle. CRAMRA is one of the examples of strict environmental protection principles operating on the premise that "Antarctica is 'closed' to all activities of exploitation until the Consultative Parties declare it 'open', area by area, by means of formal consensus

762 Professor Wolfrum expresses his doubt on the effects of the new "environmental institutions" if they do not operate on the basis of consensus. See Wolfrum (note 479), at 87.

763 "Working Paper submitted by the Delegation of Chile", XV ATCM/WP/7. http://www.ats.aq/documents/ATCM15/fr/ATCM15_fr004_e.pdf

based on environmental criteria". Meanwhile, the proposal suggested the collecting of principles and general rules applicable to all activities in Antarctica.

The Chilean proposal emphasizes the fundamental role of the ATS, as providing the basic rules and principles for all human activity in Antarctica. Moreover, it introduced an evaluation of the environmental impact of all human activities, such as scientific research, tourism, navigation and the installation of new stations. The purpose of this evaluation was to avoid: "a significant degradation of the atmospheric, earthly and marine environment of Antarctic and its dependent and related ecosystem in the face of which effective restoration is not possible within a reasonable period of time; a degradation, or significant risk, in areas of special biological, scientific, aesthetic or wilderness importance; negative effects on regional and global climatic patterns." Moreover, it emphasized the different responsibilities of the enterprises, individuals, respective states and the ATCPs in protecting the Antarctic environment.

The Chilean proposal envisages a second part of the comprehensive measures dedicated to the examination, evaluation and validity of the measures approved and in force. Recommendation XIV-2 was deemed an appropriate basis for examining the environmental impact assessment of activities, ruling out those not compatible with the principles and objectives of the comprehensive measures. It also proposes an "early warning system" on the possible environmental effects of activities.

A third element of the proposed comprehensive measures addresses the additional measures regarding the modalities for transferring waste and garbage, waste disposal, installation of antennas and electromagnetic interferences, and negative consequence of scientific research. Furthermore, these additional measures must apply to "tourism and non-governmental activities in Antarctica". In order to strengthen environmental protection, this section of the comprehensive measures ultimately settled upon should list some prohibited activities.

The next and final part of the proposed comprehensive measures addresses additional considerations. In order to fulfill environmental obligations, some monitoring schemes should be established, such as an "early warning system" and follow-up procedures for activities. State parties should report annually to the ATCPs on their compliance with comprehensive measures. Moreover, the agenda for every ATCM should have an item "Comprehensive Measures".

Completely opposite to the French-Australian proposal, the Chilean proposal tries to exhaust the competences of the ATCMs, although with the same purpose of environmental protection.[764] The Chilean proposal suggested that a Special Consultative Meeting should be convened in 1990 in order to achieve the set of principles listed in the paper and to commence the work outlined above. Under such a system, CRAMRA would not be excluded but would rather remain one of the major components of the ATS.[765]

764 Wolfrum (note 479), at 87.
765 Wolfrum (note 479), at 87.

3. *US Proposal – Comprehensive Measures*

The US paper, namely "Comprehensive Measures for the Protection of the Antarctic Environment and Dependent and Associated Ecosystems",[766] is based on the view that the parties to the Antarctic Treaty have a "shared commitment" to ensure that all human activities in Antarctica are controlled by an effective, integrated and comprehensive set of measures to avoid adverse impacts upon the Antarctic environment.

The US paper outlines a program of work for the consideration and adoption of comprehensive measures. The program is intended to begin with a delineation of standards, procedures and activities. It also provides a list of appraisal criteria considered when developing these standards, procedures and activities. Furthermore, the US believes that as a result of this appraisal, all parties to the Antarctic Treaty should establish concrete objectives for the immediate and longer-term future.

The work relating to comprehensive measures emphasizes three major areas: a) ensuring effective standards and procedures for environmental protection regarding all activities in Antarctica; b) establishing clear and enforceable obligations; and c) elaborating effective mechanisms to implement the standards and procedures.

Like the Chilean paper, the US proposal agrees that environmental protection measures should be based on the ATS and all its components, and that the ATCMs are the proper forum for elaboration of these measures.

4. *New Zealand Working Paper*

The New Zealand "Working Paper on the Protection of the Antarctic Environment"[767] affirms that the Antarctic Treaty, through the Recommendations adopted pursuant to Article IX of the Antarctic Treaty, has played an important role in minimizing the impact of human activities on the Antarctic environment. However, there are "some significant gaps in the system", and the implementation of some components of the ATS has not "always been up to the desired standards". New Zealand believes that it is necessary to improve integration and coordination within the ATS, and it is also necessary to establish an integrated regime of protection for the Antarctic environment. However, the new regime should build on and strengthen previous achievements, and should not seek "to duplicate or to replace" existing measures.

The New Zealand paper highlights a number of general measures that apply to all human activity in Antarctica as well as a series of issues which need to be specifically addressed, such as "measures specific to the conduct and support of scientific research activities", "tourism" and "uses of Antarctic ice".

---

766 XV ATCM/WP/8, "Comprehensive Measures for the Protection of the Antarctic Environment and Dependent and Associated Ecosystems, submitted by the United States".

767 XV ATCM/WP/4, "Working Paper on the Protection of the Antarctic Environment Submitted by New Zealand".

As to proposed institutions, New Zealand emphasizes that in order to increase the Antarctic Treaty's efficiency and to coordinate the relationship within the ATS, it is important to establish a permanent infrastructure or secretariat for the Antarctic Treaty. The secretariat, according to the New Zealand recommendation, should be a "small cost-efficient" unit that has "in addition to its other functions, specific environmental responsibilities and expertise". Moreover, New Zealand highly recommended a "Standing Environmental Working Group", whose function would be to give advice on environmental issues to ATCMs when undertaking policy decisions. Additionally, New Zealand considers that the precedent established by CRAMRA for inspections being carried out on behalf of an institution could be applied to a broader range of human activities that may impact on the environment. Similarly, the New Zealand paper also proposed the convening of a meeting on environmental protection for 1990.

5. *Swedish Paper – Proposed Outline of a Draft Recommendation*

The Swedish paper, titled "Comprehensive Measures for the Protection of the Antarctic Environment and Dependent and Associated Ecosystem",[768] studied all four of the other proposal papers and attempted to identify their common ground and elements in order to find possible elements for a recommendation on further work. The Swedish document outlines a draft Recommendation, which should start with a statement of the need for further protection of the Antarctic environment.

The Swedish paper recognizes the principles of environmental protection embodied in the Antarctic Treaty as well as in the Recommendations and the adopted Conventions occurring within the framework of the ATS. It suggested that a consultative meeting on environmental issues should be convened in 1990. In the substantive body of the proposed Recommendation, contained four important elements, a) the need to define more exactly the environmental concerns relating to activities in Antarctica; b) the need to define human activities that cause environmental harm, c) the need to develop further basic standards for all human activities; d) proposals for action to be considered in the short and long term.

6. *Special Consultative Meeting in 1990*

The important role of environmental protection was reflected in the fact that 21 out of the 22 Recommendations adopted in total at the XVth ATCM addressed environmental issues. These measures relate to, *inter alia*, waste disposal;[769] prevention,

768 XV ATCM/WP/14, "Comprehensive Measures for the Protection of the Antarctic Environment and Dependent and Associated Ecosystem: Proposed Outline of a Draft Recommendation by the Swedish Delegation".

769 Recommendation XV-3.

control and responses to marine pollution;[770] environmental monitoring;[771] and new sites of protected Antarctic areas.[772] These measures served as the preparations for the negotiations of a Special Consultative Meeting in 1990 to consider comprehensive measures for Antarctic environmental protection, which was met with widespread support in the XVth ATCM.

In accordance with Recommendation XV-1, the Eleventh Antarctic Treaty Special Consultative Meeting was opened in 1990 to explore and discuss proposals relating to the comprehensive environmental protection of Antarctica and its dependent and associated ecosystems. The second session took place in Madrid from April 22 to 30, from June 17 to 22 and from October 3 to 4, 1991, and was attended by all Consultative Parties, parties to the Antarctic Treaty, and international governmental organizations as well as non-governmental organizations[773] who were present as observers.

A common theme of the debate on environmental protection was the number of relevant measures already adopted under the ATS providing a level support for moving forward within the system.[774] The XIth Special Antarctic Treaty Consultative Meeting (SATCM) discussed the various proposals and concluded that there was unanimous support for a new international instrument for the protection of the Antarctic environment and its dependent and associated ecosystems.[775] Moreover, there were some delegations that attempted to convene a meeting in 1990 to elaborate the protocol on liability, which would be designed to enhance the protection of the Antarctic environment, provided for under Article 8(7) of the CRAMRA, as called for in the Final Act of the Fourth SATCM;[776] while others expressed their strong expectation for a permanent ban on mining activities in Antarctica. After the first round of objections by various nations critical of the French-Australian proposal, there were various forms of diplomatic actions: Australia and France attempted to obtain support from other nations through bilateral negotiations; the US, UK and Norway were still intent on following CRAMRA;

770 Recommendation XV-4.
771 Recommendation XV-5.
772 Recommendation XV-6.
773 The international governmental and non-governmental organizations include Antarctic and Southern Ocean Coalition (ASOC), Scientific Committee on Antarctic Research (SCAR), Commission for the Conservation of Antarctic Marine Living Resources Commission of the European Communities (CEC), Intergovernmental Oceanographic Commission (IOC), World Meteorological Organisation (WMO), International Union for the Conservation of Nature and Natural Resources (IUCN).
774 Final Report of the XV ATCM, Paris, October 9–20, 1989, paragraph 55.
775 Interim Report of the XI SATCM, Vina del Mar, 19 November to 6 December 1990, paragraph 18.
776 Final Report of the XV ATCM, Paris, October 9–20, 1989, paragraph 58.

other states, especially New Zealand, suggested a compromise approach to devise a comprehensive environmental protection regime.[777]

As a result of the deliberations, the XIth SATCM finally concluded the Protocol on Environmental Protection to the Antarctic Treaty (hereafter the Protocol) and four Annexes to the Protocol, which form an integral part thereof, namely: Annex I comprising an Environmental Impact Assessment, Annex II on Conservation of Antarctic Fauna and Flora, Annex III on Waste Disposal and Waste Management, Annex IV on Prevention of Marine Pollution. The Protocol requires that the parties protect the Antarctic environment and ecosystems, and it designates Antarctica as a natural reservation to be used only for the purposes of peace and science; moreover, it prohibits mining activities as long as the Protocol is effective. According to the Protocol, a widespread willingness to consider a "lengthy prohibition or moratorium",[778] to be set in place for 50 years, has been identified.

777 Davis (note 747), at 41.

778 Interim Report of the XI SATCM, Vina del Mar, 19 November to 6 December 1990, paragraph 20.

Part 3: CRAMRA – A Dead Convention?

Subsequent to Australia and France's having explicitly announced their intention not to sign CRAMRA, and especially after a moratorium on mining activities had been identified by the Protocol, environmentalists hailed a great victory in having protected Antarctica from mining and delightedly deemed the minerals convention already dead. However, it is difficult to agree with this opinion from the perspective of both theoretical analysis and the actual statutes. Firstly, the moratorium on mining will, according to the Protocol, be reviewed 50 years after its having entered into force. The result of that review remains to be seen. Secondly, also according to the Protocol, the moratorium could be replaced by agreed and binding regulatory measures. Thirdly, although the international environmental protection movement is still flourishing, the gradual decrease of resources is an undoubted reality, especially in terms of fuel and energy, which might drive us toward exploiting new forms of resources or virgin area. Past "achievements" do not guarantee that environmental concerns will always have the upper hand over economic benefits in a field where politicians ultimately make the decisions. Unfortunately, as Dr. Beck has stated, "politicians tend to be long on environmental rhetoric but short on action".[779] Nonetheless, to some extent, the function of CRAMRA has been diminished in the vigorous battle for establishing Antarctica as a symbol of environmental protection. The following section shall undertake to objectively evaluate CRAMRA.

A. Critiques of CRAMRA

I. CRAMRA, a Disaster or a Protector of the Environment?

The most heated debate surrounding CRAMRA has been its effects on Antarctic environmental concerns. On one hand, environmentalists desire that no mining activities ever take place in Antarctica, and they see CRAMRA as a "starter's gun for miners". Although the minerals convention imposes strict criteria for such actions, environmentalists have been dissatisfied with the continuing possibilities for mining CRAMRA provides, however small. The common fear of the devastating effects of industrial activity stems from the deterioration of the environment seen all over the inhabited part of the world, which impels environmentalists to prevent the last pristine continent on earth from suffering the same fate. They maintain that there is no guarantee at all that States will agree upon a no-mining option in future negotiations. Moreover, they argue that absent any international regulation of Antarctic mineral resource activities, exploring Antarctic mineral resources

779 Beck, Peter J., "The Future of the Antarctic Region: Is This Great White Continent Turning Green?", presented at the 33rd Annual Convention of the International Studies Association, Atlanta, Georgia, USA, (April 1992), at 20.

would be an extremely risky business venture and thus no one would ever engage in such action. However, CRAMRA reduces the risk of mining ventures and offers a legal framework to secure the required tenure, inspiring competition among mining interests to invest large sums of money.

On the other hand, Mr. Beeby has rejected the above criticism, asserting that CRAMRA provides "rigorous environmental protection criteria and safeguards" based on "the most stringent safeguards ever negotiated in an international treaty".[780] Professor Wolfrum also claims that CRAMRA, to some extent, "should be regarded as an environmental protection scheme rather than as an explorational and developmental scheme".[781]

Firstly, the adoption of CRAMRA does not automatically initiate mineral resource activities in Antarctic, since the Final Act of the Fourth Special Antarctic Treaty Consultative Meeting on Antarctic Mineral Resources reiterates and goes further[782] than the prohibition of Recommendation IX-1[783] for any mineral activities in the period preceding the entry into force of the Convention. The Final Act distinguishes prospecting activities from scientific activities. The former is defined with the aim of "identifying areas of mineral resource potential for possible exploration and development", "not including dredging and excavations... or drilling ... to depths not exceeding 25 meters".[784] Moreover, the Final Act extends the scope of the moratorium from landmass, under Recommendation IX-1, to marine areas of Antarctica.

Secondly, one of main purposes of the minerals convention is to ensure that potential mining activities are carried out in a responsible and environmentally safe manner.[785] To this purpose, CRAMRA sets up strict procedures before any area could be identified for possible exploration and development, requiring a consensus decision with full account of all factors including environmental impact. The identification of an area does not constitute permission for any exploration and development project within it. For further decision-making, a comprehensive expert environmental and technical evaluation must always be undertaken.

---

780 Beck, Peter J., "Convention on the Regulation of Antarctic Mineral Resource Activities: A Major Addition to the Antarctic Treaty System", 25 Polar Record 19 (1989), at 25.

781 Wolfrum, Rüdiger, *The Convention on the Regulation of Antarctic Mineral Resource Activities: An Attempt to Break New Ground*, (Heidelberg: Springer-Verlag, 1991), at 2.

782 This point is from: Francioni, Francesco, "Resource Sharing in Antarctica: For whose Benefit?", 1 Eur. J. Int'l L. 258 (1990), at 261.

783 Recommendation IX-1 sets forth the requirement: "to refrain from all exploration and exploitation of Antarctic mineral resources while making progress towards the timely adoption of an agreed regime concerning Antarctic mineral resource activities".

784 Article 1 paragraph 8, CRAMRA.

785 Article 2 paragraph 3, CRAMRA.

Thirdly, in accord with the various stages of mining activities, CRAMRA regulates environmental protection at a range of levels relevant to the stage of activity. At the prospecting stage,[786] the sponsoring state has to certify its operator's financial and technical ability to comply with the provisions of the Convention and has to provide an assessment of possible environmental impacts. The operator has to ensure the removal of all installations and equipment and to rehabilitate the site at the end of the prospecting activities. Moreover, all prospecting activities must comply with the general principles of the Convention.[787]

At the exploration and development stage, CRAMRA provides for a regime encompassing authorizations, permits, controls and supervision, a regime that is "one of the most demanding and sophisticated multilateral arrangements ever developed in the field of international environmental law".[788] The party applying for permission to identify an area for possible exploration and development must specify the resource(s) being sought, give a detailed description of the physical and environmental characteristics, indicate the planned scale and methods, and provide a detailed assessment of expected environmental impact.[789] Applications are considered by the Advisory Committee and Special Meeting of Parties before being later decided by the Commission. The latter must "give special weight to the conclusion of the Special Meeting of Parties" and take full account of the views and conclusions of the Advisory Committee.[790] Maximum protection of the Antarctic environment is provided by the requirement for a decision by consensus on the identification of an area for possible exploration and development.[791] Once the area is identified, a Regulatory Committee is to be established to set up guidelines for potential mining activities within the area.[792] The Regulatory Committee must then elaborate the "most important instrument in the whole mineral regime": a specifically designed management scheme,[793] which should cover, *inter alia*, "measures and procedures for the protection of the Antarctic environment and dependent and associated ecosystems, including methods, activities and undertakings by the Operator to minimize environmental risks and damage".[794]

Fourthly, CRAMRA should establish certain sensitive zones within which mining activities never take place.[795] There are three kinds of protected zones: a) zones designated under Article IX of the Antarctic Treaty, including Specially Protected Areas (SPA), Sites of Special Scientific Interest (SSSI) and other protected areas; b) protected zones, for historical, ecological, environmental, scientific or other reasons,

786 Article 37, CRAMRA.
787 Such principles are regulated in Chapter I of CRAMRA, e.g. Articles 2, 3, 4 and 8.
788 Francioni (note 782), at 262.
789 Article 39, CRAMRA.
790 Article 41 paragraph 1, CRAMRA.
791 Articles 42 paragraph 2 and 22 paragraph 2(C), CRAMRA.
792 Article 43, CRAMRA.
793 Francioni (note 782), at 263.
794 Article 47(b), CRAMRA.
795 Article 13, CRAMRA.

designated by the Antarctic Mineral Resources Commission; and c) buffer zones, which are also created by the Mineral Commission. Within the first kind of zones, mineral resources activities shall be automatically prohibited. This effectively removes those identified areas from consideration for potential exploration and development, and does not require Mineral Commission affirmation. For pending designations, the prohibition will apply to those mineral resource activities which prejudice the designated purpose. The Mineral Commission is authorized to designate protected areas for the second category, additional to those referred to in the first category, and to determine prohibitions or restrictions of mineral resource activities in such areas. Moreover, the provision especially mentions that when the Commission identifies an area for possible exploration and development, it "shall consider whether to restrict or prohibit" mining activities "in any area".[796] Thus, prohibition or restriction might be imposed not only in designated protected areas but also areas which are later set aside. In other words, the Commission can prohibit or restrict mineral resource activities all over CRAMRA's area of application, independent of the question whether it is designated as a "protected area". This design, one of the most effective provisions for environmental protection, might expand, theoretically, the no-mining option to the whole of Antarctica, both landmass and seas. The protection of environmental and ecosystems is neither a simple task nor one which could be completely fulfilled in a limited area. For this reason, CRAMRA elaborates a "buffer zone", adjacent to a protected area. Within these areas the Commission could decide to prohibit or restrict mining activities for the purpose of preserving areas of special biological, scientific, historic, aesthetic or wilderness significance. Full exercise of these provisions by various institutions under CRAMRA, could comprehensively protect and preserve the Antarctic environment and its dependent or associated ecosystem.

Bluntly put, environmentalists reject CRAMRA as their goal of prohibiting mining activities in the Antarctic cannot be satisfied by such a minerals convention. Yet CRAMRA is a convention aimed at controlling potential adverse effects of future mining activities on the Antarctic environment, rather than attempting to completely prohibit such activities. This makes criticism of CRAMRA somewhat unreasonable – minerals conventions are designed to provide a mining opportunity. To date, the environmental arguments put forward appear more appropriate as a form of political propaganda rather than a legal proposition. Actually, CRAMRA does make great efforts for the protection of the Antarctic environment and its ecosystems.

II. Lack of Clear Terminology

CRAMRA is not a perfect document: there is an abundance of vague terms, relating to environmental assessment and management. This has created the fear that the minerals convention may remain a paper tiger to be flexibly employed and interpreted at a future time to benefit exploration under commercial or political pressure.

796 Article 13 paragraph 3, CRAMRA.

Such anxiety is quite reasonable and may prove justified. However, the fact is that when the ATCPs negotiated the minerals regime in the 1980s, no prior exploitation had taken place, as appropriate technologies did not exist at that time, and quite possibly still do not exist today. Thus, the ATCPs faced an extremely difficult challenge in attempting to conclude a detailed regime regulating future activities. It is also impractical to envisage, on the basis of existing knowledge, an effective regime for managing the uncertain environmental impacts of a hypothetical technology.

CRAMRA in light of the lack of technology and therefore experience of mineral exploitation in Antarctica regulates that the region should remain closed to all mineral resource activities, until such a time as it can opened in a manner based upon adequate information. This information must demonstrate no negative impact to the Antarctic environment and its dependent or associated ecosystems[797], and must be accompanied by adequate regulations.

Firstly, CRAMRA requires a liability protocol, to be adopted by the Commission by consensus and designed to enhance the protection of the Antarctic environment and ecosystems.[798] No exploration or development shall take place until the liability protocol is in force and has been adopted by the applicant. In other words, after CRAMRA enters into force, mineral resource activities would still not be allowed unless all members of the Commission have agreed upon a liability protocol, which must enter into force according to the same procedure as does CRAMRA and be accepted by parties who intend to conduct such activities.

Secondly, the Commission, as a functionary organ under CRAMRA, is authorized to adopt measures for the protection of the Antarctic environment and its ecosystems, to adopt measures for the promotion of technical practice, and to produce a handbook of such measures.[799] These measures could be concluded at any time, if necessary, and would apply to all operators in all three stages, i.e. prospecting, exploration and development.

Thirdly, after identifying an area for possible exploration and development, the relevant Regulatory Committee should adopt guidelines identifying the general requirements for further mineral resources activities in its area of competence.[800] The general guidelines should be applicable to all operators who want to carry out exploration and development in the relevant areas and cover, *inter alia* measures and procedures for environmental and technical requirements. Moreover, for an individual operator, the Management Scheme will elaborate the specific terms and conditions for exploration and development activities. The ultimate direction should be consistent with the general guidelines, which provide the minimal requirements on relevant items. Thus, a Management Scheme might modify levels of environment protection and other conditions to incorporate knowledge and techniques at that time.

797 Article 4, CRAMR.
798 Article 8 paragraph 7(a) and (b), CRAMRA.
799 Article 21 pargraph 1(c), CRAMRA.
800 Article 43 paragraph 3 in connection with Article 47, CRAMRA.

In summary, in order to manage operators' mineral resource activities, there would pursuant to CRAMRA be a series of regulations, measures adopted by ATCPs or by the Commission, general guidelines adopted by Regulatory Committees, and particular Management Schemes agreed upon both by sponsoring States and member(s) of the respective Regulatory Committee. These regulations would improve gradually, consistent with increased knowledge and scientific and technological developments. Environmental issues, of course, would be mainly addressed within this gradually maturing regime. It is not impossible to envisage that there would ultimately be a strong and effective regime based on good scientific and environment practices.

III. UN-International Participation

The Antarctic Treaty was submitted to the UN General Assembly based on the doctrine of common heritage and the argument that all the members of UN should participate in the administration of Antarctica. CRAMRA, concluded by the ATCPs – a club of self-interested governments– cannot, as such, represent the interests of humanity as a whole. Such criticism has been expressed by a series of UN resolutions,[801] which declare that a minerals regime should not be legal until all members of the international community can fully participate in its negotiations. These resolutions express the idea that ATCPs' decision and conventions cannot directly bind third parties. This point is correct according to international law. According to the Vienna Convention on the Law of Treaties, generally speaking, "a treaty does not create either obligations or rights for a third state without its consent".[802] This provision has been confirmed by case law.[803] However, if the parties to the treaty have intended the provision to be the means of establishing the obligation, and if the third party has expressly accepted the obligation, then the obligation arises for the third party.[804] Thus, it is not by virtue of the treaty itself that the obligation is imposed on third states, but rather by the latter's acceptance. A similar requirement has been imposed for the rights of a third state; however, the third state must exercise this right according to the conditions provided for in the treaty or established in conformity with the treaty.[805] Finally, if a rule contained in a treaty has been recognized as a customary rule of international

---

801 Such as UN. Doc. A/41/88, December 4, 1986; UN. Doc. A/42/46, November 30,1987; UN. Doc. A/43/83, December 7, 1988; and UN. Doc. A/C.1/44/L.69, November 20, 1989.

802 Article 34, Vienna Convention on the Law of Treaties.

803 Cases relate to third states' obligations: the Free Zones cases, Permanent Court of International Justice, Ser. A/B, No. 46, at 41; the River Oder case, Permanent Court of International Justice, Ser. A, No. 23, pp. 19–22; and Island of Palmas arbitration, the Permanent Court of Arbitration. The North Sea Continental shelf case relates to treaty rights of a third state.

804 Article 35, Vienna Convention on the Law of Treaties.

805 Article 36, Vienna Convention on the Law of Treaties.

law, it would bind a third state without any restrictions.[806] Therefore, the Antarctic Treaty and other components of ATS are normally invalid for third states, unless some of the principles involved have become customary international law.

When the International Law Commission negotiated the Convention of Treaties, another exception to *pacta tertiis nec nocent nec prosunt* was considered, "objective regimes" whose obligations and rights are valid *erga omnes.* "Objective regimes" are "treaties for the neutralization or demilitarization of particular territories or areas", and "treaties providing for freedom of navigation in international rivers or maritime waterways".[807] The Antarctic Treaty was mentioned during the negotiation process as an example of such a treaty.[808] However, ultimately the concept was not included in the Convention of Treaties.[809] Nevertheless, the Antarctic Treaty has been valid for 50 years, even if not entirely, at least principles of neutralization and demilitarization of the applicable area have become customary international laws.

However, transferring the issues in Antarctica, especially the minerals regime issue, to the UN forum might not be a wise approach; rather, it would provoke sensitive discussions of territorial claims and the legal status of Antarctica, and consequently potentially disrupt the current peaceful order. Moreover, practice has shown that ATCPs have been the best forum for solving Antarctic issues between different groups of states, claimants and non-claimants, especially when new problems arise such as the preservation and exploitation of living and non-living resources.

The next question is whether it is legitimate that the ATCPs have authority to manage and explore Antarctic mineral resources. The Antarctic Treaty powers do not create an exclusive club as it includes both claimants and non-claimants and developed and developing states. The number of ATCPs is now 28, more than twice its original number. Some developing states have obtained positions as ATCPs, such as Brazil, China, India, Peru and Ukraine. This fact might, to some extent, reconcile the dissatisfaction with the ATS emanating from the substantial third world camp and prove that acceding to the Antarctic Treaty could easily solve the question of international participation in the management of Antarctica. Of course, the scale of ATCPs is still relatively small compared to the world as a whole. However, there

806 Article 38, Vienna Convention on the Law of Treaties.

807 United Nations, *Year Book of the International Law Commission*, Vol. II, 1966, p. 231.

808 Ibidem.

809 Although the notion of "objective regimes" has not been included in the Convention on Treaties, it has obtained support from some scholars. For example, McNair regards "dispositive or real treaties" and "constitutive or semi-legislative" treaties as having *erga omnes* effects. The former are treaties which create or transfer or recognize the existence of certain permanent rights of a territorial character. The latter are treaties of a public law character, such as treaties relating to neutralization or demilitarization. Rousseau also believes there is an exception to the general *pacta tertiis* rule in the case of treaties creating "objective regimes". See: Sinclair, Ian, *The Vienna Convention on the Law of Treaties*, (2nd Ed., Manchester University Press, 1984), at 104.

is no doubt that the status of ATS and ATCPs have been gradually recognized by the international community.

Secondly, it is more important to judge the ATS in terms of actual effects in preserving the interests of the international community in Antarctica than in terms of numeric participation.[810] The most important success of the Antarctic Treaty has been to freeze issues of sovereignty. This has kept Antarctica free from international conflicts on complicated territorial claims. Now more than 50 years along, the Antarctic Treaty and other components of the ATS have played a crucial role in preserving the continent for peaceful use, in promoting scientific research in Antarctica, in conserving the Antarctic environment and its ecosystems, and in encouraging international cooperation.

Thirdly, the rationality of questioning the authority of ATCPs is highly doubtful, as this arose only from the initiation of the mineral regime negotiations, prior to which ATCPs had concluded many measures, resolutions and even conventions. It is worth asking why other components of the ATS have not been critiqued as seriously as CRAMRA, given they were all concluded by the same organ and apply to similar areas. The exploitation of mineral resources is such a controversial matter since it directly relates to the issues of sovereignty in Antarctica. The mechanism by which the Antarctic mineral resource activities would be licensed was the crucial question which might affect the legal status of Antarctica. The non-parties to the Antarctic Treaty had eagerly expected to join the negotiations procedure, even though the final text of CRAMRA would not have been significantly changed by international participation, as the resulting procedural justice would play an important role in increasing the transparency of the proposed minerals regime and in eliminating misunderstanding and reconciling confrontations between the ATCPs and the international community.

The rationale behind the issues of legitimacy does not focus on which is the more appropriate organ to conclude the minerals regime, the ATCPs or the UN, but rather whether the existing minerals convention adopted by the ATCPs has the capacity to satisfy external interests in Antarctica through specific provisions on access to resources, and the sharing of benefits, information and decision-making procedures.[811]

IV. Inadequate Protection of Developing States

Pressure placed on CRAMRA comes partly from the developing states, which have not been sufficiently considered in the negotiations nor effectively protected in the final text of the minerals convention. This flaw in CRAMRA is based on the fact that there were indeed quite few developing states involved in the negotiations process. When the IVth SATCP started in 1982, there were (excluding the claimant states of Argentina and Chile who possess different interests to other developing states)

810 Francioni (note 782), at 266.
811 Ibidem.

only two developing states, Poland and South Africa, who attended the meetings as consultative parties. It is not a surprise that the special needs of developing states were ignored given such a low developing-developed state ratio at the ATCPs.

Firstly, not all parties to CRAMRA obtain positions in the decision-making organs, the Commission and the Regulatory Committees. Full membership depends on consultative party status, which requires satisfying the "substantial activity" criterion. Normally, operations in Antarctica are costly and risky, a huge burden for less wealthy developing states. Thus, the structure of the Commission does not benefit developing states in achieving their objectives. Some remedies for this obstacle were envisaged, e.g. non-consultative parties could take part as observers in the meetings of the Commission and Regulatory Committees; and they are involved in the Advisory Committee and the Special Meeting of Parties, both advisory organs. Moreover, the Commission might open to a special non-consultative party if it is actively engaged in substantial research in the applicable area of CRAMRA or if it is a sponsoring state during a relevant Management Scheme in force.[812] These provisions might be useful for specific developing states interested in Antarctic mineral resources activities, but they offer quite limited functions for developing states as a whole.

Secondly, developing states have not been sufficiently protected in the structures of Regulatory Committees. According to the Article 29(3)(b) of CRAMRA, at least three developing states should be "adequately and equitably" involved in each Regulatory Committee, considering the "overall balance between developed and developing country members of the Commission". Argentina and Chile hold both claimant status and developing state status. Thus, generally, only one seat will be reserved for non-claimant developing states. Of course, it is possible to assign more than one position for normal developing states; however, developing states universally lack confidence in such an assumption.

Thirdly, the decision-making process in Regulatory Committees only requires at least one developing state be involved in the two-thirds majority under Article 32 of CRAMRA.[813] Such a regulation has difficulty satisfying developing states' interests as one of the two claimant developing states may agree with a Regulatory Committee decision thus satisfying the regulation, if not the other developing states interests.

It cannot be denied that CRAMRA provides some opportunities for developing states to have access to mineral resources. Firstly, cooperation and international participation, including in particular developing states, has been explicitly required by CRAMRA, although without practical and effective measures.[814] Secondly, when there are competing applications, resolutions should respect the broadest

812 Article 18 paragraph 2 and 3, CRAMRA.

813 Final Act of the IVth SATCM

814 Article 6, Article 41 paragraph 1(d) and Article 44 paragraph 1(e), CRAMRA.

participation among interested parties, in particular, developing states regardless of whether they are ATCPs.[815]

V. Disposition of Surplus

After all the administrative and operational costs of the regime have been met it is still not clear who has a right to whatever could be categorized as surplus. The answer could be found in Article 35(7)(b), which in its "cryptic and awkward language"[816] requires that "the interests of the members of Regulatory Committees having the most direct interest in the matter in relation to the areas in question are respected in any disposition of that surplus". The members who have the "most direct interest in the matter" are the respective claimants in the identified areas.

The provision could be read as a covert recognition of a claimant state's right to a special share of revenues, which might be further construed as a recognition of territorial claim. If a provision of CRAMRA provides a remedy to this disposition of surplus problem, it must be the requirement for a three-quarters majority of the Commission in decision making; however, this criterion is not the strictest in CRAMRA nor in the Commission regime. This provision reflects the complicated reality of Antarctica and the compromise outcome of the conflicts between claimants and non-claimants.

Claimants insisted on their territorial sovereign rights during the negotiations and declared that sovereignty "will not be spent".[817] Claimants initially supposed their ownership of substantive management power in their claimed sectors. However, to achieve consensus among ATCPs, only special economic benefits were recognized in the final text of the Convention, meaning that claimants might assert a special entitlement to the economic benefits of exploration and development activities carried out in their claimed areas.

Even though the disposition of surplus might be inconsistent with the Antarctic Treaty, it is on occasion wise to sacrifice innate principles in order to achieve compromise. It is hard to image a minerals regime which disregards the interests of claimants. Moreover, much like the procedure of the making the Antarctic Treaty, if ATCPs – in the tense international atmosphere during the mineral negotiations – had failed to agree among themselves, this matter probably would have been "taken out of their hands".[818]

CRAMRA could be seen as an environmental treaty intended to protect the environment to the greatest extent possible, although nobody knows its real effects on environmental protection. While international participation in the negotiation did exclude some groups, is not really a serious challenge to the validity of either the CRAMRA or the authority of ATCPs having concluded the minerals convention.

815 Article 43(2)(e), CRAMRA.
816 Francioni (note 782), at 264.
817 Beck (note 780), at 28.
818 Maddox, John, "Antarctic Mining Regime at Risk", 307 Nature 105 (January 1984).

Having said that, the absence of protection for developing states definitely brought enormous objection from the international community. The important lesson to be learned from the CRAMRA process therefore is the need for greater consideration of external parties for future minerals regimes.

B. Merits of CRAMRA

CRAMRA itself is not a failed convention; on the contrary, as evaluated by Professor Wolfrum: "CRAMRA will serve as model for the future development of international law concerning the cooperation of states in the administration of areas beyond national jurisdiction or with disputed jurisdiction".[819] Moreover, it is a good example of ATCPs capacity to resolve emerging issues related to Antarctica through peaceful measures – namely comprehensive negotiating – which enhances the crucial role of the ATS in Antarctic matters.

I. CRAMRA and the Reconciliation of Internal Tensions

Mineral resources activity is a sensitive question relating to the territorial sovereignty of Antarctica. On one hand, claimants used their territorial claims to seize authority in regulating such activities. On the other hand, non-claimants obviously rejected national control of Antarctic mineral resource activities. In order to get wider support, CRAMRA balances various interests through its series of organs and procedural safeguards.

Firstly, powers under CRAMRA have been distributed into two decision-making organs, the Commission and Regulatory Committee. The former is the functionary organ, which possesses general competence and includes each ATCP "on the date when this Convention was opened for signature".[820] The latter possesses special competence with each Committee having ten members including four claimants, four non-claimants and the US and USSR. The purpose of each Regulatory Committee is to obtain support from claimants by providing them particular privileges. Article 46 provides that the Management Scheme, which "prescribe[s] the specific terms and conditions for exploration and development of mineral resources",[821] should "have recourse to" the sponsoring State and respective claimant(s), on whose claimed sectors proposed mineral resource activities are to be carried out. This provision reveals that the ultimate operational mechanism of the minerals convention is held by both the respective claimant(s) and sponsoring State, but especially by the former. If there are obstacles preventing a respective claimant(s) and sponsoring State from achieving agreement, proposed mining activities become very difficult. Thus,

819 Wolfrum (note 781), at 2.
820 Article 18 paragraph 2, CRAMRA.
821 Article 47, CRAMRA.

it is the claimants who finally determine whether there will be mineral resource activities or not.

In order to limit the claimant's huge privileges, non-claimants have made great effort to establish a system of "checks and balances" which strive to control mining activities as much as possible.[822] Firstly, the Commission is authorized to elaborate "any additional associated conditions" for mineral resource activities when it decides to identify an area.[823] For example, this provision enables the Commission to establish rules to restrict the discretionary powers of the relevant Regulatory Committee.

Secondly, Article 49 of CRAMRA provides a review procedure. Any member of the Commission or of a Regulatory Committee may, within one month, request the Commission to review a Regulatory Committee's approval of a Management Scheme or issuance of a development permit. Where this review application is well founded, the Regulatory Committee would be asked to reconsider its decision. Having said this, this review design is too weak to exercise effective function, reducing the reconsideration procedure to a role based more on political pressure than one of legal restriction.

Thirdly, CRAMRA calls for the establishment of an effective system of inspection of Antarctic mineral resource activities. This also provides observers designated under the Antarctic Treaty with the ability to inspect mineral resource activities,[824] and to be be designated by any member of the Commission, the Commission or a relevant Regulatory Committee.[825] Moreover, aerial inspection may be carried out at any time.[826] No exploration or development activities are allowed unless an effective system of inspection has been established in the identified area.[827] The minerals convention does not explicitly express which organ has the right and obligation to elaborate the inspection regimes, although it is safe to assume that both the Commission and the relevant Regulatory Committee are authorized to do so.

Fourthly, a compulsory dispute settlement process provides another mechanism to reconcile the conflicts among states. Generally, any dispute concerning the interpretation or application of CRAMRA, a measure in effect pursuant to this convention or a Management Scheme shall be solved under the settlement procedure provided in the Convention, although this compulsory requirement is softened by the exclusion of some categories of disputes.[828] CRAMRA not only provides an opportunity for judicial review of Management Schemes, but also envisages the establishment of additional procedures for third-party settlement of disputes regarding a Regulatory Committee's action concerning a decision to decline a Management

822 Francioni (note 782), at 264.
823 Article 41 paragraph 1(e), CRAMRA.
824 Article 11, CRAMRA.
825 Article 12 paragraph 1, CRAMRA.
826 Article 12 paragraph 2, CRAMRA.
827 Article 12 paragraph 8, CRAMRA.
828 Article 58, CRAMRA.

Scheme or a development permit or concerning a decision to suspend, modify or cancel a Management Scheme or to impose monetary penalties.[829]

Fifthly, the decision-making process of Regulatory Committees might weaken the claimants' privileges. A Management Scheme, concluded by a relevant claimant(s) and a sponsoring State, or a development permit must be approved by a two-thirds majority of the members of the relevant Regulatory Committee, with this majority to include a simple majority of claimants and a simple majority of non-claimants.[830]

II. CRAMRA and Part XI of UNCLOS

Before CRAMRA was concluded, there were lively discussions about the relationship between the proposed mineral regime and the existing and pending UNCLOS. Discussions circled around matters including whether maritime zones and a continental shelf exist in the Antarctic, how to delimit the deep seabed and the continental shelf of Antarctica, and whether the common heritage principles would be applied, among others. All these questions are based on the sensitive issue of territorial claims, i.e. the legal status of Antarctica. CRAMRA deals with these complicated questions in a subtle way.

Firstly, CRAMRA continued to sidestep the territorial issue and did not address the delimitation of the maritime zones. It regulated that all mineral resource activities within the applicable area, including landmass and maritime areas, should be controlled by the minerals convention. Article 5 of CRAMRA[831] resolves disputes between claimants and non-claimants regarding conflicts of interest by ensuring that the minerals convention is applied to all mineral resource activities taking place on the Antarctic continent and "in the seabed and subsoil of adjacent offshore areas up to the deep seabed".

---

829 Article 59, CRAMRA.

830 Article 32, CRAMRA.

831 Article 5 Area of Application reads:
1. This Convention shall, subject to paragraphs 2, 3 and 4 below, apply to the Antarctic Treaty area.
2. Without prejudice to the responsibilities of the Antarctic Treaty Consultative Parties under the Antarctic Treaty and measures pursuant to it, the Parties agree that this Convention shall regulate Antarctic mineral resource activities which take place on the continent of Antarctica and all Antarctic islands, including all ice shelves, south of 60 deg. south latitude and in the seabed and subsoil of adjacent offshore areas up to the deep seabed.
3. For the purposes of this Convention 'deep seabed' means the seabed and subsoil beyond the geographic extent of the continental shelf as the term continental shelf is defined in accordance with international law.
4. Nothing in this Article shall be construed as limiting the application of other Articles of this Convention in so far as they relate to possible impacts outside the area referred to in paragraphs 1 and 2 above, including impacts on dependent or on associated ecosystems.

Secondly, Article 5 also resolves the dispute between, on one side, the opinion that ISA has jurisdiction over the whole of the deep seabed and, on the other, the argument that the Antarctic is a special area which is under the regime of the ATS and that ISA should thus be excluded in the Antarctic Treaty area, i.e. south of 60°south latitude.[832] It generally regulates that the applicable scope is the Antarctic Treaty area; however, to the extent seabed and subsoil are beyond the continental shelf in "accor[d] with international law", those areas fall under the jurisdiction of international law. This regulation ensures the unification of the ATS, on one hand, and respect for international law on the other hand, perfectly reflecting the flexibility of the ATS.

III. CRAMRA and the International Community

As has been discussed above, the international community has paid considerable attention to Antarctic and especially to the mineral regime negotiations since the 1980s. CRAMRA does make some efforts to improve international cooperation in respect to the mineral regime.

As a general principle, ATCPs should take into account the "interests of the international community as a whole" when they consider mineral resources activities.[833] Moreover, all parties, both ATCPs and non-ATCPs to CRAMRA, should improve their level of cooperation and encourage international participation in Antarctic mineral resource activities.[834] These principles determine that the international community should be respected during the entire mining activities process. In addition to cooperation between the parties and other international actors, CRAMRA also establishes cooperation with international organs, e.g. the Commission should cooperate with the UN and its relevant specialized agencies, with the International Union for the Conservation of Nature and Natural Resources and with other relevant international organs, including NGOs with a special interest in Antarctica.[835]

With regard to practical provisions for international cooperation, the Commission may accord observer status in the Commission and the Advisory Committee to international organizations if it is appropriate and helpful for the institution in question. Observers of the Commission or the Advisory Committee automatically acquire observers' positions at a Special Meeting of Parties.[836] Further accommodating the interests of the international community, there is a provision making public the records of the Commissions' meeting and decisions as well as their information, notifications and submitted reports,[837] and also Regulatory Committees' decisions

832 Wolfrum (note 781), at 33.
833 Article 2 paragraph 3(g), CRAMRA.
834 Article 6, CRAMRA.
835 Article 34 paragraph 2 and 3, CRAMRA.
836 Article 34 paragraph 4, CRAMRA.
837 Article 21 paragraph 4, CRAMRA.

and Management Schemes in force.[838] Although this level of transparency might not affect the final decision – as these are items normally published after decisions have been made – the international community will know what has happened in respect of Antarctica and will have the opportunity to advise and make comments. Moreover, the Commission and Regulatory Committees with decision-making authority will have to consider the reaction of the international community as well as their own reputations and can be expected to – more or less – exercise prudent actions, although the pervasive effects of these transparency measures are debatable.

C. Underlying Reasons

Like all treaties, CRAMRA is not perfect. Both advocates and critics could provide adequate reasons in support of their opinions. However, the fate of CRAMRA is so unpromising that much attention has been paid to the Convention *per se* rather than the deeply embedded reasons behind it. The minerals convention has been sidelined by virtue of two claimant states, Australia and France, having explicitly expressed their unwillingness to sign it in the name of protection of Antarctica environment and its ecosystems. Furthermore, the Protocol suspends mining activities in Antarctic for 50 years. Are all these actions really for the admirable reason of environmental protection?

I. Political reasons

Australia has declared that the "Antarctic environment is extremely fragile and critically important to the whole global ecosystem", and that "mining in Antarctica will always be dangerous, and could be catastrophically so".[839] Thus, the country was the first to decide not to sign CRAMRA and believed that its action would be supported and commended by most parts of the international community. Moreover, were this position to prevail, the world would have to recall that it was Australia that stood in the vanguard to protect Antarctica for all mankind. Thus Australia would reap important political benefits instead of potential economical revenues. "As a middle power", Australia praises itself for having "le[d] efforts to liberalize world trade in the Uruguay Round; initiating moves to establish close economic cooperation in [its] Asia Pacific region; hosting the recent Government-Industry Conference against Chemical Weapons; actively exerting international pressure on the abhorrent system of apartheid"[840]; and here attempts to play the same "leading role" in the Antarctic environmental protection movement. "With that attitude and that

838 Article 31 paragrah 3, CRAMRA.

839 Hawke, Bob, "Speech on the theme 'Antarctica's future: Continuity or Change' at the 16th national conference of the Australian Institute of International Affairs in Hobart on November 18, 1989", 60 Australian Foreign Affairs and Trade- Monthly Record 627 (November 1989), at 627.

840 Hawke (note 839), at 628.

aspiration we approach these vital decisions that must be made about the further of Antarctica", Australia firmly maintains that Antarctica should be left unexploited.

It is ironic that France was playing the role of a second trumpet for a mining free movement designed to effectively protect the environment in Antarctica, all while building an airstrip on that continent. The French scientific station, Dumont d'Urville, was constructed in 1956 and was a year-round base located in its claimed sector of Adelie Land. Since it was generally surrounded by packed ice over the sea, the station could operate only in January and February. In 1969, France considered building an airstrip to give access to the base. Without any environmental assessment, the plan was approved by the French government in 1981 and began in 1987. The proposed location of the air base is a chain of five small islands. To build a 1,100-meter runway, three islands were flattened and linked using the extra soil to fill the shallow waters between them. This airstrip was completed in 1994. Unfortunately, it was struck by a violent storm before its first use and was never used. The construction, especially blasting activities, would have seriously affected Antarctic environment and its ecosystems, e.g. the population of Adelie penguins who were living on the islands which were subject to blasting. Other species, such as emperor penguins, giant petrels, cape pigeon, snow petrel and Wilson's storm petrel have suffered to different extents. This story as regards France might reflect the phenomenon whereby governments consider environmental protection when it appears necessary but ignore such protection when there is something else of greater necessity to be considered.

When Australia and France led the mining free camp in the late 1980s, it is hard to image that their ultimate purpose was really to protect Antarctic from damage. Various factors have to be considered by governments when they are making policy decisions. In fact, the most critical and basic reason why some of the claimants chose not to sign CRAMRA was that it did not fully protect their interests. CRAMRA neither explicitly recognizes sovereign rights, especially royalties, in respect of mining activities, nor entrusts a claimant with the right to manage activities which take place in its claimed area.[841] Environmental protection is therefore merely one of the most convenient, popularly acceptable, and thus safe arguments that provides as a lofty and idealistic cloak to cover true political and economic intentions. Australia, claiming the biggest sectors constituting 42 percent of the Antarctic continent, had the most to lose if mineral resource activities had taken place under CRAMRA. The former Australian finance minister, Paul Keating, asserted: "Signature (of CRAMRA) would mean we would in effect concede our economic claims over Antarctic for virtually nothing."[842]

841 Redgwell, Catherine, "Environmental Protection in Antarctica: the 1991 Protocol", 43 Int'l & Comp. L. Q. (1994), at 600.

842 Hansard, Australian Parliamentary Debates: Senate, 3 May 1989, p. 1649, as quoted in Beck, Peter J., "The Antarctic Resource Conventions Implemented: Consequences

II. Economic and Technological Reasons

CRAMRA, a compromise between the claimants and non-claimants, attempts to protect claimants' privileges through disposition of surplus and the creation of Management Schemes. However, some claimants were satisfied by such regulations. They complained that the minerals convention did not "contain any provision conferring upon claimant states the right to claim royalties for minerals extracted from their territory".[843] Additionally, at the time the convention was to be signed, neither technology nor economic conditions were mature enough to enable costly and risky mineral resource activities to be carried out. No state was in a position to exploit Antarctica; making its postponement at the time somewhat immaterial. Moreover, the CRAMRA regulation excluding prospecting from scientific research activities may have restricted activities underway at that time, such as geological research.

CRAMRA was concluded as a good example of international cooperation and of dealing with new issues through international forums, i.e. Antarctic Treaty Consultative Meetings. However, it has ultimately failed to enter into force because of intense pressure from UN, incisive criticisms of environmentalists, complicated considerations of various national interests, and because of its own fiscal and technical deficiencies. Having said this, it is hard to determine the ultimate fate of the Convention. The limited resources in exploited continents are being rapidly consumed, begging the question of how long Antarctica can be kept mining free.

D. CRAMRA and the Environmental Protocol

There is no doubt that the CRAMRA process has provided a successful precedent for both the ATS and international law, proving that disputes on territorial sovereignty claims were not an insurmountable barrier to the process of establishing a legal regime on economic usage of area in question.[844] However, it was still a challenge for CRAMRA to satisfy all the complex interests groups in Antarctica, especially relating to accommodate potential economic benefits, which might be abundant should mineral activities eventuate. The bumpy fate of the mineral Convention was finally determined by Australia and France who refused to sign and ratify the Convention, with the result that their claimant State status prevented the Convention entering into force. Following the adoption of the environmental Protocol, it seemed as though the issue of Antarctic mineral activities may ultimately be possible, where previously such activities had been prohibited by the Protocol. A thorough analysis of CRAMRA and the Protocol might provide a clearer answer.

---

for the Sovereignty Issues", in Jorgensen-Dahl, Arnfinn / Ostreng Willy, (eds.), *The Antarctic Treaty System in World Politics* (Palgrave Macmillan, July 1991), p. 254.

843 Redgwell, Catherine, "Antarctica: Wilderness Park or Eldorado Postponed?", 1 Environmental Politics 137 (1992), at 140.

844 Wolfrum (note 781), at 84.

I. Background

In terms of the negotiating periods for these two documents, CRAMRA took nine times longer to complete from the initial informal discussion at the XIth ATCM, than the Protocol, which focused on mineral activities in Antarctica as a question completely new to ATCPs and international society as a whole. There were neither adequate evidences of Antarctic mineral resources nor appropriate technology or acknowledgement of Antarctic mineral activities, limiting ATCPs to trying to elaborate a regime to accommodate complicated internal and external accommodations for a potential future issue. All the interested parties, including States and international organizations, wanted to engage in and influence the process towards their own interests, such as benefit-sharing, operation or participation in mineral activities, environmental protection, and others. The intensive discussions that took place resulted in what were innovative principles at the time as a means of resolving competitions of interests, such as precautionary measures, and institutions such as Regulatory Committees. On the other hand, the Protocol took only two years to be adopted, which indicates that true innovations would be hard to find in this rapidly produced document.

CRAMRA was criticized for its "behind closed door" negotiations, with NGOs in particular not receiving any formal status during the negotiating sessions. Conversely, the Protocol negotiations included NGOs, such as the Antarctic Southern Ocean Coalition (ASOC), the International Union for the Conservation of Nature (IUCN), the World Tourism Organization as having a formal and substantial role. Moreover, some NGOs' representative sat on national delegations, such as the Wilderness Society in the US delegation, the Fondation Cousteau in the French delegation, the German Overseas Institute in German delegation, and Greenpeace in the delegations of New Zealand, Australia and Denmark.[845] ASOC capitalized on its formal position at the negotiating sessions to submit documents which clearly enunciated NOGs' objectives on environmental protection,[846] and several of these objectives were reflected in the final text of the Protocol.[847] Generally speaking, the

845 Blay, S.K.N., "New Trends in the Protection of the Antarctic Environment: the 1991 Madrid Protocol", 86 A.J.I.L. 377 (April 1992), at 397.

846 ASOC submitted following working papers: Antarctic Protected Areas, XVI ATCM/IP063, http://www.ats.aq/documents/ATCM16/ip/ATCM16_ip063_e.pdf; the Regulation of tourism in Antarctica, XVI ATCM/IP077, http://www.ats.aq/documents/ATCM16/ip/ATCM16_ip077_e.pdf; the Place of Science on an Environmentally Regulated Continent, XVI ATCM/IP079, http://www.ats.aq/documents/ATCM16/ip/ATCM16_ip079_e.pdf; and Draft initial environmental evaluation: Removal of World Park Base, Cape Evans, Ross Island 1991/92, XVI ATCM/IP094, http://www.ats.aq/documents/ATCM16/ip/ATCM16_ip094_e.pdf.

847 Annex V on protected areas was adopted at the XVI ATCM, Recommendation XVI-10, Doc. XVI ATCM/WP 57, (October 18, 1991). http://www.ats.aq/documents/ATCM16/wp/ATCM16_wp057_e.pdf .

creation of regulations and institutions becomes more complicated in direct correlation with the number of participants in the negotiations. This may explain why, the standards and procedures provided in the Protocol are less explicit.

Two oil spill incidents in 1989 brought about the downfall of CRAMRA and the initial discussion on an Antarctic environmental protection convention. On January 28, 1989, *Bahia Paraiso*, an Argentine Polar Transporter ran aground near Janus and DeLaca islands in Antarctica, and diesel fuel and other petroleum products leaked into the sea. On March 24, 1989, an oil tanker *Exxon Valdez* ran aground in Prince William Sound and spilled hundreds of thousands of barrels of crude oil in Alaska. These accidents strengthened the idea that CRAMRA might result in environmental damage to Antarctica, although it would permit mineral activities in the last virgin continent, Antarctica definitely needed a very robust environmental protection regime.

The Protocol was designed as a framework to set up basic regulations managing all human activities in Antarctic and differs from CRAMRA only in terms of managing mineral activities. It is a supplement to the Antarctic Treaty and as such, neither modifies nor amends the Treaty.[848] Therefore, any rights and obligations under other instruments in force within the ATS could be derogated by the Protocol, and its six attached Annexes, which refer to Environmental Impact Assessment, Conservation of Antarctic Fauna and Flora, Waste Disposal and Waste Management, Prevention of Marine Pollution, Area Protection and Management and Liability Arising from Environmental Emergencies. Additional Annexes may also be adopted.

II. Mining Issue

The radical difference between CRAMRA and the Protocol is that Article 7 of the Protocol, which has become a central provision, unconditionally prohibits any non-scientific activity relating to mineral resources in Antarctica. This prohibition can be amended or waived in two ways. Firstly, according to Article XII (1) of the Antarctic Treaty, if, at any time, all contracting parties unanimously agree to amend the prohibition on mining. Although this sounds simple, in practice it is almost impossible to achieve such unanimity, as each contracting party is also granted a veto on amendment or modification of the mining prohibition. Secondly, after a period of fifty years, the mining prohibition may be amended at a review conference. Any amendment may enter into force if ratified by three-quarters of all ATCPs, including all ATCPs at the time the Protocol was adopted, effectively also granting each of the original ATCP a veto over the mining issue. Moreover, the Protocol requires that any amendment of the mining issue should be accompanied by a binding legal regime on Antarctic mineral resources activities, which grants ATCPs another veto to block mineral activities. On this point, the Protocol is effectively on par with

848 The Protocol, Article 4.

CRAMRA, and its moratorium on mineral activities until an appropriate regulatory regime has entered into force.[849]

III. Principles

The Protocol transplants some of the principles of CRAMRA and applies them to all human activities in Antarctica rather than only to mineral activities. Firstly, all parties to the Protocol have obligations to protect the Antarctic environment and its dependent and associated ecosystems, and treat Antarctica as "a natural reserve, devoted to peace and science".[850] In CRAMRA, similar principles were described as "Antarctica shall continue forever to be used exclusively for peaceful purposes and not become the scene or object of international discord",[851] and ATCPs has "special responsibility... for the protection of the environment and the need to protect the Antarctic environment and dependent and associated ecosystems..."[852] The difference between these two documents is that the Protocol raises the value of Antarctica "as an area for the conduct of scientific research, in particular research essential to understanding the global environment" to the same high level as environmental protection, which means that such research has been granted priority over all other human activities in Antarctica. However, scientific research activities should be restricted when it could cause adverse effects on environmental protection.

According to the Protocol, achieving the goal of comprehensive protection of Antarctic environment and dependent and associated ecosystems, requires adverse impacts to be limited by thoroughly planning all human activities and how they are conducted. The specific rules on how to plan and conduct human activities are mainly from Article 4 of CRAMRA, titled "Principles Concerning Judgments on Antarctic Mineral Resource Activities".[853] Article 4 paragraph 4 of CRAMRA, requires that mineral activities may only take place when sufficient information is available

849 Final Act of the Fourth Special Antarctic Treaty Consultative Meeting on Antarctic Mineral Resource Activities. Wolfrum (note 781), at p88.

850 The Protocol, Article 2.

851 Article 2 paragraph 1, CRAMRA.

852 Article 2 paragraph 3(a), CRAMRA.

853 The Protocol, Article 3 paragraph 2 (b) specifies that planned and conducted activities should avoid adverse effects on climate or weather patterns, which is similar to CRAMRA, Article 4, paragraph 3; significant adverse effects on air or water quality, which is the same as CRAMRA, Article 4, paragraph 2 (a); significant changes in the atmospheric, terrestrial (including aquatic), glacial or marine environments, which is similar to CRAMRA, Article 4, paragraph 2 (b); detrimental changes in the distribution, abundance or productivity of species or populations of species of fauna and flora, which is similar to CRAMRA, Article 4, paragraph 2 (c); further jeopardy to endangered or threatened species or populations of such species, same as CRAMRA, Article 4, paragraph 2 (d); degradation of, or substantial risk to, areas of biological, scientific, historic, aesthetic or wilderness significance, almost the same as CRAMRA, Article 4, paragraph 2 (e).

and provides an outline against which all human activities should be planned and conducted. Sufficient information in this instance is considered to be information sufficient to allow prior assessments and informed judgments about possible impacts on the Antarctic environment and dependent and associated ecosystems.[854]

IV. Environmental Impact Assessment

Article 8 of the Protocol requires parties to assess the environmental impact of proposed activities before deciding to undertake activities in the Antarctic Treaty area. The impacts of proposed activities are divided into three sorts, less than a minor transitory impact, a minor or transitory impact, and more than a minor or transitory impact.[855] Annex I, as an integral part of the Protocol, further sets up several rules to guide such assessment. The Annex regulates that when an activity is determined as having less than a minor transitory impact, it may preced. However, it also regulates that the original assessment of a proposed activity should be in accordance with appropriate national procedure.[856] The problem is that different States have a range of procedures and standards, making it a uniform standard of "less than a minor or transitory impact" impossible. Therefore, the environmental impact assessment system has been built without a firm basis, which definitely undermines its effectiveness.

When an activity is assessed as having minor or transitory impact or higher, an Initial Environmental Evaluation or additional Comprehensive Environmental Evaluation should be prepared by the relevant party.[857] The Annex only provides some contents of evaluation,[858] and the substantive provisions of the Protocol and the Annex remain silent on what constitutes a "minor or transitory impact", or which standards or procedures should be followed.

---

854 These judgments include: the scope of the activity, including its area, duration and intensity; the cumulative impacts of the activity, both by itself and in combination with other activities in the Antarctic Treaty area, which is similar to CRAMRA, Article 4, paragraph 5; whether the activity will detrimentally affect any other activity in the Antarctic Treaty area; whether technology and procedures are available to provide for environmentally safe operations, which is similar with Article 4, paragraph 4 (a); whether the capacity exists to monitor key environmental parameters and ecosystem components so as to identify and provide early warning of any adverse effects of the activity and to provide for such modification of operating procedures as may be necessary in light of the results of monitoring or increased knowledge of the Antarctic environment and dependent and associated ecosystems, which is similar to Article 4, paragraph 4 (b); and whether the capacity exists to respond promptly and effectively to accidents, particularly those with potential environmental effects, which is similar to Article 4, paragraph 4 (b).

855 Article 8 paragraph 1, the Protocol.

856 Article 1 paragraph 1, Annex I, the Protocol.

857 Article 2 and Article 3, Annex I, the Protocol.

858 Article 2, Article 3, Annex I, the Protocol.

Another weakness of the evaluation system lies in its decision making process. Article 4 of the Annex indicates only that decisions on a proposed activity should proceed and should be based on the Comprehensive Environmental Evaluation and other relevant considerations, while providing no instruction as to who has the right to make this decision. Either the ATCM or the proposed State could decide, as Article 3 paragraph 5 appears not to grant veto power to the ATCM, stating that "[n]o final decision shall be taken... unless there has been an opportunity for consideration of the draft Comprehensive Environmental Evaluation by the Antarctic Treaty Consultative Meeting on the advice of the Committee [for Environmental Protection]".[859] Such vague regulations suggest that the Protocol is ensuring Antarctic environmental protection in part by limiting operational ability.

In comparison with the Protocol, CRAMRA provides more practical regulations on environmental impact assessment, which is not defined in an independent provision, but rather as part of all procedures of mineral activities. Generally speaking, Antarctic mineral activity is dependent on an assessment of it potential to impact on the Antarctic environment and its ecosystems,[860] and on global or regional climate or weather patterns.[861]

Before commencing prospecting activities, the sponsoring State must provide an assessment of possible environmental and other impacts as a part of notifying the Commission.[862] Any member of the Commission may question this notification and request that the Commission take appropriate action,[863] including determining maximum drilling depths, and placing restrictions or prohibitions on prospecting.[864] It is worth mentioning that CRAMRA grants membership of the Commission to States that were not ATCPs at the time the Convention was opened for signature, while that party is actively engaged in environmental research in general, and in conducting the assessments mentioned above in particular.[865] This opens the field of assessments to a broader and more expert body, who may affect the final decision on prospecting applications.

The Commission must identify an area for potential exploration and development based on the application submitted by the party seeking to undertake said activity. This application must include a detailed environmental assessment[866], and a comprehensive environmental and technical assessment carried out by the Advisory Committee.[867] When carrying out its comprehensive environmental and technical assessment of the proposed actions, the Advisory Committee must consider, inter

859 Blay (note 845), at 392.
860 Article 4 paragraph 2, CRAMRA.
861 Article 4 paragraph 3, CRAMRA.
862 Article 37 paragraph 7 (d), CRAMRA.
863 Article 38 paragraph 1, CRAMRA.
864 Article 21 paragraph 1 (e), CRAMRA.
865 Article 18 paragraph 2 (b), CRAMRA.
866 Article 44 paragraph 2 (b) and Article 53, paragraph 2 (b), CRAMRA.
867 Article 26 paragraph 4, CRAMRA.

alia: a) the adequacy of existing information; b) the nature, extent, duration and intensity of direct environmental impacts resulting from the proposed activity; c) possible indirect impacts; d) means and alternatives to reduce direct or indirect impacts; e) cumulative impacts of the proposed activity; f) capacity to respond effectively to accidents with potential environmental effects; g) the environmental significance of unavoidable impacts; and h) the probability of accidents and their environmental consequences.[868] These specific concerns have been adopted into the Protocol as elements of a Comprehensive Environmental Evaluation.[869] The Advisory Committee's conclusion should be fully considered by the Commission, before making the final decision on an identified area. Similarly, all members of the Commission were granted a veto on the decision of such area, allowing the Commission to identify an area for possible exploration and development if a consensus of its members could be achieved.[870]

All applications were to be examined by the Regulatory Committee. Exploration applications could be rejected at any time during the process of consideration[871], however development applications, could be refused by the Regulatory Committee by modifying the Management Scheme.[872]

This analysis appears to show that the environmental impact assessment in CRAMRA would have been more operational than that contained in the Protocol, although the mineral convention does not provide a standard for "adverse effects", the specific procedures and explicit power of the Commission and the Regulatory Committees would compensate for the absence of uniformity. Moreover, the Advisory Committee's function in checking professional opinions and advice on the environmental impact of proposed activities', arguably makes the final decision more scientific.

V. Responsibility & Liability

CRAMRA has clearly indicated that a Party State was the one responsible for planning mineral activities in Antarctica.[873] In contrast, the Protocol answered this most vital question in an obscure way. Article 8, paragraph 2 of the Protocol regulates that each party must apply the environmental assessment procedure to planning processes leading to decisions about any scientific research programmes, tourism and all other governmental and non-governmental activities in Antarctica, including associated logistic support activities. Practically identical to Article 7 of CRAMRA, Article 13 of the Protocol also regulates that each Party shall take appropriate measures within its competence to ensure compliance. This raises the same question of

868 Article 26 paragraph 4, CRAMRA.
869 The Protocol, Annex I, Article 3(2).
870 Article 22 paragraph 2 (c) and Article 41, paragraph 2, CRAMRA.
871 Article 45 paragraph 2, CRAMRA.
872 Article 54 paragraph 4, CRAMRA.
873 Article 7, CRAMRA.

whether competence is based on personal jurisdiction, or on territorial jurisdiction, or both. Jurisdiction is a sensitive topic, and causes fierce competition among States – not an easy situation to deal with in a document that manages all human activities in a specific area. It is therefore, it is not hard to understand why the Protocol left this question open. However, Annex IV of the Protocol namely, Prevention of Marine Pollution, was prepared to make a clear statement, providing that responsibility falls upon the flag State or the port State.[874] It could be concluded therefore that the home State of the operator who plans to undertake activities in Antarctica, has the power and the respective management obligation in the planning process.

As a great improvement, Annex VI of the Protocol namely, Liability Arising from Environmental Emergencies, might provide some clues to clarify the question of liability under the Protocol. Firstly, each Party must require its operators to take prompt and effective response action to environmental emergencies arising from their activities.[875] Where the operator does not respond, the Party of the operator or other Parties are encouraged, but not obliged, to take such action. Under CRAMRA, the Sponsoring State is liable for any damage that would not have occurred had that State carried out its obligations under the Convention.[876] Secondly, under the Annex, the operator, be it a State or non-State operator, who fails to respond promptly and effectively to environmental emergencies arising from its activities, shall pay the cost of response action taken by other Parties.[877] Thirdly, the Party State is liable for environmental emergencies caused by its warships, naval auxiliaries, or other ships or aircraft owned or operated by it and being used on governmental non-commercial service, when it fails to provide for prompt and effective response action.[878] Fourthly, liability in the Annex will only cover the cost of response actions taken by Parties, which is much more limited than the liability in CRAMRA, which includes damage to the Antarctic environment, and/or dependent or associated ecosystems, loss of or impairment to an established use of dependent or associated ecosystems, loss of or damage to property of a third Party or loss of life or personal injury to a third Party, and reimbursement of reasonable costs incurred by necessary response action.[879] Fifthly, liability is reduced where a Party has taken appropriate measures within its competence, including the adoption of laws and regulations, administrative actions and enforcement measures to ensure compliance with the Annex.[880] This extent of relief provided by this regulation is hard to define, although it is clear that it lessens a Party's liability for monitoring. It does however imply that Parties have obligations for activities of their non-State operators. Finally, Article 12 of the Annex echoes

---

874 Article 2 Annex IV of the Protocol.
875 Article 5 Annex VI of the Protocol.
876 Article 8 paragraph 3 (a), CRAMRA.
877 Article 6 paragraph 1 Annex VI of the Protocol.
878 Article 6 paragraph 4 Annex VI of the Protocol.
879 Article 8 paragraph 2, CRAMRA.
880 Article 10 Annex VI of the Protocol.

the CRAMRA requirement that a fund be established to reimburse reasonable and justified costs incurred by a Party or Parties taking response action.

In comparison with CRAMRA, which set up an explicit system of responsibility and liability for potential Antarctic mineral activities, the Protocol, despite its Annex dedicated to the arduous task of protecting the Antarctic environment and its ecosystem against all human activities, still needs to draw up more comprehensive rules or Annexes to illustrate and regulate these vital issues.

VI. Institutions, Inspection & Dispute Settlement

The Protocol does not establish any organ with real law-making or implementation power, instead creating an advisory organ, that includes all parties to the Protocol, the Committee for Environmental Protection.[881] It is the ATCM that defines general policy for Antarctic environmental protection and adopts measures under Article IX of Antarctic Treaty.[882] In contrast, CRAMRA built up a set of bodies, some of which, such as the Commission and Regulatory Committees, were independent and had the power to adopt measures and implement the Convention.

Furthermore, the Protocol lacks a full inspection system which may undermine its effectiveness. Article 14 of the Protocol imitates CRAMRA's two-tier inspection system, granting observers designated under Article VII of the Antarctic Treaty the right to inspect. The difference between CRAMRA and the Protocol is that, the former does not impose any restrictions on that inspection,[883] while the latter limits the inspection one in accordance with the purposes of promoting "the protection of the Antarctic environment and dependent and associated ecosystems, and to ensure compliance with this Protocol". Secondly, both CRAMRA and the Protocol design new measures for designation of observers. CRAMRA provides very clear rules, under which observers may be designated by the Commission, or any of its members or the relevant Regulatory Body.[884] Conversely, the Protocol does not have a clear rule allowing ATCMs to nominate observers.

Finally, in comparison to that offered under CRAMRA, the Protocol regulates a restricted system for the settlement of disputes, limiting the mandatory authority of the International Court of Justice and the Arbitral Tribunal to the settlement of disputes concerning the interpretation or application of the prohibition of mineral resource activities (Article 7), environmental impact assessments (Article 8) and

881 Article 11, The Protocol, the function of the Committee is listed in Article 12.

882 Article 10, The Protocol.

883 Article 11, CRAMRA, which reads: "All stations, installations and equipment, in the Antarctic Treaty area, relating to Antarctic mineral resource activities, as well as ships and aircraft supporting such activities at points of discharging or embarking cargoes or personnel at such stations and installations, shall be open at all times to inspection by observers designated under Article VII of the Antarctic Treaty for the purposes of that Treaty."

884 Article 12 paragraph 1, CRAMRA.

emergency responses Action (Article 15). Provisions in the Protocol concerning dispute settlement procedures and the operation of the Arbitral Tribunal are nearly identical to those in CRAMRA. The Protocol only allows the procedure to be initiated among State Parties while CRAMRA would have provided an opportunity for an operator to initiate proceedings against a Regulatory Committee.[885]

In conclusion, the Protocol adopts some characteristics of CRAMRA and applies them to all human activities in Antarctica. The comparative weaknesses of the Protocol is obvious, it lacks explicit definitions, standards and operational procedures, which restrict its effectiveness. On the other hand, specific and detailed rules could apply to special activities, where CRAMRA would only have been able to regulate mineral activities, but not other human activities. ATCPs would therefore be required to create additional Annexes rendering the Protocol a more complete, and therefore more effective environmental protection system. In this process, CRAMRA, with its high standard of environmental protection, must serve as the baseline for future elaboration of any regulations.

E. Influence of CRAMRA on the ISA

CRAMRA provides a model for international society to govern mining activities in areas beyond national jurisdiction or in areas between contested by two or more States. The Regulations on Prospecting and Exploration for Polymetallic Nodules in the Area (hereafter Regulations) elaborated by the International Seabed Authority (ISA), as its legitimate heir, is heavily influenced by CRAMRA in many respects, especially regarding the protection of the environment, for which CRAMRA has provided a high standard of principles and mechanisms. Moreover, while these two documents both deal with human mining activities which would be carried out beyond national jurisdictions, at the same time, both have to balance different the interests of various States groups (claimant States versus non-claimant States for CRAMRA; developing States versus developed States and pioneer investors versus the others), the Regulations adopt similar processes for mining activities and institutional systems as CRAMRA.

I. Impacts on Environmental Protection

1. *The Precautionary Approach*

One significant character of the Regulations is the "precautionary approach" adopted for environmental protection. The original Regulations, adopted in 2000, only mentioned this approach one in Regulation 31(2), however, the Amendment of the Regulations repeated it two more times in Regulation 2(2) and 5(1). While the precautionary principle formally emerged in international treaties in the 1980s, the concept of "*Vorsorgeprinzip*", which literally translates as the precautionary

885 Article 59 in connection with Article 41, paragraph 1(f), CRAMRA.

principle, had already been used in a number of German laws in the 1970s.[886] By the time this approach had been adopted by the Rio Declaration on Environment and Development (Hereafter Rio Declaration)[887] as a core principle, it had become an intrinsic part of international environmental law.[888] In the Rio Declaration, the precautionary approach is read in principle 15 as: "In order to protect the environment, the precautionary approach shall be widely applied by States according to their capabilities. Where there are threats of serious or irreversible damage, lack of full scientific certainty shall not be used as a reason for postponing cost-effective measures to prevent environmental degradation", which is transplanted *in toto* into the Regulations[889]. The precautionary approach attempts to bridge the gap between the preservation of the environment and ecosystems and scientific uncertainty surrounding the potential damage arising from the exercise of human activities. The approach has now been incorporated into numerous international documents.[890] Although there is neither a uniform understanding of the meaning of the precautionary principle nor a set of criteria for its execution,[891] some consistency has arisen and, in a well-known definition, it has come to be phrased as: "When an

886 Cameron, James / Abouchar, Juli, "The Precautionary Principle: A Fundamental Principle of Law and Policy for the Protection of the Global Environment", 14 B.U. Int'l & Comp. L. Rev. 1 (1991), at 6; Gardiner, Stephen M., "A Core Precautionary Principle", 14 The Journal of Political Philosophy 33 (2006), at 34–35. Use in German legislation includes inter alia: Bundesimmissionsschutzgesetz (1974), Article 5, §1, No. 2; Atomgesetz, Article 7, § 2, No. 3; and Gesetz über die Umweltverträglichkeitsprüfung.

887 Report of the United Nations Conference on Environment and Development, Rio de Janeiro, 3–14 June 1992, vol. I, Resolutions adopted by the Conference, Resolution 1, annex I. at: http://www.un.org/documents/ga/conf151/aconf15126-1annex1.htm.

888 Wolfrum, Rüdiger, "Precautionary Principle", in Beurier, Jean-Pierre, Kiss, Alexander, and Mahmoudi, Said (ed.), *New Technologies and Law of the Marine Environment*, (Kluwer International Law, 2000), p. 203.

889 Regulation 2(2) and 31(2) of the Regulations claims that "a precautionary approach, as reflected in principle 15 of the Rio Declaration on Environment and Development" shall be applied by relevant mining actors.

890 The precautionary principle, or its underlying rationale, has been incorporated into: the 1992 Rio Declaration, the 1992 Convention on the Protection of the Marine Environment of the North-East Atlantic (OSPAR Convention), the 1992 Convention on the Protection of the Marine Environment of the Baltic Sea (Helsinki Convention), the 1995 Agreement for the Implementation of the United Nations Conventions on the Law of the Sea of 10 December 1982 relating to the Conservation and Management of Straddling Fish Stocks and Highly Migratory Fish Stocks (UN Fish Stocks Agreement), the 1996 London Protocol to the Convention on the Prevention of Pollution by Dumping of Wastes, the 2000 Cartagena Protocol on Biosafety, and the 2001 Stockholm Convention on Persistent Organic Pollutants.

891 For more detailed explication see: Wang, Runyu, The Precautionary Principle in Maritime Affairs, 10 WMU Journal of Maritime Affairs 143 (2011), at 146–148.

activity raises threats of harm to human health or the environment, precautionary measures should be taken even if some cause and effect relationships are not fully established scientifically."[892] Notwithstanding that the precautionary approach was not explicitly mentioned by CRAMRA, it does contain a clear expression of the spirit of this approach.

According to principle 15 of the Rio Declaration, the precautionary approach includes four main components: a) under which level of damage the approach should be applied; b) the scientific criterion as being uncertainty of the relationship between a given activity and a given effect; c) the requirement for decision-makers to respond to the given activity; and d) the party who wants to carry out an activity must prove that the activity is not harmful to environment.[893] However, the critical problem of this approach is that it is vague enough to define a damage is "serious and irreversible", a definition which requires individual case-by-case consideration.

In the 1970s, ATCPs had recognized that even though "adequate scientific data concerning the harmful environmental effects of activities related to the exploration and exploitation of Antarctic mineral resources, should they occur, are not yet available". In light of this, ATCPs further concluded that they had a special responsibility to ensure that "any activities in Antarctica, including commercial exploration and exploitation in the future, should they occur, should not become the cause of (...) danger to the unique Antarctic environment (...) or be otherwise contrary to the principles or purposes of the Antarctic Treaty".[894] The adoption of a precautionary approach and a voluntary moratorium for the exploration and exploitation of Antarctic mineral resources were an issue of basic consensus during the negotiating of CRAMRA, and similar expressions could be found in its Preamble. These include: firstly, "recognizing that Antarctic mineral resource activities could adversely affect the Antarctic environment or dependent or associated ecosystems"; and secondly, "believing that the protection of the Antarctic environment and dependent and associated ecosystems must be a basic consideration in decisions taken on possible Antarctic mineral resource activities". Thus, CRAMRA accepted that mining activities would harm Antarctic environment and regulations and that appropriate remedies should be established. Article 4 (4) in particular elaborates a precautionary approach stating that no Antarctic mining activities should take place if suitable technology and procedures, capacity to monitor key environmental parameters and ecosystem components and capacity to response effectively to accidents were not available. In other words, where the science was uncertain, the onus of proof that

---

892 Wingspread Statement on the Precautionary Principle, the Wingspread Conference Centre, Racine, Wisconsin, January 1998, at: http://www.gdrc.org/u-gov/precaution-3.html.

893 Wolfrum (note 888), at 209; Manson, Neil A., "Formulating the Precautionary Principle", 24 Environmental Ethics 263 (2002), at 265; Wang (note 891), at 149–154.

894 Preamble of Recommendation XI-1, Report of the Ninth Consultative Meeting, London, 19 September – 7 October, 1977.
http://www.ats.aq/documents/ATCM9/fr/ATCM9_fr001_e.pdf.

activities would not cause adverse effects on Antarctic environment rested on the party seeking to undertake Antarctic mining activities. Moreover, Article 4 (1), (2) and (3) enhance the precautionary approach by providing supplementary provisions. Firstly, as stated above, any decisions relating to Antarctic mineral resource activities should be made based on adequate information about their possible impacts, and no such activities could take place in the absence of this information. Secondly, the party seeking to undertake Antarctic mineral resource activity must provide proof that potential mineral activities would not cause "a) significant adverse effects on air and water quality; b) significant changes in atmospheric, terrestrial or marine environments; c) significant changes in the distribution, abundance or productivity of populations of species of fauna or flora; d) further jeopardy to endangered or threatened species or populations of such species; or e) degradation of, or substantial risk to, areas of special biological, scientific, historic, aesthetic or wilderness significance." Inadequate evidence of any of the above aspects would impede the implementation of proposed mineral resource activities. Thirdly, potential actors should prove their activities would not cause significant adverse effects on global or regional climate or weather patterns.

CRAMRA has the same weakness as the precautionary approach elaborated in Rio Declaration, with its vague definition of "significant adverse effects". However, given the difficulty in foreseeing the state of the Antarctic environment many years into the future, this flexibility is not unreasonable, allowing for the line between significant and non-significant harm to be determined on a case by case basis.

As a precedent of international mining convention, CRAMRA provides a high standard of environmental protection not only for the Regulations but also for general international environmental laws, such as Rio Declaration. CRAMRA did not restrict the application of the precautionary approach to environmental protection, but extended it to ecosystems and to scientific, historic, aesthetic considerations. This approach is also prevalent in the Regulations, which claims that no further prospecting or exploration shall take place should any human remains of an archaeological or historical nature be found in the exploration area.[895]

2. Protection and Preservation of the Marine environment

The Regulations claim that a potential prospector must notify the Authority of its intention to engage in prospecting, *inter alia* including data relevant to the protection and preservation of the marine environment.[896] CRAMRA required that the sponsoring State should provide an assessment of the possible environmental and other impacts of the prospecting.[897] Under the Regulations, after the notification has been approved, the prospector must take measures to minimize or eliminate

895 Regulation 35, the Regulations.
896 Regulation 3 (4) (d) (iii), the Regulations.
897 Article 37 paragraph 7(d), CRAMRA.

adverse environmental impacts from prospecting, cooperate with the Authority's monitoring and evaluation of potential impacts on marine environment, and immediately notify of any incident arising from prospecting.[898] These requirements are taken directly from Article37 (7)(e) of CRAMRA, which demands that a Sponsoring State should describe measures, including monitoring programs, to avoid harmful environmental consequences, provide measures for the event of any accident as well as contingency plans for evacuation in an emergency. The difference between these documents is that, in CRAMRA all these demands were preconditions for potential prospecting activities, while these requirements are only triggered by the approval of notification under the Regulations. Therefore, it is clear that CRAMRA had a much stricter standard of environmental protection than the Regulations during prospecting.

For the exploration phrase, the Regulations require that each application for approval of a plan of work for exploration shall include a description of the applicant's financial and technical ability to respond to any incident or activity which causes serious harm to the marine environment,[899] which could be found in Article 44(2)(iv) of CRAMRA. Similar to CRAMRA,[900] an applicant for exploration for Polymetallic Nodules shall also submit an assessment of the potential impact of the proposed activity on the marine environment.[901] Moreover, the Regulations require that applicants submit a description of proposed measures for the prevention, reduction and control of pollution and other hazards and possible impacts to the marine environment.[902] Although this point is not explicitly mentioned in CRAMRA, the Management Scheme, which would have guided exploration and development of the mineral resources in relevant area, did cover *inter alia* measures and procedures for environmental protection including methods, activities and undertakings by the Operator to minimize environmental risks and damage.[903] Comparing these two designs, the former provides freedom for an applicant to decide what environmental actions would be taken without imposing a uniform standard, while the latter provided mandatory requirements for all applicants, proposing mineral resources activities in Antarctica. The innovation of the Regulations with regard to exploration is that each applicant must submit a description of the program for oceanographic and environmental baseline studies to enable an assessment of the potential environmental impact of the proposed activities.[904]

An external improvement made by the Regulations is that protection and preservation of the marine environment was highlighted in a separate Part V. In this part, with the exception of regulations regarding to the coastal States' rights, the other

---

898 Regulation 5, the Regualtions.
899 Regulation12 (10)(c), the Regulations.
900 Article 44 paragraph 2(b)(iii), CRAMRA.
901 Regulation 18 (c), the Regulations.
902 Regulation 18 (d), the Regulations.
903 Article 47(b), CRAMRA.
904 Regulation 18 (b), the Regulations.

provisions repeat or provide additional detail for previous provisions. Firstly, the Authority is authorized to establish and review environmental rules, regulations and procedures to ensure effective protection for the marine environment from harmful effects arising from activities in the Area.[905] In CRAMRA, both the Commission and Regulatory Committees could elaborate measures for the protection of the Antarctic environment.[906] Secondly, Part V reiterates that the precautionary approach should be applied to activities in the Area;[907] and that each contractor must undertake the necessary measures to prevent, reduce and control pollution and other hazards to the marine environment which may arise from its activities;[908] that each contractor shall gather environmental baseline data and establish that baseline;[909] and that contractors must immediately notify the Secretary-General of the discovery of any human remains of an archaeological or historical nature in the exploration area, or any object or site of a similar nature and its location.[910] Thirdly, this Part adds to the Council's ability to issue emergency orders to prevent serious harm to the marine environment arising out of activities in the Area.[911] It is clear that Part V only initiates the process of marine environmental protection in mining activities: some provisions require additional detail, such as the application of precautionary approach, the determination of what constitutes a significant adverse effect in the marine environment, explicit measures to prevent, reduce and control pollution and other hazards to the marine environment, and the responsibility of operators whose activities could cause environmental damages. Compared to the Regulations, CRAMRA provided higher standards of environmental protection, and a thorough inspection[912] and monitoring system[913] to implement these standards along with detailed provisions of liability, which were intended to be completed by a separate Protocol,[914] as well as comprehensive dispute settlement regime[915]. CRAMRA clearly supplied not only fertile ground for the Regulations with regard to the protection of the environment, some of which was absorbed by the latter, but will also certainly affect ISA in its further detailed provisions for marine environmental protection for mining activities in the Area.

905 Regulation 31(1), CRAMRA.
906 Article 21 paragraph 1(c), Article 47 (b), CRAMRA.
907 Regulation 31 (2), the Regulations.
908 Regulation 31 (5), the Regulations.
909 Regulation 32, the Regualtions.
910 Regulation 35, the Regulations.
911 Regulation 33, the Regulations.
912 Article 12, CRAMRA.
913 Article 51 and 52, CRAMRA.
914 Article 8, CRAMRA.
915 Chapter VI, and Annex for an Arbitral Tribunal, CRAMRA.

II. Impacts on Institutions

Before analyzing institutions of the ISA, it is necessarily to introduce some basic information about the current legal regime for the Area, which is contained in Part XI and Annex III of the UNCLOS and in the Agreement Relating to the Implementation of Part XI of the United Nations Convention on the Law of the Sea of 10 December 1982 (the 1994 Agreement). Institutions, through which the regime of the Area could be implemented, are primarily established by the Part XI of the UNCLOS. Firstly, the International Seabed Authority (ISA), which came into being immediately when UNCLOS entered into force on 16 November 1994, is the organization through which State parties organize and control activities in the Area.[916] Secondly, ISA comprises three principal organs[917]: an Assembly, which consists of all the members of the Authority, a Council,[918] which consist of 36 members of the Authority elected by the Assembly,[919] and a Secretariat. In addition, a Legal and Technical Commission and an Economic Planning commission, as support organs of the Council, are provided for by UNCLOS.[920] The Assembly is granted power to elaborate the general policies of the ISA in conformity with the relevant provisions of UNCLOS on any question or matter within the competence of the Authority.[921] Relating to prospecting and exploration for polymetallic nodules in the Area, it has the right to approve the Regulations and amendments to the Regulations.

The Council is the executive organ of the Authority and has the power to establish the specific policies to be pursued by the Authority on any question or matter within the competence of the Authority.[922] In polymetallic nodules mining activities, the Council has power to adopt the regulations and to amend them, as such, it combines the functions of the Commission and Regulatory Committees of CRAMRA. Firstly, the Council may reject exploitation activities in an area, based on the risk of serious harm to the marine environment.[923] This function could be found in Article 21 (1)(b) and 13 of the CRAMRA, which grant the Commission of CRAMRA the right designate protected areas in which Antarctic mineral resource activities should be prohibited or restricted for historic, ecological, environmental, scientific or other reasons. Secondly, just as the Regulatory Committees had right to consider applications for possible exploration and development permits,[924] the Council may also determine whether an applicant is financially and technically capable of carrying out the proposed plan of work for exploration and of fulfilling

916 Article 157, UNCLOS.
917 Article 158, UNCLOS.
918 Article 159, UNCLOS.
919 Article 161, UNCLOS.
920 Article 163, UNCLOS.
921 Article 160 (1), UNCLOS.
922 Article 162 (1), UNCLOS.
923 Regulation 2 (3), and Regulation 4 (3), the Regulations.
924 Article 31 paragraph 1(b), Article 45, 46, and 54, CRAMRA.

its financial obligations to the ISA,[925] and consider and approve plans of work for exploration.[926] Thirdly, the Council may withdraw a preference or priority of a contractor who has an approved plan of work for exploration, if the contractor has failed to comply with the requirements.[927] Similarly, had an operator failed to comply with CRAMRA or relevant measures of a Management Scheme, the Regulatory Committee was able cancel the exploration or development permit. In addition, under CRAMRA, the Regulatory Committees could also modify, suspend or cancel the Management Scheme, or impose a monetary penalty.[928] Comparing these two regimes, the Regulatory Committees were granted more choices than the Council to deal with inappropriate mining activities. Fourthly, the Council has specific powers to implement the Regulations, such as deferring the schedule of relinquishment,[929] extension for the plan of work for exploration,[930] issuing emergency orders and carrying measures by itself to prevent, contain and minimize serious harm or threat of serious harm to the marine environment,[931] making determinations regarding discoveries of human remains and objects and sites of archaeological or historical nature,[932] modifying or withdrawing recommendations of the Legal and Technical Commission,[933] reviewing the Regulations[934] or the fees for applications for plans of work for exploration.[935] Most of the above provisions could be found in CRAMRA, in which functions of the Commission and Regulatory Committees were regulated more broadly and more systematically and explicitly than in the Regulations. Under the Regulations, the Council can only exercise those limited and fragmented functions to manage exploration activities in the Area. The Council should look to CRAMRA to enrich its capacity, for example in aspects of inspection, monitoring and dispute settlement.

The Legal and Technical Commission (hereafter the Commission) is a technical branch of the Council containing 15 members, elected in their personal capacity

925 Regulation 12 (1), and Regulation 18(e), the Regulations.
926 Regulation 22, the Regulations.
927 Regulation 24 (2), the Regulations.
928 Article 51 paragraph 3, CRAMRA.
929 Regulation 25 (2), the Regulations. According to the Regulations, the contractor shall gradually relinquish portions of the area allocated to it to revert to the Area: by end of the third year from the date of the contract 20 percent of the area should be relinquished; by end of the fifth year an addition 10 percent should be relinquished; and by end of the eighth year an additional 20 percent should be relinquished.
930 Regulation 26 (2), the Regulations.
931 Regulation 33 (6) and (7), the Regulations.
932 Regulaiton 35, the Regulaitons.
933 Regulation 39 (2), the Regulations.
934 Regulation 42, the Regulations.
935 Regulation 19 (2), the Regulations.

as legal and technical experts in ocean mining.[936] Its main function is to provide recommendations to the. For example, the Council should consider the recommendations of the Commission, when it designates a reserved area;[937] or determines effective protection of the marine environment;[938] when it takes emergency response measures;[939] and when it adopts and applies provisional amendments to the Regulations.[940] Similarly, CRAMRA established an Advisory Committee, to provide advice to the Commission and Regulatory Committees on the scientific, technical and environmental aspects of Antarctic mineral resource activities as required.[941] The Regulations do however allow the Commission to determine whether to recommend the proposed plan of work for exploration to the Council or not[942], while the Council can only consider recommended proposals.[943] In addition, the Commission may from time to time provide contractors with technical or administrative recommendations.[944] This ability to provide guidance is somewhat undermined by a lack of clear legal status for such guidance, or the consequences for contractors who ignore or violate such guidance.

III. Impacts on the Phases of mining activities

As the name of the Regulations has showed, it manages two phases, i.e. prospecting and exploration of polymetallic Nodules in the Area; this does not include an exploitation or development phase. Under the influence of CRAMRA, the Regulations set out procedures and requirements for notification for prospecting activities; it requires a prospector to submit an annual report to the Secretary-General of the ISA, taken from Article 37(8)(c) of CRAMRA; it requires a prospector to notify the ISA of "any incident arising from prospecting which causes serious harm to the marine environment, taken from Article 37(7)(3). In addition, both documents clarify that prospectors have no exclusive rights and no rights to resources.[945]

For the exploration phase, Regulation 18 requires applicants to submit data and information for approval of a plan of work for exploration, which is influenced by Article 39 of CRAMRA. In particular, the Regulations require a preliminary environmental assessment of the possible impact of the proposed exploration activities on

936 Lodge, W Michael, "International Seabed Authority's Regulations on Prospecting and Exploration for Polymetallic Nodules in the Area", 20 Journal of Energy & Natural Resource Law 270 (2002), at 275.
937 Regulation 16 (1) and (2), the Regulations.
938 Regulation 31 (1), (2) and (3), the Regulations.
939 Regulation 33 (5), the Regulations.
940 Regulation 42 (3), the Regulations.
941 Article 26, CRAMRA.
942 Regulation 21(5) and (6), the Regulations.
943 Annex to the 1994 Agreement, section 3 paragraph 11.
944 Regulation 39 (1), the Regulations.
945 Regulation 2(4), the Regulations. Article 37 paragraph 1, CRAMRA.

the marine environment, reflecting Article 39(2)(e) of CRAMRA, which requires any notification contain a detailed assessment of the environmental and other impacts of potential mining activities in Antarctica.

IV. Impacts on Confidential Nature of Date and Information

Regulations on commercial confidentiality in CRAMRA provide the Regulations a mechanism for dealing with commercial data and information. Firstly, CRAMRA did not require that commercial data be disclosed under normal circumstances,[946] unless for scientific or environmental purposes.[947] Similarly, the Regulations recognize that data and information, should be considered confidential with the exception of three specific situations[948], but declares that data and information required for the formulation of rules, regulations and procedures concerning protection and preservation of the marine environment and safety by the Authority shall not be deemed confidential.[949] Secondly, CRAMRA declared that data and information lost commercial value 10 years after collection, unless otherwise certified, as well as requiring regular reviews to check whether or not such data and information should be released.[950] The Regulations similarly impose a sunset clause of 10 years after the submission of data and information or the expiration of the contract for exploration, whichever is the later, and every five years thereafter, data and information should be reviewed to determine whether it should remain confidential.[951] Even though the explicit expression in those two documents are not exactly the same, the spirit of confidentiality of commercial data has been adopted by the Regulations, which also improve and develop relevant rules on time following the expiration of the contract of exploration,[952] on waiver of confidentiality,[953] and on procedures to ensure confidentially[954].

Based on above analysis, it is clear that CRAMRA provided valuable experience for the ISA to elaborate regulations on mining activities. Particularly in respect to marine environmental protection, the high environmental standards in CRAMRA have significantly influenced the formulation of the Regulations on Prospecting and

---

946 Article 37 paragraph 10, CRAMRA.

947 Article 37 paragraph 11, CRAMRA.

948 Regulation 36 (1), the Regulations. Data and information are not confidential when a) it is generally known or publicly available; b) it has been previously made available by the owner to others without an obligation concerning its confidentiality; c) it is already in the possession of the Authority with no obligation concerning its confidentiality.

949 Regulation 36 (2), the Regulations.

950 Article 37 paragraph 12, CRAMRA.

951 Regulation 36 (4), the Regulations.

952 Regulation 36 (5), the Regulations.

953 Regulation 36 (6), the Regulations

954 Regulation 37, CRAMRA.

Exploration for Polymetallic Nodules in the Area. This proves that CRAMRA has already become a model for the development of international law regarding environmental protection in common spaces or spaces beyond national jurisdiction.[955] Moreover, CRAMRA's influence can also be found in the designation of institutions and phases of mining activities, and in the protection of commercial data and information. Although CRAMRA failed to enter into force, one cannot really say that is completely dead and gone, its valuable intrinsic nature was a success and continues to work effectively in managing mining activities in the Area.

955 Wolfrum (note 781), at 89.

Part 4: Protection of the Antarctic Environment and Ecosystem

A. Management of Antarctic Living Resources before the Protocol

The original motive for man to go to Antarctica was economic interest, such as seal hunting which began in the last quarter of the eighteenth century and whaling from the early 1900s. Such uncontrolled explorations ultimately threatened some Antarctic living resources with extinction, notably the blue whale in 1930s.[956] The persistent decline in the number of whales gained international attention, and the International Convention for the Regulation of Whaling (the Whaling Convention) was signed in Washington DC on 2 December 1946. To protect the Antarctic environment and its living resources, the Antarctic Treaty alone is definitely not enough. Therefore, ATCPs have been engaging in a long process to develop a series of instruments to conserve and manage Antarctic living resources. In 1964, the Agreed Measures on the Conservation of Antarctic Fauna and Flora was approved as Recommendation III-VIII to the Antarctic Treaty and was described as a "treaty within a treaty"[957]. In order to achieve the principles and objectives of the Antarctic Treaty, it designated the area south of 60° South Latitude as a "Special Conservation Area", in which human activities related to Antarctic plants and animals were to be restricted.[958] As a traditional target of hunting activities, Antarctic seals were almost extinct in the nineteenth and early twentieth centuries.[959] In 1972, a component of the ATS, the Convention for the Conservation of Antarctic Seals was adopted. The next endangered Antarctic living species is krill, which is vital to the Antarctic ecosystem. In 1970s, the harvesting of this resource rose from less than 10,000 tons in 1972 to approximately 140,000 tons in 1977–1978.[960] In order to avoid over-exploitation of certain Antarctic living resources, the Antarctic Treaty

956 Couratier, Josyane, "The Regime for the Conservation of Antarctica's Living Resources", in Vicuña, Francisco Orrego, *Antarctic Resources Policy: Scientific, Legal, and Political Issues*, (Cambridge University Press, 1983), 139–148 at 139.

957 Heap, John A. / Holdgate, Martin W., "The Antarctic Treaty System as an Environmental Mechanism: An Approach to Environmental Issues", in Polar Research Board (ed.), *The Antarctic Treaty System as An Assessment: An approach to Environmental Issues*, (US National Reesarch council, 1986), 195–210 at 203.

958 Beck, Peter J., "The Antarctic Resources Convention Implemented: Consequences for the Sovereignty Issue", in Jørgensen-Dahl, Arnfinn / Østreng Willy (eds.), *The Antarctic Treaty System in World Politics* (Palgrave Macmillan, 1991), Chapter 15 at 242.

959 Couratier (note 956), at 141.

960 Couratier (note 956), at 140.

Consultative Meetings (ATCMs) started in 1978 to build a detailed, binding regime on conservation of Antarctic living resources and finally concluded the Convention on the Conservation of Antarctic Marine Living Resources (CCAMLR) in 1980.

I. Antarctic Treaty

It is fair to say that the parties to the Antarctic Treaty did not pay much attention to the issue of environmental protection when it was adopted in 1959,[961] although it did put the spotlight on management of a living resource. Article IX (1)(f) states that the ATCPs shall meet periodically to consult issues relating to "preservation and conservation of living resources in Antarctica". At that time, the living resources in Antarctica were the prevailing value of this continent.[962] Although this provision refers only to living resources of the Antarctica, an intrinsic link between these resources and physical environment could provide the ground to elaborate further regulations for the protection of the Antarctic environment, with the preservation of environment being the precondition for the preservation and conservation of living resources.[963]

II. International Convention for the Regulation of Whaling

The Whaling Convention,[964] signed on 2 December 1946 and taking effect on 10 November 1948, intends to "provide for the proper conservation of whale stocks and thus make possible the orderly development of the whaling industry".[965] The Whaling Convention comprises a framework convention and a Schedule of rules (the Schedule), which provides measures to govern the conduct of whaling throughout the world, including, among other things, listing protected species; the specification of areas as whale sanctuaries; limitations on the number and size of whales which

---

961 The Antarctic Treaty does not explicitly mention environmental protection. An idea states that Article I, peaceful use of Antarctica, could be a basic environmental safeguard. However, it is hard to say that the prohibition of military use equals to environmental protection, while not only military actions but also other human activities, such as tourism or scientific research and so on, could damage the fragile Antarctic environment. Conversely, Article V, prohibition of nuclear explosions and the disposal of nuclear waste on Antarctica, provide clearer basis for environmental protection. However, this provision is too narrow to preserve and conserve the overall Antarctic environment.

962 Blay, S.K.N., "New Trends in the Protocol of the Antarctic Environment: the 1991 Madrid Protocol", 86 A.J.I.L. 377, (Apr., 1992), at 379,

963 Ibidem.

964 The Whaling Convention is a successor to the International Agreement for the Regulation of Whaling, signed in London on 8 June 1937. However, the pioneer treaty was not effective in protecting all whale species from over-fishing.

965 Preamble to the Whaling Convention.

may be taken; open and closed seasons; and areas for whaling.[966] Moreover, the Whaling Convention has made a huge contribution by establishing the International Whaling Commission (IWC) to keep the measures laid down in the Schedule to the Convention under review and revised as necessary. There are 89 member nations of the IWC, including almost all the Antarctic Treaty members.[967] The IWC has four main committees – the Scientific Committee, the Technical Committee, the Finance and Administration Committee, and the Conservation Committee, the latter being a new committee which first met in 2004.

1. *The moratorium on commercial whaling*

The IWC, authorized by the Whaling Convention, can amend the Schedule, and the amendment is effective for all member States, except where a member State presents its objection to such amendment within 90 days of notification of the decision.[968] This mechanism is often criticized as rendering the IWC "toothless"; however, its function in making the Whaling Convention more acceptable and ease resistance from States is clear. Additionally, a member State has the right to withdraw from the Convention,[969] whereupon it is no longer bound by any of the regulations.

In 1982, a majority vote of the IWC decided to limit catches for all commercial whaling and to set the level to zero, the so-called "moratorium" which entered into force for the 1986 coastal and 1985/86 pelagic seasons and thereafter.[970] As this moratorium of commercial whaling should be kept under review, restricted commercial whaling may be permitted based upon an improved scientific knowledge in the future. The IWC Scientific Committee embarked on a review of the status of whale stocks and developed management objectives and procedures. It spent over eight years developing a Revised Management Procedure (RMP), to provide a method to set safe catch limits for certain whale stocks, i.e. seeking a balance between the risk to individual stocks and the need for a yield. Although the IWC accepted and endorsed it in 1994, it has yet to be implemented. This is due to the IWC

966 The Schedule of the Whaling Convention, as amended by the Commission at the 63rd Annual Meeting Jersey, Channel Islands, July 2011.

967 All members of the Antarctic Treaty have membership in the IWC, except Ukraine, at http://www.iwcoffice.org/commission/iwcmain.htm#conservation.

968 Article V, the Whaling Convention.

969 Article XI, the Whaling Convention.

970 Paragraph 10(e), the Schedule of the Whaling Convention. The governments of Japan, Norway, Peru and the (then) USSR lodged objection to paragraph 10(e). For all other member states this paragraph came into force on 3 February 1983. Peru withdrew its objection on 22 July 1983. The government of Japan withdrew its objections with effect from 1 May 1987 with respect to commercial pelagic whaling; from 1 October 1987 with respect to commercial coastal whaling for minke and Bryde's whales; and from 1 April 1988 with respect to commercial coastal sperm whaling. The objections of Norway and the (now) Russian Federation not having been withdrawn, the paragraph is not binding upon these governments.

decision to implement an inspection and observation scheme, the so-called Revised Management Scheme (RMS), to ensure that catch limits are not exceeded before the RMP is implemented and the moratorium on commercial whaling lifted. The RMS will include not only the scientific aspects such as the RMP, but also a number of non-scientific issues, including inspection and enforcement and, perhaps, rules on humane killing techniques. The RMP has set an important example for wildlife resource management. It takes into account inevitable scientific uncertainty and requires only relatively simple-to-obtain information, such as knowledge of the population size, past and present catches, and stock identity.[971]

2. *Whaling under Scientific Permit*

The killing of whales is still conducted in the name of science. Article VIII of the Whaling Convention allows any member State to grant permits to its nationals to kill whales for purposes of scientific research. As a response to the moratorium on commercial whaling which came into effect subsequent to 1986, Japan, Norway and Iceland have issued scientific permits to their national whalers, although Norway has since ceased the practice. From 1985 to 2010, there have been more than 14,000 whales killed under the scientific permits.[972]

Among these permits issued by different States, Japan's "JARPA" program, which lasted for 16 years and finished in the 2004/05 Antarctic season, was partly conducted in Antarctic waters.[973] Research activity in this program comprised a randomized lethal catch, which met with strong protest in the international community. Although "[t]he results of the JARPA program, while not required for management under the RMP, have the potential to improve management of minke whales in the Southern Hemisphere",[974] the IWC adoption of the Southern Ocean Sanctuary[975] at the 46th Annual Meeting in 1994, may render information on improving management of whaling in that region unnecessary. Nevertheless, the government of Japan

---

971 For more information about the RMP see: http://www.iwcoffice.org/conservation/rmp.htm#topofpage.

972 http://www.iwcoffice.org/conservation/table_permit.htm.

973 The JARPA programme was also conducted in the North Pacific.

974 International Whaling Commission, "Report of the Intercessional Working Group to Review Data and Results from Special Permit Research on Minke Whales in the Antarctic", in the 48 Reports of the International Whaling Commission 377 (1998), at 389, a PDF version can be found at http://www.iwcoffice.org/_documents/conservation/RIWC-48-377-412.pdf. However, JARPA was criticized in that "none of the goals of JARPA 1 had been reached, and that the results of the JARPA 1 program are not required for management under the RMP", see the Resolution 2007–1, "Resolution on Japan", made at the 59th IWC Annual Meeting in 2007, at http://www.iwcoffice.org/meetings/resolutions/resolution2007.htm.

975 The Southern Ocean Sanctuary is regulated in the paragraph 7(b) of the Schedule to the Whaling Convention.

lodged an objection with respect to minke whales, which has been condemned on many occasions by the IWC.[976]

Following the first JARPA program, Japan commenced a new large-scale Antarctic program – JARPA II – during the austral summer of 2005/06. Under JARPA II, 850±10% Antarctic minke whales, 50 humpback whales and 50 fin whales are to be killed annually for the proposed full-scale research.[977] The whaling activity of the government of Japan has encountered opposition within the IWC and its Scientific Committee, both expressing extreme doubt as to the value of this research. In 2005, the IWC strongly urged Japan "to withdraw its JARPA II proposal or to revise it so that any information needed to meet the stated objectives of the proposal is obtained using non-lethal means."[978] In 2007, the IWC called again upon Japan to refrain from issuing a permit for JARPA II.[979] However, Japan has not stopped issuing permits and not stopped whaling in the Southern Ocean.

3. *Southern Ocean Sanctuary*

Because the terrible weather and natural conditions prevented humans from exploring and exploiting resources in the Antarctic waters, the area has been regarded as a paradise for marine mammals. Thus, in 1938, the first Whaling Sanctuary was established in the Antarctic, south of 40°S between longitudes 70°W and 160°W, in an area where commercial whaling had hitherto not been prosecuted. This sanctuary was continued by the IWC until 1955, when the pressure of catches on the rest of the Antarctic whaling grounds impelled the opening of this sector.[980] In 1994, at its 46th Annual Meeting the IWC adopted another area as a sanctuary in the Antarctic, the Southern Ocean Sanctuary (SOS),[981] in which commercial whaling is prohibited. This prohibition was to be reviewed 10 years after its initial adoption. Moreover, in

---

976 The IWC has passed Resolutions urging Japan not to issue a permit for these catches. Such as Resolution 2001–7, "Resolution on Southern Hemisphere Minke Whales and Special Permit Whaling", at: http://www.iwcoffice.org/meetings/resolutions/resolution2001.htm#7; Resolution 2003–3, "Resolution on Southern Hemisphere Minke Whales and Special Permit Whaling", at: http://www.iwcoffice.org/meetings/resolutions/resolution2003.htm#3, and etc.

977 See: http://www.iwcoffice.org/conservation/permits.htm#notes.

978 The Resolution 2005–1, "Resolution on JARPA II", which was passed by a count of 30 votes to 27 votes with 1 abstention, at, http://www.iwcoffice.org/meetings/resolutions/resolution2005.htm#1.

979 The Resolution 2007–1, "Resolution on JARPA", which was passed by 40 votes in favour, 2 votes against and 1 abstention; 27 States did not participate in the vote, at http://www.iwcoffice.org/meetings/resolutions/Resolution2007-1.pdf.

980 International Whaling Commission, "Report of the Scientific Committee", 6 Report of the International Whaling Commission 17 (1955).

981 The northern boundary of this Sanctuary follows the 40°S parallel of latitude except in the Indian Ocean sector where it joins the southern boundary of that sanctuary

1979, the IWC established the India Ocean Sanctuary, whose duration has since been extended twice. Two additional proposals for the establishment of sanctuaries in the South Atlantic and South Pacific have failed to gain the required three-quarters majorities at the 56th IWC Annual Meeting in 2004.

The primary objective of the SOS was to "contribute to the rehabilitation of the Antarctic marine ecosystem by reinforcing and complementing other measures for the conservation of whales and the regulation of whaling, in particular by the protection of all Southern Hemisphere species and populations of baleen whales and sperm whales in their feeding grounds".[982] The SOS has been criticized as being based on vague goals that are difficult to measure, lacking an executive method for its operation and also lacking an effective monitoring system to ensure that its objectives are being met.[983] The SOS does not effectively stop harvesting whales and does not satisfy generally accepted criteria for reserve design and operation.[984]

Moreover, it has been claimed that the sanctuary design is ultra vires and without effect, as the decision to establish the SOS does not meet the requirements of Article V(2) of the Whaling Convention, which specifies that any IWC amendments of the Schedule shall (a) "be such as are necessary to carry out the objectives and purposes of this Convention and to provide for the conservation, development, and optimum utilization of the whale resources"; (b) shall be "based on scientific findings"; and (c) shall "take into consideration the interests of the consumers of whale products and the whaling industry".[985] According to this opinion, the SOS is unnecessary to the objectives of the Whaling Convention as other conservation measures, i.e. the moratorium on commercial whaling and the RMP, have been rigorously tested for prudent application. Moreover, the current moratorium on commercial whaling and future implementation of the RMP makes the SOS unnecessary. Finally, the SOS does not meet the requirement of being based on scientific findings. In all, the adoption of the SOS by the IWC was said to be "in excess of the commission's authority and a violation of the treaty".[986]

---

at 55°S, and around South America and into the South Pacific where the boundary is at 60°S.

982 International Whaling Commission, "Chairman's Report of the Forty-Fourth Meeting, Appendix 4. Resolution on a Sanctuary in the Southern Hemisphere", in 43 Report of the International Whaling Commission 41 (1993).

983 Zacharias, Mark A. / Gerber, Leah R. / Hyrenbach, K. David, "Review of the Southern Ocean Sanctuary: Marine Protected Areas in the context of the International Whaling Commission Sanctuary Programme", 8 Journal of Cetacean Research and Management 1 (2006), at 9.

984 Botsford, Louis W. / Micheli, Fiorenza / Hastings, Alan, "Principles for the Design of Marine Reserves", 13 Ecological Applications Supplement 25 (2003).

985 See Burke, William T., "Memorandum of Opinion on the Legality of the Designation of the Southern Ocean Sanctuary", in 27 Ocean Dev. & Int'l L. 315 (1996).

986 See Burke, William T., "Legal Aspects of the IWC Decision on the Southern Ocean Sanctuary", in 28 Ocean Dev. & Int'l L. 313 (1996), at 314.

Although the Whaling Convention did not impose regulations specifically for Antarctica, it was an important step toward achieving protection of Antarctic cetaceans. At the 63rd IWC in July 2011, the catch limits in the Southern hemisphere for the 2011/2012 pelagic season and the 2012 coastal season were set to zero, which came into effect on 23 January 2012.[987] Thus, the IWC generally prohibits whaling activities in Antarctica. Moreover, in the mid-1970s, the Scientific Committee established a research program known as the "International Decades of Cetacean Research", which was a series of international cruises in the Antarctic with the aim of obtaining estimates of the Southern Hemisphere minke whale population. The cruises started in 1978/79 and ceased in 1996/97 and were replaced by the currently operating Southern Ocean Whale and Ecosystem Research (SOWER) program. These programs have provided a considerable amount of information on the distribution and the abundance of cetaceans in the waters south of 60°S.

III. Agreed Measures for the Conservation of Antarctic Fauna and Flora

Since adopting the "Measures for the Protection of Living Resources" in 1961,[988] the ATCPs have long been involved in improvement of the protection of the native fauna and flora of Antarctica. They realized the "urgent need for measures to conserve" the Antarctic living resources and "to protect them from uncontrolled destruction or interference by man".[989] Thereafter, many other measures dealing with related issues have been concluded in subsequent ATCMs. One of these, the Agreed Measures for

987 In Tables 1, 2 and 3 of the Schedule of the Whaling Convention, which has been amended by the IWC at the 63rd Annual Meeting, in 2011, at http://www.iwcoffice.org/commission/schedule.htm. The original footnote of Tables 1, 2, and 3 reads: "The Government of the Czech Republic lodged an objection within the prescribed period to the amendments to the Schedule arising from the 63rd Annual Meeting of the Commission, i.e. changes to the dates of the pelagic and coastal whaling seasons given in paragraphs 11 and 12 and Tables 1, 2 and 3. For all other Contracting Governments, these provisions came into force on 23 January 2012. The Czech Republic lodged similar objections to the Schedule amendments arising from the 60th, 61st and 62nd Annual Meetings of the Commission. These objections were withdrawn by the Government of the Czech Republic on 8 June 2011."

988 "Measures for the Protection of Living Resources", annexed to Recommendation ATCM I-VIII, adopted in Canberra, in 1961, at: http://www.ats.aq/devAS/info_measures_listitem.aspx?lang=e&id=9. Since July 1, 2011, the Measures have been designated as no longer current by the Decesion 1, 2011, ATCM XXXIV- CEP XIV.

989 "Measures for the Protection of Living Resources", at http://www.ats.aq/devAS/info_measures_listitem.aspx?lang=e&id=9.

the Conservation of Antarctic Fauna and Flora[990] adopted in 1964 at the third ATCM was the "most comprehensive instrument for wildlife conservation",[991] attempting to protect any animal from being harmed, captured or molested.[992] Moreover, the Agreed Measures includes a preamble, application zone, definitions, final clauses and annexes, giving this instrument convention-like qualities.

In order to implement the principles and purposes of the Antarctic Treaty (hereafter the Treaty), the Agreed Measures applies to the same geographic area as the Treaty area,[993] and designates the Treaty area as a "Special Conservation Area".[994] In the application area, both native animals (mammals and birds) and plants, with the exception of whales, are protected.[995] According to the rules on the preservation of native animals it is prohibited to kill, wound, capture or molest any native animal without a permit for restricted purposes.[996] To achieve these goals, the "Special Protected Species" (SPSs)[997] and "Special Protected Areas" (SPAs)[998] were established. To date, only one SPS (the Ross seal)[999] and 29 SPAs have been designated[1000]. In order

990 The Agreed Measures for the Conservation of Antarctic Fauna and Flora annexed to Recommendation III-VIII was adopted in 1964, Brussels. It entered into force on 1 November 1982 when Japan submitted its approval. For more information see at: http://www.ats.aq/devAS/info_measures_listitem.aspx?lang=e&id=35.

991 Joyner, Christopher C., "Protection of the Antarctic Environment: Rethinking the Problems and Prospects", 19 Cornell Int'l L. J. 259 (1986), at 266.

992 Article VI (1), the Agreed Measures.

993 Article 1, the Agreed Measures.

994 The Preamble of the Agreed Measures.

995 Article II, the Agreed Measures. Whales were already protected by a special convention in 1946. Moreover, Article I of the Agreed Measures guarantees the rights of the high seas under international law. Consequently, neither mammals on the high seas nor any fish is covered in the regime of the Agreed Measures.

996 Articles V and VI, the Agreed Measures. The purposes include providing indispensable food and securing specimens.

997 Article VI (5), (6), (7), the Agreed Measures.

998 Article VIII, the Agreed Measures.

999 There were previously two kinds of SPSs: Fur seals and Ross seals were specified in the Agree Measures. In 2006, ATCM adopted a Measure XXIX-IV, which decided to remove the words "'All species of the genus Arctocephalus, fur seals" from Appendix A to Annex II to the Protocol on Environmental Protection to the Antarctic Treaty to the Antarctic Treaty, because SCAR has determined that "the Antarctic fur seal (Arctocephalus gazella) and the Sub-Antarctic fur seal (Arctocephalus tropicalis) are no longer at significant risk of extinction". This Measure became effective on 23 June 2007. See: http://www.ats.aq/devAS/info_measures_listitem.aspx?lang=e&id=433.

1000 In 1991 the ATCPs adopted the Protocol on Environmental Protection to the Antarctic Treaty to the Antarctic Treaty to impel the environmental protection in Antarctica. Annex V to the Protocol, which was adopted at ATCM XVI under Recommendation XVI-10, provides two new site designations: Antarctic Specially Protected Areas (ASPAs) and Antarctic Specially Managed Areas (ASMAs).

to preserve the normal living conditions for any native mammal or bird, the Agreed Measures require minimizing harmful interference, which is specified as allowing dogs to run free, the flying, landing or driving of vehicles close to animal populations, the use of explosives or firearms close to animal populations, and any disturbance of animals during their breeding periods.[1001] Moreover, the Agreed Measures prohibits the introduction of non-native species into the Treaty Area.[1002]

The Agreed Measures is the first comprehensive instrument to preserve the living resources of Antarctica, and it drew a framework for the draft of Annex II to the Protocol. However, the Measures have only limited applicable area and weak implementation mechanisms: the Agreed Measures are an internal instrument of the ATS, which is only binding on the contracting parties of the Antarctic Treaty, and the rules stating how their objectives and principles should be met are not clear.

IV. Convention for the Conservation of Antarctic Seals (CCAS)

In 1964, a Norwegian expedition visited Antarctica and took 850 seals.[1003] This action spurred ATCPs to safeguard against any possible resumption of uncontrolled sealing in the Antarctic. The Convention for the Conservation of Antarctic Seals (Seals Convention, CCAS) was concluded in 1972 and entered into force in 1978. As of 2012, there are 16 contracting parties to the Convention.[1004]

The Seals Convention applies to the seas south of 60°S and covers all species of seals.[1005] Contracting states may not kill or to capture species of seals,[1006] unless for specific purposes, such as providing indispensable food for men or dogs; providing for scientific research; or providing specimens for museums, educational or cultural institutions.[1007] However, it is forbidden to kill or capture seals in the closed season, which is from 1 March to 31 August,[1008] in a reserve (seal breeding area) or in a site

According to Annex V to the Protocol, on entry into force of Annex V, all SPAs and Site of Special Scientific Interest (SSSIs) should become ASPAs. Annex V entered into force in 2002. The number of the SPAs is counted as the areas which are built based on the Agreed Measures. See http://www.ats.aq/documents/recatt/att093_e.pdf.

1001 Article VII, the Agreed Measures.

1002 Article IX, the Agreed Measures.

1003 Headland, R.K., *Chronological List of Antarctic Expeditions and Related Historical Events* (Cambridge: Cambridge University Press, 1989). Cited from Harris, Colin M., / Meadows, Janice, "Environmental Management in Antarctica: Instruments and Institutions", 25 Marine Pollution Bulletin 239 (1992), at 242.

1004 The contracting parties are: Argentina, Australia, Belgium, Brazil, Canada, Chile, France, Germany, Italy, Japan, Norway, Poland, Russian Federation, South Africa, United Kingdom and United States.

1005 Article 1(1) and (2), CCAS.

1006 Article 2(1), CCAS.

1007 Article 4(1), CCAS.

1008 Paragraph 3 of Annex of CCAS.

of long-term scientific research.[1009] The Convention also defines six sealing zones, each of which is closed for a year in rotation.[1010]

CCAS does not establish new machinery, and SCAR plays the role of a functional institution, including making assessments of populations and stimulating the development of a necessary information base for those populations.[1011] However, the contracting parties may develop a commission with inspection and control mechanisms, and a scientific advisory committee "at any time after commercial sealing has begun".[1012] The text also includes provisions governing future scenarios, which form "most salient feature" of this convention.[1013] Although commercial sealing had not resumed at the time of the writing of CCAS, some measures were adopted regarding potential future hunting.[1014]

The fact that there was no commercial harvesting of seals at the time when the Seals Convention was promulgated, its protection of a species prior to the existence of a commercial threat, constituted a preventative measure.[1015] To date, no commercial seal harvesting has occurred in the Antarctic.

V. Convention on the Conservation of Antarctic Marine Living Resources

The extraordinarily simple Antarctic food chain means that if one part of the ecosystem is harmed, all other parts will quite possibly suffer as well, making regulation of the Antarctic ecosystem as a whole crucial.[1016] Moreover, the interest in Antarctic mineral resources that arose in the ATCMs in 1970s, gave some urgency to the conclusion of a regulatory instrument before exploitation posed a serious threat.[1017] Therefore, in February 1978 the negotiation of the Convention for the Conservation of Antarctic Marine Living Resources (CCAMLR) began; two years later, in 1980, the Convention was concluded, entering into force two years later in 1982. The

---

1009 Paragraph 5 of Annex of CCAS.

1010 Paragraph 4 of Annex of CCAS.

1011 Article 5(4), CCAS.

1012 Article 6(1), CCAS.

1013 Couratier (note 956), at 144.

1014 Paragraph 7 of Annex of CCAS.

1015 Gulland, John A., "The Antarctic Treaty System as a Resource Management Mechanism", in Triggs, Gillian D. (ed.), *The Antarctic Treaty Regime: Environment and Resources* (Cambridge: Cambridge University Press, 1987), 116–127, at 120.

1016 Gulland, J. A., indicates that: "the real concern which led to the establishment of CCAMLR [the Commission] was less for the krill or fish in themselves, than for the impact that a large-scale krill fishery might have on those species that feed on krill". In Gulland, (note 1015), at 120. Also see Couratier (note 956).

1017 Zegers, Fernando, "The Canberra Convention: Objectives and Political Aspects of Its Negotiation", in Vicuña, Francisco Orrego (ed.), *Antarctic Resource Policy: Scientific, Legal and Political Issues*, (Cambridge: Cambridge University Press, 1983), 149–156 at 151–154.

Convention was a milestone in the evolution of the Antarctic Treaty System (ATS), strengthening and also broadening the system as detailed below.

1. *The Application Area*

CCAMLR applies to a scope beyond the Antarctic Treaty zone up to the Antarctic Convergence,[1018] which is based not on artificial boundaries, but rather on the krill's natural habitat.[1019] In other words, it applies to the area between the Antarctic continent in the south and the Antarctic Polar Front in the north (at about 50° South Latitude).[1020]

The broad applicable area reflects a hallmark of the convention, the ecosystem approach.[1021] As the krill fishery mostly impacts krill predators,[1022] and as prior fisheries management dealt solely with target species and ignored dependent and related species, it lacked adequate methodologies.[1023] This made it necessary to adopt an entire ecosystem approach, which also considers the conservation of birds,[1024] rather than adopting the single-species approach inherent in most fishery agreements.[1025] The applicable area of the Antarctic Convergence is a safeguard for achieving the Convention's objective of protecting all marine resources. The natural habitat of Antarctic marine species means that all species can be found in the Antarctic Convergence; this regime allows for all ecological parameters to be

1018 Article 1(1), CCAMLR. The Antarctic Convergence is a line varying between 60°S and 45°S, which extends the CCAMLR area beyond the limits to which the Antarctic Treaty applies.

1019 See Shapley, Deborah, *The Seventh Continent: Antarctica in a Resource Age*, (Washington D.C.: Johns Hopkins University Press, 1985), at 151–152.

1020 The Polar Front is the zone where colder, fresher waters flowing north from the Antarctic meet the warmer, saltier waters flowing south from the Atlantic, Indian and Pacific Oceans. See: Commission for the Conservation of Antarctic Marine Living Resources, *CCAMLR's Management of the Antarctic*, p. 4.

1021 CCAMLR is the first fishery convention to apply to a whole marine ecosystem. See Harris (note 1003), at 243.

1022 In the Antarctic Peninsula and on the South Orkney Island, krill fishery is usually undertaken in areas within 100 kilometers of predator colonies during breeding season, pre-moult periods of adults and post-fledging periods when animals have a limited foraging range. See Kock, K.H., "Fishing and Conservation in Southern Water", 30 Polar Record 3 (1994).

1023 Powell, D.L., "Scientific and Economic Considerations Relating to the Conservation of Marine Living Resources in Antarctica", in Vicuña, Francisco Orrego (ed.) *Antarctic Resources Policy: Scientific, Legal, and Political Issues*, (Cambridge: Cambridge University Press, 1983), 111–118, at 112.

1024 Article 1(2), CCAMLR.

1025 Cordonnery, Laurence, "Environment Protection in Antarctica: Drawing Lessons from the CCAMLR Model for the Implementation of the Madrid Protocol", in 29 Ocean Dev. & Int'l L. 125 (1998), at 126.

considered. Therefore, CCAMLR not only broadens the Antarctic Treaty's area but also improves the practical capacity of international environmental law.

Nevertheless, the area of CCAMLR does overlap EEZs of sub-Antarctic islands which are the undisputed territory of certain States. This problem has been solved by the Statement of the Chairman of the Conference on the Conservation of Antarctic Marine Living Resources. It recognizes the existence of state sovereignty on the waters adjacent to Kerguelen and Crozet under the French jurisdiction and in waters around other sub-Antarctic islands within the convention's application area. The Statement provides four approaches: Firstly, measures adopted by coastal states prior to the entry into force of the Convention remain in force after the Convention becomes effective. Secondly, once the convention comes into force, its Commission should examine the conservation needs of the marine living resources in the general area under national jurisdiction, and the respective coastal State must decide whether or not the waters in question should be included in the area of application of any specific conservation measure under consideration. Thirdly, the specific conservation measures are binding, if adopted by consensus and agreed to by the relevant coastal State. Fourthly, the relevant coastal states have to enforce both national measures and measures adopted by the Commission of CCAMLR. This approach has achieved a compromise on the assignment of jurisdictions on conservation of marine living resources between coastal States and the Convention. Notably, no rejection of the Statement was recorded, meaning the participants reached consensus on this issue at that time.

2. *Participants of CCAMLR*

CCAMLR is open to accession by non-members of the Antarctic Treaty. There are two approaches to access to CCAMLR: accession to the convention or joining its institutions. Any State interested in research or harvesting activities of Antarctic marine living resources, or in regional economic integration organization may accede to the convention.[1026] Moreover, any State engaged in research or harvesting activities of Antarctic marine living resources that has acceded to CCAMLR could also be a member of the Commission.[1027] The difference between the two types of accession lies in the level of research or harvesting activities undertaken: merely acceding to the convention only requires interest in such activities; however, membership of the Commission requires active engagement.[1028]

All contracting parties, especially non-members of the Antarctic Treaty, must comply with the principles and purposes of the Antarctic Treaty;[1029] with Article IV and VI of the Antarctic Treaty in their relations with each other, regardless of

1026 Article XXIX, CCAMLR.
1027 Article VII paragraph 2 (b), CCAMLR.
1028 Couratier (note 956), at 146.
1029 Article III, CCAMLR.

whether they are parties to that Treaty or not;[1030] and to respect the status of the ATCPs, the Agreed Measures, the Whaling Convention, CCAS, and other measures recommended by the ATCPs in fulfilment of their responsibility.[1031] Consequently, CCAMLR, to some extent, broadens the effectiveness of the ATS.

3. *Conservation Measures*

Article II of CCAMLR provides two concepts for the conservation of Antarctic marine living resources, the precautionary approach and the ecosystem approach. According to the precautionary approach, management decisions should seek to minimise long-term adverse effects, and where information is uncertain, for example insufficient data of the actual size of exploited stocks, or targeted new stocks.[1032] Moreover, CCAMLR aims to protect not only the targeted species, but also dependent and related Antarctic marine living resources, ultimately, using the "ecosystem approach" to protect the entire ecosystem.[1033] In this approach, the level of interdependence between the harvested species and other species in the system is the key consideration,[1034] which is why the area of application of the convention is extended to the Antarctic Convergence[1035], making CCAMLR a successful example of ecosystem approach in fishery management.

4. *Evaluation of CCAMLR*

CCAMLR represents a compromise between conservation states and fishing states. While it permits the "rational use" of marine living resources, it lays down three principles to be respected: maintenance of substantial populations; maintenance of ecological relationships between harvested, dependent and related populations; and prevention of changes in the ecosystem which are not potentially reversible over two or three decades.[1036] To achieve the purposes of conserving Antarctic marine living resources, CCAMLR creates a permanent structure including a commission, a scientific committee, an executive secretary, a secretariat and a system

---

1030 Article IV, CCAMLR.

1031 Article V and VI, CCAMLR.

1032 de la Mare, W.K., "CCAMLR's Management Task and the Definition of its Operational Objectives", in Kock, Karl-Hermann (ed.), *Understanding CCAMLR's Approach to Management*, (http://www.ccamlr.org/en/system/files/am-all.pdf, 2000) pp. 8–10, at p. 9; and Commission for the Conservation of Antarctic Marine Living Resources, *CCAMLR's Management of the Antarctic*, p. 6.

1033 De la Mare (note 1032), at 9; Commission for the Conservation of Antarctic Marine Living Resources, *CCAMLR's Management of the Antarctic*, p. 6; and Zegers (note 1017), at 151.

1034 Joyner, Christopher C., "Review Article: the Evolving Antarctic Legal Regime", 83 A.J.I.L. 605 (1989), at 606.

1035 Zegers (note 1017), at.151.

1036 Article II (2), CCAMLR.

of observation and inspection.[1037] These institutional arrangements, as the first attempt of the ATS to empower a subsidiary organization to further specific Treaty aims, provides a precedent for both the ATS and international environmental law.

According to CCAMLR, the Commission for the Conservation of Antarctic Marine Living Resources should facilitate research; compile, ensure, analyse, disseminate and publish related data on populations and harvest; and identify conservation needs and formulate appropriate measures.[1038] The Commission is supported by a consultative body, a Scientific Committee, which provides a setting for consultation and cooperation in regards to the conservation of marine living resources. Based on scientific information and advice submitted by the Scientific Committee,[1039] these conservation measures should be reviewed and adjusted at each annual meeting of the Commission. All the measures taken by the Commission regulating the opening and closing of areas and seasons for fishing, imposing total allowable catch numbers for harvested species and implementing other restrictions[1040] are binding upon all members of the Convention.[1041] These measures have played an important role in ensuring a rational exploitation of Antarctic marine resources.[1042] The Commission also adopts non-binding resolutions to complement the measures, and contracting parties are encouraged to implement each resolution.[1043]

Unlike CCAS, CCAMLR is designed to be executed by its Commission through adoption of measures and the supervision of their application, by its Scientific Committee through provision of the necessary scientific assessment and information and recommendations for the appropriate measures, and by a system of control and inspection. As such, it has been recognized that CCAMLR has created a new model for conserving ecosystems, which is different in both substance and form from traditional fishery agreements.[1044]

1037 Articles VII, VIII and IX of CCAMRL are about the Commission; Articles XIV-XVI are on the Scientific Committee; Article XIII relates to the Secretariat; and Article XXIV addresses the system of observation and inspection.

1038 Article IX (1), CCAMLR.

1039 Darry Powell, a former executive secretary of the Commission, noted:" The Commission has interpreted the Convention as requiring management decisions to flow directly from scientific advice. This interpretation places a responsibility on the Commission to ensure that the information necessary for the implementation of the Convention is gathered." Powell, Darry, "Antarctic Marine Living Resources and CCAMLR", in Herr, R.A. and others (eds.) *Antarctica's Future Continuity or Change?* (Hobart: Australian Institute of International Affairs, 1990), 61–70 at 69.

1040 Article IX (2), CCAMLR.

1041 Article IX (6), CCAMLR.

1042 These measures refer to fishing tools and methods, minimization of incidental mortality, fishing seasons and areas, by-catch limits, exploratory measures for special fish species, and the like. See at http://www.ccamlr.org/en/conservation-and-management/browse-conservation-measures.

1043 http://www.ccamlr.org/en/conservation-and-management/conservation-measures.

1044 See Zegers (note 1017), at 151.

However, there are still some obstacles to the enforcement of this Convention. Firstly, the provision that each party implement the Convention within its own competence creates ambiguity in its enforcement procedures.[1045] Secondly, with CCAMLR, sovereignty issues of Antarctica, just like other instruments of the ATS, are an obsession. The model of "agreement of disagreement" makes situations unclear when a party attempts to enforce its domestic measures of conservation, or when two parties claim jurisdiction over the same area.[1046] Thirdly, some crucial provisions of this convention, such as Article II (3) (a), (b) and (c),[1047] are less precise, with the outcome that they can be used only as guidelines and cannot be strictly implemented. Therefore, during the first several years, the Scientific Committee – lacking detailed scientific information – did not make recommendations, and the Commission did not adopt conservation measures.[1048] Fourthly, there are weaknesses in the observation and inspection system which were exploited during the first seven years of CCAMLR's approach leading to criticism of its as "work[ing] largely to benefit fishing nations".[1049] This system, intended to bolster the Commission's enforcement capacity, permits nationally designated observers and inspectors to carry out inspections and observation of fishing or research vessels of another state.[1050] However, inspectors should report to their national States; and the States in turn report to the Commission.[1051] This led to problems, firstly in ensuring an inspector was accepted by a foreign vessel and, secondly, in preventing monitoring by other states from interfering with other monitoring activities. The CCAMLR System of Inspection (Inspection System), established in 1989, may provide some answers, including the procedures for the designation of inspectors; the rights and responsibilities of inspectors; procedures for boarding and inspection; inspection reporting and the procedures for flag State prosecutions; and sanctions.[1052] Inspectors designated by

1045 Article XXI, CCAMLR.

1046 Triggs, Gillian D., *International Law and Australian Sovereignty in Antarctica*, (Sydney: Lega Books Ltd., 1986), at 197.

1047 Article II(3) of CCAMLR reads as follow: ...
(a) prevention of decrease in the size of any harvested population to levels below those which ensure its stable recruitment. ...
(b) maintenance of the ecological relationships between harvested, dependent and related populations of Antarctic marine living resources ...
(c) prevention of changes or minimisation of the risk of changes in the marine ecosystem...

1048 Heap, J.A., "Has CCAMLR Worked? Management Policies and Ecological Needs", in Jorgensen-Dahl, A. / Østreng, W. (ed.), *The Antarctic Treaty System in World Politics* (Basingstocke: Macmillan Academic and Professional, 1991), pp. 43–53.

1049 Joyner, Christopher C., *Antarctica and the Law of the Sea*, (Dordrecht: Martious Nijhoff Publishers, 1992), p. 246.

1050 Article XXIV(2)(b), CCAMLR.

1051 Article XXIV(2)(c), CCAMLR.

1052 http://www.ccamlr.org/en/document/publications/ccamlr-system-inspection.

contracting states should register with the Commission[1053] and are entitled to board a fishing or research vessel of any contracting party[1054] and to inspect catch gear and fishery activities[1055]. This appears to grant inspectors a mandatory right to carry out their duties, complemented by the requirement that a flag State should launch, prosecute and impose sanctions severe enough to negate any economic benefit arising from illegal activities[1056] once a violation of conservation measures has been proven.[1057] Additionally, the flag State should ensure that its illegal vessel conducts no further fishing activities until the sanctions are paid.[1058] Actual operation of the Inspection System was undertaken during the 1989/90 season with inspectors being designated by Argentina, Chile, the United States and the Soviet Union.[1059]

CCAMLR is confronted with a crucial problem common to other international environmental laws. Namely, it is the States who must recognize and automatically fulfill conservation requirements for the Antarctic environment and its ecosystems. Although the birth of the Protocol was seen as a new chance to combine and improve the existing mechanisms preserving Antarctica from destruction by human activities, it is hard to satisfy such a lofty expectation.

Generally however, CCAMLR does set up a successful regime for the exploitation of Antarctic marine living resources. Its ultimate success demonstrates the importance of international cooperation when dealing with Antarctic issues. An impressive innovation was the extension of the CCAMLR beyond that of the Antarctic Treaty, providing the ATCPs with precious experience dealing with allocation of jurisdiction between the convention and the coastal States for future policy on other Antarctic issues. CCAMLR demonstrates that the Antarctic Treaty System is quite flexible, imaginative, and capable of managing emerging issues in Antarctica, and could be a precedent in this regard for future negotiations on a regime for the exploration and exploitation of Antarctica's mineral resources.[1060]

B. The Protocol on Environmental Protection to the Antarctic Treaty

In the 1980s, ATCPs focused on issues related to Antarctic mineral resource exploitation and initiated the negotiations surrounding a minerals convention. This step was accompanied by serious debates both in- and outside the Antarctic Treaty System. Within the ATS, sovereignty concerns were at the core of the struggle between the

1053 Paragraph II of the CCAMLR System of Inspection.
1054 Paragraph III of the CCAMLR System of Inspection.
1055 Paragraph VI of the CCAMLR System of Inspection.
1056 Paragraph XIII of the CCAMLR System of Inspection.
1057 Paragraph XI of the CCAMLR System of Inspection.
1058 Paragraph XIV of the CCAMLR System of Inspection.
1059 Cordonnery (note 1025), at 137.
1060 Zegers (note 1017), at 154–155.

claimant states and non-claimant states.[1061] External pressure was imposed on the ATS by both developing states working within the UN forum and international environmental organizations. The former insisted that Antarctica should be treated as *res communis*, and the latter believed that mineral resources activities were incompatible with the conservation of the environment.[1062] In order to alleviate this pressure, the ATCPs decided to commence the negotiation of a general instrument concerning Antarctic environmental protection.[1063]

During negotiations for the new environmental regime, there was a significant debate regarding the coordination and compatibility of the norms for the proposed instrument and the prior ATS. The environmental regime may effectively limit some human activities in Antarctica which might have been recognized by other components of the ATS, raising conflicts of competence between these instruments. These environmental, political, international and internal pressures, led the XV ATCM in Paris in 1989, to convene a special Antarctic Treaty Consultative Meeting – the XI SATCM – to consider all proposals related to the comprehensive protection of the Antarctic environment and its dependent and associated ecosystems. Recommendation XV-1 provided the mandate for the XI SATCM,[1064] which established two working groups to assess the new environmental instrument.

1061 Some claimant states have difficulty accepting that CRAMRA did not recognize royalties and competence over activities on territory subject to sovereign claims. However, some of them ultimately accept the legitimacy of CRAMRA (for example, Chilean Minister of Foreign Affairs J. Berguno noted that in the end CRAMRA dealt successfully with these difficulties in achieving a fair balance between the interests of claimants and non-claimants. See: Berguno, J., "The Antarctic Park: The Issue of Environmental Protection", in Verhoeven et al. (eds.), *The Antarctic Environment and International Law* (London: Graham and Trotman Limited, 1992), 103–107 at 104. Others, such as Australia and France, chose to abandon it.

1062 It was accepted by many states that Antarctic environmental protection would not be effective unless Antarctic mineral activities were prohibited and CRAMRA were not adopted. See Vicuña, Francisco Orrego, "The Protocol on Environmental Protection to the Antarctic Treaty: Questions of Effectiveness", in 17 Geo. Int'l Envtl. L. Rev. 1 (1994–1995), at 1–2.

1063 Recommendation XV-1, in 1989.

1064 Recommendation XV-1 sets out five interrelated aspects of environmental protection: 1) Formulating principles for Antarctic environmental protection; 2) reviewing the existing body of environmental protection measures for Antarctica in order to identify those to be strengthened or otherwise improved and gaps where supplemental measures should be elaborated; 3) strengthening the legally binding and enforceable nature of obligations to protect the Antarctic environment; 4) establishing new institutional arrangements to underpin effective implementation of Antarctic environmental protection measures; and 5) promoting environmental assessment and scientific research and monitoring and exploring the role of an information and data base related to effective environmental management in Antarctica.

Working Group I focused on the structures and components of the proposed system of protection for Antarctica, and discussed a range of proposals submitted by various states. Australia, Belgium, France and Italy put forward an indicative draft of a convention for the comprehensive protection of the Antarctic environment;[1065] New Zealand put forward a detailed draft protocol to the Antarctic Treaty;[1066] Argentina, Norway, the UK, the US, and Uruguay put forward an outline proposal[1067] calling for a protocol with annexes to the Antarctic Treaty, with the UK submitting a supplementary daft protocol[1068] and the USA submitting a draft protocol plus four specific annexes;[1069] India put forward an outline approach to comprehensive measures.[1070]

Working Group II focused on elaborating details of the four annexes to the Protocol and laid the groundwork for progress in 1991 on additional annexes. The four annexes cover environmental impact assessment, waste disposal, marine pollution and conservation of Antarctic fauna and flora. The work of this group was based primarily on the New Zealand draft protocol and on the draft annexes submitted by the USA as well as a working paper on environmental impact assessment tabled collectively by Australia, Belgium, France and Italy.

At the last session of the XI SATM, there had been widespread support for the Protocol with annexes approach. The measures contained in the annexes touch upon matters regulated in the ATS, thereby elevating the status of annexes to the Protocol to the level of the Treaty itself. There is also flexibility to add more annexes in

---

1065 Australia, Belgium, France and Italy, "Indicative Draft of a Convention for the Comprehensive Protection of the Antarctic Environment", XI Special Antarctic Treaty Consultative Meeting, November 18, 1990. http://www.ats.aq/documents/SATCM11_1/wp/SATCM11_1_wp001_e.pdf.

1066 New Zealand, "Draft Protocol to the Antarctic Treaty on Environmental Protection", XI Special Antarctic Treaty Consultative Meeting, November 18, 1990. http://www.ats.aq/documents/SATCM11_1/wp/SATCM11_1_wp002_e.pdf.

1067 Argentina, Norway, United Kingdom, United States of America and Uruguay, "Comprehensive Measures for the Protection of the Antarctic Environment and its Dependent and Associated Ecosystems – Outline of a Protocol Supplementing the Antarctic Treaty", XI Special Antarctic Treaty Consultative Meeting, November 18, 1990. http://www.ats.aq/documents/SATCM11_1/wp/SATCM11_1_wp005_e.pdf.

1068 United Kingdom, "Provisions for A Protocol Supplementing the Antarctic Treaty", XI Special Antarctic Treaty Consultative Meeting, November 18, 1990. At: http://www.ats.aq/documents/SATCM11_1/wp/SATCM11_1_wp003_e.pdf.

1069 United States of America, "Protocol Supplementing the Antarctic Treaty", XI Special Antarctic Treaty Consultative Meeting, November 18, 1990. At: http://www.ats.aq/documents/SATCM11_1/wp/SATCM11_1_wp004_e.pdf.

1070 India, "Comprehensive Measures for the Protection of the Antarctic Environment and Dependent and Associated Ecosystems", XI Special Antarctic Treaty Consultative Meeting, November 18, 1990. At: http://www.ats.aq/documents/SATCM11_1/wp/SATCM11_1_wp007_e.pdf.

future, a common approach in international environmental treaties.[1071] To minimize the possible conflicts, the Protocol to the Antarctic Treaty with attached annexes was adopted as a supplementary protocol which acknowledges the validity of the preceding ATS at the XVI ATCM .[1072]

Within the Protocol, Articles 4 and 5 intend to maintain the "integrity and the continuing" status of the prior ATS.[1073] Article 4 affirms that the Protocol is a supplement and not a modification or amendment to the Antarctic Treaty. It also stipulates that the new environmental regime cannot derogate from the rights and obligations of its parties under the other components of ATS, including CCAMLR, CCAS and measures adopted pursuant to the Treaty. Article 5 provides a procedure for consultation and cooperation among the parties to the Protocol, parties to other components of the ATS, and among the respective institutions associated with these components. The ultimate purpose of consultation and cooperation is to ensure the achievement of the objectives and principles of the Protocol without impairing the objectives envisaged by other instruments of the ATS. These two provisions present a modest position for the environmental regime in the ATS, not allowing environmental protection to impair the right of widespread and established activities undertaken in accordance with other components of the ATS. However, a contradiction exists between the non-prejudice clause and the principles clause. The Antarctic Treaty and ATS instruments grant freedom of access to marine resources, tourism and scientific research, however, while Article 4 exempts these rights from more restrictive limitation under the Protocol, Article 3 limits all these activities to those consistent with the principles of environmental protection. In particular, Article 3(4) stipulates that all the activities mentioned above be modified, suspended or cancelled if they cause or might cause inappropriate impact on the Antarctic environmental or its ecosystems. Any completely unregulated human activity in Antarctica would affect its environment and ecosystems. Thus Article 5, as a follow-up to Article 4, requires that the environmental protection objective of the Protocol should be achieved despite conflicts between the Protocol and other components of the ATS, which will require ATCPs to reach compromises through consultation and cooperation.

I. General Rules

1. Objects and Principles

One of the most controversial issues in the negotiation of the Protocol was how to balance protection of the Antarctic environment with the intrinsic value of

1071 Redgwell, Catherine, "Antarctica: Wilderness Park or Eldorado Postponed?", 1 Environmental Politics 137 (1992), at 140.

1072 Francioni, Francesco, "Madrid Protocol on the Protection of the Antarctic Environment", 28 Tex. Int'l L. J. 47 (1993), at 55.

1073 Ibidem.

Antarctica as an area for the conduct of scientific research, with such research inevitably impacting the Antarctic environment. On one hand, any activities, including scientific research, will affect the unique Antarctic environment and ecosystems to some degree. On the other hand, the Antarctic scientific community will not accept an approach of comprehensive environmental protection that unduly limits and restricts research activities. The Scientific Committee on Antarctic Research (SCAR) advocates controls that should be customised to reflect the degree and extent of the likely impact of scientific and logistic support activities.[1074]

The Protocol deals with this issue by recognizing the importance of the protection of the Antarctic environment and its ecosystems, and by designating Antarctica as "a natural reserve, devoted to peace and science".[1075] The term "natural reserve, devoted to peace and science" is not a common international concept and lacks any substantial meaning. Moreover, the Protocol neither provides a detailed explanation of the concept nor does it regulate specific rights or obligations, if any, that are intended to flow from other provisions. Thus, this designation appears more politically symbolic than a legal concept.[1076]

In order to preserve the environment, the Protocol articulates certain principles with which all human activities must comply. These principles are listed in Article 3 and are the heart of the Protocol.[1077] Article 3 paragraph 1 sets forth the essential principle: activities in Antarctica must be planned and conducted to account for the protection of the Antarctic environment and ecosystems and the intrinsic value of Antarctica, including its wilderness and aesthetic values and its value as an area for the conduct of scientific research, in particular research essential to understanding the global environment. This principle imposes two fundamental considerations: environmental protection and intrinsic value protection. Moreover, it divides the intrinsic value of Antarctica into two components, wilderness and aesthetic values on one side and value as an area for scientific research on the other, and specifies this research should be related to understanding the global environment. It is worth

---

1074 Kimball, Lee A., "Report on Antarctica", (Washington, D.C., World Resources Institute, January 1991), (copy at SPRI), p. 6.

1075 Article 2, the Protocol.

1076 Watts, Arthur, *International Law and the Antarctic Treaty System* (Cambridge: Grotius Publications Limited, 1992), p. 277; Kammerer, Jörn-Axel, "The Protocol on Environmental Protection to the Antarctic Treaty", 45 Law and State 68 81992), at 69; and Redgwell, Catherine, "Environmental Protection in Antarctica: the 1991 Protocol", 43 Int'l & Comp. L.Q. 599 (1994), at 606. However, one scholar believes this designation has a legal influence on the Protocol that limits the various freedoms of use. It expresses that Antarctica, henceforth, becomes a zone controlled and protected in its entirety, not only in relation to the specially protected zone adopted by the prior ATS, but also in relation to the ecological, peaceful and scientific objectives of the Treaty. See: Francioni (note 1072), at 61–62.

1077 Watts (note 1076), at 277; and Redgwell (note 1076), at 606.

noting that all of these "fundamental considerations" are found in Article 2(3) of CRAMRA.

This principle provides a fundamental dimension to the relationship established between environmental protection and scientific research. As scientific research ranks as one of the intrinsic values of Antarctica, it has no greater status than Antarctic "wilderness and aesthetic values". Therefore, when considering scientific research in Antarctica, other intrinsic values of Antarctica cannot be ignored, meaning all activities must be consistent with the goal of environmental protection and then other intrinsic values of Antarctica. This has the significant effect of eliminating the dominant position of scientific research in Antarctic matters, and thus its free and unlimited exercise.

Paragraph 2 of Article 3 of the Protocol requires that in order to achieve the general goal of protecting the Antarctic environment and intrinsic values, any activity undertaken must be planned and conducted in such a way as to limit its adverse impact on the Antarctic environment. This is not a new principle; it can be derived from Principle 21 of the 1972 Declaration of the United Nations Conference on the Human Environment[1078] and was confirmed in 1982 in the World Charter of Nature[1079].

Generally speaking, adverse impact on the Antarctic environment and its ecosystems should, according to this provision, be limited. The degree of this limitation however, is not regulated. The provision does include six specific consequences of planned and conducted activities to be avoided, including adverse effects on climate or weather patterns; significant adverse effects on air or water quality; significant changes in the atmospheric, terrestrial, glacial or marine environments; detrimental changes in the distribution, abundance or productivity of species or populations of species of fauna and flora; further jeopardy to endangered or threatened species or populations of such species; and degradation of or substantial risk to areas of

---

1078 United Nations Conference on the Human Environment, UN. Doc. A/conf.48.14 (1972). Principle 21 reads: "States have, in accordance with the Charter of the United Nations and the principles of international law, the sovereign right to exploit their own resources pursuant to their own environmental policies, and the responsibility to ensure that activities within their jurisdiction or control do not cause damage to the environment of other States or of areas beyond the limits of national jurisdiction." http://www.unep.org/Documents.Multilingual/Default.asp?documentid=97&articleid=1503

1079 World Charter for Nature was adopted by the UN. General Assembly, on October 28, 1982, UN. Doc. A/RES/37/7. Part I, paragraph 4, reads: "4. Ecosystems and organisms, as well as the land, marine and atmospheric resources that are utilized by man, shall be managed to achieve and maintain optimum sustainable productivity, but not in such a way as to endanger the integrity of those other ecosystems or species with which they coexist." http://www.un.org/documents/ga/res/37/a37r007.htm

biological, scientific, historic, aesthetic or wilderness significance.[1080] This list is transplanted from CRAMRA Article 4(2): which uses exactly same terminology, e.g. "air and water quality", "atmospheric, terrestrial or marine environments", "fauna or flora" and "biological, scientific, historic, aesthetic or wildness significance".[1081] There are differences between these two instruments however; the principles in CRAMRA provide the basis of whether to proceed with minerals activity, while the Protocol establishes them as the basis for both planning and conducting activities. The relationship of these principles in these two documents creates a level of uncertainty as to the exact meaning of the terminology used, such as "significant", "detrimental" and "substantial". These uncertainties have already attracted serious criticism from environmentalists.

This criticism rightly raises the question of how to determine which kind of activity would cause these adverse effects. In response, a regime of advance assessment of proposed activities has been envisaged which require prior assessment of environmental impact and regular and effective environmental monitoring for any proposed activity in the Antarctic.

Subparagraph (c) of Article 3 imposes the positive obligation of possessing enough knowledge to make a judgment on the likely environmental consequences of a proposed activity. Theoretically, if the level of knowledge associated with an activity does not sufficiently allow for the completion of an environment assessment, the activity should be excluded. Besides these prior assessments, it is also necessary to make informed judgments about proposed activities. The judgments should be concluded in consideration of cumulative impacts, potential conflicts of use, the availability of appropriate technologies, environmental monitoring, and emergency response capabilities. This provision also stems from Article 4 of CRAMRA, which prescribes stricter requirements for permitting a mineral resource activity. Mineral activities would not be permissible unless the capacity to monitor key environmental parameters and ecosystem components exists to identify any adverse effects. The Protocol on the other hand, only requires informed judgments to take account of "whether" the capacity to monitor exists. In other words, CRAMRA attempts to establish a threshold of knowledge and technology for the activities, which becomes a mandatory condition for any proposed mineral resource activities,

1080 Article 3(2), the Protocol.

1081 CRAMRA Article 4(2) reads: "No Antarctic mineral resource activity shall take place [until it is determined] that the activity in question would not cause: a) significant adverse effects on air and water quality; b) significant changes in atmospheric, terrestrial or marine environments; c) significant changes in the distribution, abundance or productivity of populations of species of fauna or flora; d) further jeopardy to endangered or threatened species or populations of such species; or e) degradation of , or substantial risk to, areas of special biological, scientific, historic, aesthetic or wilderness significance."

while the Protocol does not explicitly or implicitly prohibit human activities even though knowledge regarding involved capacities is lacking.[1082]

Subparagraphs (d) and (e) require that permitted activities should be regularly and effectively monitored. This monitoring is aimed not only at verifying the predicted impacts but also at facilitating early detection of potential unforeseen effects, both within or outside the Antarctic Treaty area.

Paragraph 3 of this Article recognizes that scientific research has priority over other activities. At first glance, this may appear to contradict[1083] the position that scientific research does not have priority over environmental protection. However, this is not the case. This provision is intended to recognize the priority of scientific research over other kinds of human activity in Antarctic.[1084] In other words, any conflicts between the intended scientific research and another kind of activity, tourism for example, would result in the latter ceding priority to the former. Nevertheless, if the scientific research were foreseen as potentially harming the Antarctic environment or its ecosystems, it would be limited or even totally prohibited according to the above principle.

This is enhanced, specifically, by paragraph 4 of Article 3, which requires certain Antarctic activities be planned in advance according to Antarctic Treaty Article VII (5) to be carried out in a manner consistent with the principles set out above. These activities include scientific research programs, tourism and all other governmental and non-governmental activities. Where these actions result in or might result in inappropriate environmental consequences, they must be modified, suspended or cancelled. This provision reaffirms the criterion of protecting the Antarctic environment and its ecosystems even for scientific research.

To sum up, the Protocol acknowledges that it is important to protect the Antarctic environment and ecosystem along with its intrinsic values; and any activity that may cause adverse effects on the Antarctic environment should be limited or avoided. Therefore, the freedom of scientific research, a traditional right in the Antarctic, has been limited in a manner consistent with criteria designed to maintain a healthy and stable environment.

1082 Redgwell (note 1076), at 608.

1083 Some scholars claim that these principles are sometimes of a competing nature. The principle which gives priority to scientific research is immediately followed with the caveat that activities such as scientific programs shall "take place in a manner consistent with the principles in the Article" and that such programs should be "modified, suspended or cancelled if they result in impact upon the Antarctic environment or dependent associated ecosystems inconsistent with those principles." Moreover, modifications to these unacceptable impacts of scientific research are not explained. See Redgwell (note 1076), at 606–607.

1084 Watts (note 1076), at 280.

2. Dispute settlement

Generally, disputes concerning the interpretation or application of the Protocol should be solved by negotiation, inquiry, mediation, conciliation, arbitration, judicial settlement or other peaceful means.[1085] In addition, the dispute settlement system in CRAMRA has been carried over into the Protocol to solve disputes arising from specific provisions, including Article 7, prohibition on minerals activities; Article 8, environmental impact assessment; Article 15, emergency response action; and Article 13, compliance with the environmental Protocol to the extent that it applies to these articles. Furthermore, the mandatory third-party judgment procedures envisioned in CRAMRA apply to specific annexes to the Protocol.

Article 19 of the Protocol allows each party to select either, or both, the International Court Justice or the Arbitral Tribunal for the settlement of disputes regarding specific Articles. Disputes will be settled according to the means chosen by the parties unless otherwise agreed. Where parties cannot agree on a means for solving these specific disputes, the dispute will be resolved according to the mandatory third-party judgment procedure at the request of any involved party.[1086]

However, the Final Act of the XI Special Antarctic Treaty Consultative Meeting states that: "With reference to Article 18, the Meeting agreed that an inquiry procedure should be elaborated to facilitate resolution of dispute concerning the interpretation or application of Article 3 with respect to activities undertaken or proposed to be undertaken in the Antarctic Treaty area."[1087] There have not yet been any further measures concluded on this topic, nor has it even been discussed. As Article 3 does not fall under the special third-party judgment procedure, in theory, disputes concerning this Article should be solved by peaceful means through consultation. Nevertheless, a third-party procedure cannot be ruled out where the parties to the dispute have so agreed. Therefore, generally speaking, the third-party judgment mechanism established in Article 19 and Article 20 of the Protocol could apply to any dispute beyond the extent of the Article 19, where agreed by the disputing parties.

3. Other Matters

Some proposals before the meeting of the Protocol suggested that the new environmental regime should go beyond the northern limits of the Antarctic Treaty area, i.e. 60° South latitude, to the Antarctic convergence, which would include the area of application defined in CCAMLR. Nevertheless, as a protocol to the Antarctic Treaty, the new regime can "neither modify nor amend that Treaty";[1088] thus it

1085 Article 18, the Protocol.
1086 Article 20(1), the Protocol.
1087 Final Act of the XIth SATCM, see http://www.ats.aq/documents/recatt/att326_e.pdf.
1088 Article 4(1), the Protocol.

applies to the "Antarctic Treaty area", i.e. the area in Article VI of that Treaty. The Protocol and its annexes apply to all human activities in Antarctica governed by the Antarctic Treaty[1089], and are open for accession by any contracting state to the Antarctic Treaty.[1090] Non-ATCPs are also encouraged to accede to this Protocol by means of withholding the position of an ATCP this step has been taken.[1091]

Parties are obliged to cooperate in the planning and conduct of activities in the Antarctic Treaty area and, to the extent possible, to share information that may be helpful to other parties.[1092] The cooperation obligation is stipulated as a list of desiderata rather than concrete legal obligations as mandatory regulations covering the wide range of all the human activities to which the Protocol applies would be difficult to craft.[1093]

Article 17 of the Protocol requires that each party report annually on steps made toward implementing the Protocol. This report should be circulated to all party states and to the Committee for Environmental Protection (hereafter the CEP) and made publicly available. A party's annual report should include details of its efforts to comply with the Protocol,[1094] contingency plans to respond to environmental emergencies in the Antarctic Treaty area,[1095] and any other notifications and information.

This publication requirement is a huge improvement, as access to environmental information is a critical element of an effective environmental protection system.[1096] It provides a double benefit by increasing the availability of documentation regarding the ATS;[1097] while also providing a somewhat disguised opportunity to monitor parties' actions in respect to Antarctic environmental protection.

II. Prohibition of Mining Action

Over the course of three sessions of the XI SATCM, which commenced in Vina del Mar, Chile, in November 1990 and continued in Madrid in April and June 1991, the critical issue for achieving consensus on the Protocol was the resolution of minerals

1089 Article 3(4), the Protocol.
1090 Article 22(2), the Protocol.
1091 Article 22(4), the Protocol.
1092 Article 6, the Protocol.
1093 Redgwell (note 1076), at 609.
1094 Article 17(1) in connection with Article 13(3), the Protocol. According to these provisions, party states have to report appropriate measures within their competence, administrative actions and enforcement measures to ensure compliance with the Protocol, and they have to make appropriate efforts to ensure that no one engages in any activities contrary to the Protocol.
1095 Article 17 in connection with Article 15, the Protocol.
1096 Redgwell (note 1076), at 612.
1097 Ibidem.

issues.[1098] Article 7 of the Protocol, a delicate compromise between a permanent ban on mineral resource activities and the regulation of such activities under the minerals convention, prohibits any activities regarding mineral resources other than scientific research. However, Article 25, which governs amendment or modification of the Protocol, provides an opportunity to unfreeze the moratorium on mining activities. According to this article any ATCP may request a conference to review the operation of the Protocol 50 years after its entry into force. Given the important status of Article 7 of the Protocol, any modification or amendment is specifically regulated by a provision under Article 25.

Paragraph 5 (a) of Article 25 stipulates, firstly, that if a modification or amendment to Article 7 is proposed at a Review Conference, it must include a binding legal regime, which is in force, on Antarctic mineral resource activities. Secondly, the binding system has to meet certain conditions, including an agreed means for determining whether, and if so, under which conditions any such activities would be acceptable. Moreover, this regime should fully safeguard the interests of all States referred to in Article IV of the Antarctic Treaty and apply the principles thereof. Thirdly, Article 7, unlike other provisions of the Protocol, can only be modified or amended after 50 years from entry into force by a Review Conference convened under the Protocol, regardless of the procedures for the modification of the Treaty set forth in Article XII (1) (a) and (b).[1099] Therefore, the precondition for revoking the prohibition of Antarctic mineral resource activities is the conclusion of a valid binding minerals regime among the ATCPs, with the procedure itself requiring consideration of the Protocol for a Review Conference.

Any proposed change to the Protocol must be adopted by a majority of parties, including three-quarters of the states which were ATCPs at the time the Protocol was adopted. It will then only come into force upon approval by three-quarters of ATCPs, including approval by all states which were ATCPs at the time of the adoption of the Protocol. Therefore, every ATCP, having such status when the Protocol was adopted, is given a "double veto" to extend the moratorium on mineral resource activity, by rejecting any modification of Article 7 or the binding legal regime on Antarctic mineral resource activities. As discussed above, mineral resource activities in the Antarctic under CRAMRA cannot be conducted until an appropriate regulatory regime has been introduced to manage them.[1100] Thus, Article 7 is nothing more

---

1098 Redgwell (note 1071), at 137–138.

1099 Paragraph 1of Article 25 reads: "without prejudice to the provisions of Article 9, this Protocol may be modified or amended at any time in accordance with the procedures set forth in Article XII (1) (a) and (b) of the Antarctic Treaty." However, Article 25(5) (a) requires that an amendment to Article 7 be proposed at a Review Conference referred to in paragraph 2 Article 25.

1100 Final Act of the Fourth Special Antarctic Treaty Consultative Meeting on Antarctic Mineral Resource Activities.

than "the institutionalization of the moratorium already agreed upon during the negotiations on CRAMRA."[1101]

However, a "walk away" provision provides huge opportunity for a waiver of the prohibition on mineral resource activities. Article 25(5) (b) permits a state to withdraw from the Protocol at any time with two years' notice, if a modification has not entered into force within three years of the date of its adoption. The effect of this provision means that the ban on mining could be lifted 55 years after the entry into force of the Protocol, or amended at any time by consensus of ATCPs in accordance with Article XII (1)(a) of the Antarctic Treaty.

In conclusion, the Protocol prohibits mineral resource activities for 50 years and imposes very strict conditions on waiving the prohibition after that period. The ban will remain in place in a range of scenarios: as long as ATCPs are unable to conclude a binding legal regime covering such activities; if the binding legal regime is not in force; if the proposal for modification or amendment cannot be adopted; or if the proposal cannot enter into force. However, if any modification or amendment regardless of its relationship to Article 7 of Protocol has not entered into force within three years of its adoption, any party to the Protocol may withdraw from the regime, effectively removing the prohibition on mining activities.

III. Environmental Impact Assessment (EIA)

Article 8 and Annex I of the Protocol represent one of the most significant environmental principles[1102] and provide the *sine qua non* of effective environmental protection – the environmental impact assessment.[1103] Article 8 of the Protocol prescribes basic principles of an EIA and Annex I[1104] stipulates specific assessment procedures.

Article 8 requires all activities requiring prior notice under Article VII (5) of the Antarctic Treaty, be assessed for their impact on the Antarctic environment and its ecosystems. It sets the threshold at activities that have a "minor or transitory

---

1101 Wolfrum, Rüdiger, *The Convention on the Regulation of Antarctic Mineral Resource Activities: An Attempt to Break New Ground*, (Heidelberg: Springer-Verlag, 1991), at 88–89.

1102 Watts (note 1076), at 281.

1103 Redgwell (note 1076), at 610–611.

1104 Annex I to the Protocol, namely "Environmental Impact Assessment" establishes Recommendation XIV-2, "Man's Impact on the Antarctic Environment: Environmental Impact Assessment", 1987. Recommendation XIV-2 builds on UNEP's "Goals and Principles on Environmental Impact Assessment" and SCAR's work. There are some other instruments referring to the EIA, including: Recommendation VIII-11, namely "Code of Conduct for Antarctic Expeditions and Station Activities" (Oslo, 1975), which was revised in 1989 by Recommendation XV-3 (Paris, 1989), and Recommendation XV-17. Additionally, the 1991 Espoo Convention on Environmental Impact Assessment in a Transboundary Context also affects the Annex I. See: Redgwell (note 1076), at 617.

impact" on the Antarctic environment or on dependent or associated ecosystems.[1105] Specifically, human activities are divided into three categories in the Antarctic Treaty area, those having: "a) less than a minor or transitory impact; b) a minor or transitory impact; c) more than a minor or transitory impact." Only activities of the first category may proceed without further approval; the latter two categories should observe the procedures in Annex I.

Under Article 8, all parties must apply the assessment procedures in the planning processes leading to decisions about any activities undertaken in the Antarctic Treaty area. The EIA therefore integrates environmental considerations into the decision-making process for any proposed activities.[1106] The assessment procedures apply to "any changes in any activity" which might "arise[] from an increase or decrease in the intensity of an existing activity, from the addition of an activity, the decommissioning of a facility, or otherwise."[1107] For some activities taken by more than one party, all parties involved must nominate a coordinator to implement the EIA procedures.[1108]

1. *Preliminary Stage and Initial Environmental Evaluation*

Annex I elaborates the procedure of the EIA. The initial assessment of a proposed activity's environmental impacts should be determined in accordance with "appropriate national procedures".[1109] If the assessment concludes a "less than a minor or transitory impact", the proposed activity may directly proceed.[1110] To satisfy EIA requirements, party states must establish "appropriate" national procedures according if they do not already exist[1111], although what is "appropriate" has been left vague. Similar issues arise regarding the standard for "minor or transitory impact" – this must be clarified in the relevant "appropriate" act(s) of national legislation. Authority to determine the results of an EIA is determined by the nation State.

Where a proposed activity is not assessed as "less than minor or transitory", the activity must go through either an Initial Environmental Evaluation (IEE) or a Comprehensive Environmental Evaluation (CEE). An IEE must be prepared (unless a CEE is pending) and contain sufficient detail to assess whether a proposed activity may have more than a minor or transitory impact. If an IEE indicates that a proposed activity is likely to have "no more than a minor or transitory impact", it may proceed. The activity must be assessed and verified by appropriate procedures, such as a monitoring

1105 Redgwell (note 1076), at 617.
1106 Evteev, Sveneld, "Antarctica and its Place in the Contemporary Environmental Movement", in Jorgensen-Dahl, Arnfinn / Ostreng Willy, (eds.), *The Antarctic Treaty System in World Politics* (Palgrave Macmillan, July 1991), 147–152 at 151.
1107 Article 8(3), the Protocol.
1108 Article 8(4), the Protocol.
1109 Article 1(1), Annex I to the Protocol.
1110 Article 1(2), Annex I to the Protocol.
1111 Watts (note 1076), at 282.

system that provides a regular and verifiable record of the impacts of the proposed activity[1112], and which are circulated to party states.[1113]

2. *Comprehensive Environmental Evaluation*

There are more detailed provisions for the third category of activities in the Antarctic Treaty area. A Comprehensive Environmental Evaluation (CEE) is required for any proposed activity likely to have more than a minor or transitory impact, and may be commenced as either the preliminary assessment or as the result of an IEE. According to the requirements and procedures laid out in Article 3 of the Annex, the CEE should be made based on a series of information and considerations and go through a consultation phase.[1114] The draft should be publicly available and circulated to all parties for comment for a period of 90 days. During this period, the parties may provide comments on the draft of the CEE. The draft must be sent to the CEP at the same time as it is circulated to the parties, at least 120 days before the next consultative meeting. Article 3 (5) provides that "no decision to proceed with a proposed activity shall be delayed ... for longer than 15 months from the date of circulation of the draft Comprehensive Environmental Evaluation"[1115] to avoid indefinite postponement of the decision[1116]. The final CEE should include all comments received, be publically available and circulated to all parties 60 days before the commencement of the proposed activity.

The CEE and "other relevant considerations" provide the basis for determining whether an activity should proceed.[1117] If so, monitoring procedures should be put in place to monitor key environmental indicators and to assess and verify the impact of the proposed activity. However, it is uncertain whether the operator, the

1112 Article 5(2), Annex I to the Protocol.

1113 Article 6(1) (c), Annex I to the Protocol.

1114 Article 3(2) of Annex I to the Protocol lists items that should be included in a CEE, including a description of the proposed activity; a description of the initial environmental reference; a description of the methods and data used to forecast the impacts of the proposed activity; estimation of the nature, extent, duration, and intensity of the likely direct impacts of the proposed activity; consideration of possible indirect or second order impacts of the proposed activity; consideration of cumulative impacts of the proposed activity; identification of measures that could be taken to minimise or mitigate the proposed activity; identification of unavoidable impacts of the proposed activity; consideration of the effects of the proposed activity on the conduct of scientific research and on other existing uses and values; an identification of gaps in knowledge and uncertainties encountered in compiling the information required under this paragraph; a non-technical summary of the information provided; information on the person or organization which prepared the CEE.

1115 Article 3(5) Annex I to the Protocol.

1116 Francioni (note 1072), at 64.

1117 Article 4, Annex I to the Protocol.

sponsoring state or the CEP is responsible for establishing this monitoring system. Moreover, all activities and processes must provide a regular and verifiable record of the impacts of the activity to facilitate decisions relating to suspension, cancellation or modification of the activity.[1118]

However, the procedures and provisions on an EIA under this annex do not apply to emergency cases relating to the safety of human life or of ships, aircraft or equipment and facilities of high value, or the protection of the environment.[1119]

3. *Conclusion*

The system of EIAs adopted in the Protocol does have some advantages. Firstly, this annex requires that all activities in the Antarctic Treaty area should be assessed before they take place. The mandatory EIA has been highly commended as "it became indispensable and conferred credibility on the system of environmental standards established in the Madrid Protocol, which, in the absence of such a procedural instrument, risked becoming into a merely cosmetic operation."[1120] Secondly, the annex provides transparency in the evaluation process.[1121] Important information must be made public, such as the national EIA procedures, an annual list of any IEE and any decisions taken in consequence thereof, significant information obtained from the monitoring system, and information referred in Article 3 (6),[1122] and a draft CEE that must also be circulated to the parties and to the CEP. The rationale behind the publicity requirement is that if an activity might cause more than minor and transitory impact to Antarctic environment, the rest of the international community has the right to be informed and comment by various methods.[1123] The CEP and any parties may provide comments, providing the opportunity to influence and operator's decisions through political pressure. Therefore, public comment is treated as the "best available means of avoiding environmentally harmful activities being undertaken".[1124]

However, the flaws of the annex are also very obvious and serious. Firstly, the annex does not indicate who is authorized to determine which level of impacts the proposed activities will cause. Secondly, there is no obvious criterion inherent in "minor and transitory impact", resulting in potentially different standards from different states. Thirdly, there is an absence of obligatory provisions to regulate actions contrary to this annex. Fourthly, it is not clear who can make decision on whether a proposed action is to proceed or not. Last but not least, the EIA system lacks an independent institutional mechanism to manage and coordinate the whole

1118 Article 5, Annex I to the Protocol.
1119 Article 7, Annex I to the Protocol.
1120 Francioni (note 1072), at 64.
1121 Francioni (note 1072), at 64–65; Redgwell (note 1076), at 620.
1122 Article 6, Annex I to the Protocol.
1123 Watts (note 1076), at 284.
1124 Watts (note 1076), at 285.

assessment procedure.[1125] Without such a mechanism, national authorities may use the EIA may in an instrumental and cosmetic manner rather than as a mechanism to protect the Antarctic environment.[1126] The publicity requirements, giving the most credibility of the EIA system, would become a means of mitigating political, rather than environmental, risk.[1127]

IV. Implementation of the Protocol

1. *Institutions*

Article 11 of the Protocol establishes a Committee for Environmental Protection (hereafter the CEP), which comprises all parties to the Protocol. Observer positions are granted to parties to the Antarctic Treaty who are not parties to the Protocol as well as to the president of SCAR and the chairman of the Scientific Committee for the CCAMLR. Other relevant scientific, environmental and technical organizations approved by the ATCM may also obtain observer positions at the CEP.

Article 12 defines the CEP as an advisory organ, without independent authority on implementation or supervision, which provides advice and recommendations on compliance with the Protocol and its annexes. However, the Protocol does not require that the members of the CEP possess specific expertise. Therefore, in carrying out its functions, the CEP should consult with the SCAR, the Scientific Committee for the CCAMLR and other relevant organizations. Moreover, the Protocol provides that the CEP is allowed to perform other functions "as may be referred to it by the ATCMs". The CEP meets every year in conjunction with the ATCM. Article 10 of the Protocol recognizes the function of the ATCMs, as defining general policy for the comprehensive protection of the Antarctic environment and its ecosystems, and adopting measures according to Article IX of the Antarctic Treaty. In addition, the ATCMs shall review and consider the work of the CEP. This contrasts strongly with CRAMRA and its complex system of institutions – the Antarctic Mineral Resource Commission, Regulatory Committees, Special Meetings of Parties, Advisory Committee and a Secretariat – for controlling mineral resource activities and to protect the Antarctic environment and its ecosystems. In particular, the Antarctic Mineral Resource Commission and Regulatory Committees are authorized to set general criteria for mineral resource activities and establish special requirements for relevant activities in specific areas as regards possible exploration and development activities.

In conclusion, under the Protocol, there is no special functionary organ to manage environmental protection in Antarctica. The ATCMs continue to play the decision-making role in the environmental protection regime, with the ATCMs

---

1125 Francioni (note 1072), at 65.

1126 Francioni (note 1072), at 65–66.

1127 Lyons, David, "Environmental Impact Assessment in Antarctica under the Protocol on Environmental Protection", 29 Polar Record 111 (1993), at 117.

working upon the CEP's advice and recommendations, as well as upon the advice of the SCAR.

2. Compliance and inspection under the Protocol

Article 13 regulates that each party to the Protocol is obliged to comply with it, through adoption of laws and regulations, administrative actions and enforcement measures. As the Antarctic is open to all peaceful uses, parties and non-parties to the Protocol, or even to the Antarctic Treaty, including international governmental and non-governmental organizations, are therefore permitted to legitimately exercise activities in Antarctica. This poses a risk, as treaties cannot bind third parties, neither the Antarctic Treaty nor the Protocol is able to limit third parties' rights, unless relevant provisions have become customary international law, making international cooperation, beyond the party States, essential to effectively protect the Antarctic environment. Thus paragraph 2 of this Article encourages each party, consistent with the UN Charter, to exert efforts to avoid and eliminate any activity contrary to the Protocol. Where a party considers an activity affects the implementation of the Protocol, it should draw the attention of all other parties to it. This is further enhanced by the ATCMs bringing any such activity to a non-party's attention. These provisions are all found in CRAMRA, which also provide that the Commission may draw the attention of all parties to any inappropriate activities carried out by any parties[1128], and elaborate a deliberate and detailed system of liability for the operator and the sponsoring States.[1129] As Professor Wolfrum indicates, the Protocol, compared to CRAMRA, leaves one of the most vital questions unresolved as it does not identify "who is responsible for planning human activities in Antarctica."[1130] Even though the Protocol provides principles and procedures to limit human activities to protect the Antarctic environment and its ecosystems, its rules on compliance are very flexible and indistinct.

The Protocol applies the inspection system detailed in the Antarctic Treaty to ensure compliance. Article 14 of the Protocol recognizes individual and collective inspection by observers, assigned under Article VII of the Antarctic Treaty. Observers may be nationals designated by any of the ATCPs, or designated by the ATCMs. The latter could be some form of institutional inspection,[1131] with inspection being carried out under special procedures established by an ATCM. The observers have the right to access to all parts of stations, installations, equipment, ships and aircraft as well as all records maintained. The final inspection report is then sent to the relevant parties, for comment. The reports and comments are to be made publicly available and circulated to all parties and to the CEP. The two-tier inspection system in the Protocol also comes from the approach taken by Article 12 of CRAMRA.

1128 Article 7(7), CRAMRA.
1129 Article 8, CRAMRA.
1130 Wolfrum (note 1101), at 91.
1131 Redgwell (note 1076), at 612.

However, the ATCMs have not yet begun to establish any institutional inspection system. Similarly, the second prong of collective institutional inspection remains to be worked out in the near future. Thus, at present the only applicable inspection system is carried out by observers either designated in accordance with Article VII of the Antarctic Treaty or designated under the consensus rule.[1132]

3. *Liability*

The attribution of liability plays an important role in making an environmental protection regime effective.[1133] Article 16 of the Protocol requires that "parties undertake to elaborate rules and procedures relating to liability for damage arising from activities taking place in the Antarctic area and covered by this Protocol", and those rules should be included in annex(es) to the Protocol.

In 1998, the XXII ATCM and the First CEP commenced negotiating the liability annex to the environmental protection protocol in accordance with Article 16 of the Protocol. The ATCM and the CEP appointed Working Group I of the ATCM to undertake further negotiation of the liability annex to the Protocol.[1134] In July 2001, Decision 3 of the XXIV ATCM and the IV CEP decided to draft texts for an annex on the liability aspects of environmental emergencies. The negotiations on the liability annex were based on the draft text proposed by the US (ATCM XXIV/WP6) and the draft text proposed by the chairman of Working Group I (ATCM XXIV/WP17),[1135] taking into account the paper produced by SCAR and the Council of Managers of National Antarctic Programs (COMNAP)[1136] in response to Resolution 5 (1999)

1132 Wolfrum (note 1101), at 92.

1133 Redgwell (note 1076), at 615.

1134 Decision 3 of ATCM XXII-CEP I (1998), at: http://www.ats.aq/devAS/ats_meetings_meeting_measure.aspx?lang=e

1135 Final Report of the Twenty-fourth Antarctic Treaty Consultative Meeting, Item 10: The Question of Liability as Referred to in Article 16 of the Protocol, pp. 17–20. At: http://www.ats.aq/documents/ATCM24/fr/ATCM24_fr001_e.pdf

1136 The Council of Managers of National Antarctic Programs (COMNAP), created in 1988, is the international association that brings together National Antarctic Programs from around the world to develop and promote best practice in managing the support of scientific research in Antarctica. Each country that is a signatory to the Antarctic Treaty 1959 normally establishes a National Antarctic Program, which has national responsibility for managing the support of scientific research in the Antarctic Treaty Area on behalf of its government and in the spirit of the Antarctic Treaty. COMNAP currently brings together the National Antarctic Programs of 28 countries from all around the world. The purpose of COMNAP is: to develop and promote best practice in managing the support of scientific research in Antarctica, by: serving as a forum to develop practices that improve the effectiveness of activities in an environmentally responsible manner; facilitating and promoting international partnerships; providing opportunities and systems for information exchange; and providing the Antarctic Treaty System with objective

(ATCM XXIV/WP14). In 2005, the XXVIII ATCM and the VIII CEP adopted Annex VI to the Protocol: Liability Arising from Environmental Emergencies. As this must still be approved by all ATCPs, and to date, only five have done so, this annex is not likely to enter into force in the short term.[1137]

a) General provisions for environmental emergencies

Article 15 of the Protocol regulates that every party should provide a prompt and effective response action to any emergency arising in the performance of scientific research, tourism and all other governmental or non-governmental activities. It also includes some specific procedures for such emergency response actions, and requires contingency plans be established to respond to incidents with potential adverse effects on the Antarctic environment or its ecosystems. Annex VI specifies the scope and requirements: Article 1 of Annex VI states that the annex applies to environmental emergencies in the Antarctic Treaty area caused by scientific research programmes, tourism and all other governmental and non-governmental activities, including associated logistic support activities, and applies to all tourist vessels and other vessels. Article 4 of the annex elaborates the components of contingency plans to be established by operators, including procedures for assessing the incident; notification procedures; identification and mobilisation of resources; response plans; training; record keeping; and demobilisations. Moreover, each party to the Protocol should establish and implement procedures for immediate notification of, and cooperative responses to environmental emergency.

b) Response action

In response to environmental emergencies arising from the activities of one operator, the following actions are required under Article 5 of this annex. First, the operator must take prompt and effective response action. Secondly, if the operator does not take prompt and effective response actions, the sponsoring state[1138] and other parties are encouraged to take such actions.

Thirdly, if other parties wish to respond to an environmental emergency, they must notify the sponsoring state of the operator and the Secretariat of the Antarctic Treaty of their intention beforehand to provide the sponsoring state with opportunity to undertake such actions itself. However, if the threat of significant and harmful impact to the Antarctic environment is imminent, other parties could

---

and practical, technical and non-political advice drawn from the National Antarctic Programs' pool of expertise. https://www.comnap.aq/

1137 These states are Finland, Peru, Poland, Spain and Sweden.

1138 Annex VI does not use the term "sponsoring state" but rather "the Party of the operator". The present author chooses the former because "sponsoring state" has been defined by CRAMRA as "the party with which an operator has a substantial and genuine link", and its meaning is specific and will not be confused with other states. See Article 1(12), CRAMRA.

reasonably take immediate response action and notify the sponsoring state and the Secretariat as soon as possible. Other parties can only respond immediately to an environmental emergency under only three situations: a) the harmful impact is imminent and the action is reasonable; b) the sponsoring state does not notify the Secretariat within a reasonable time that it will take a responsible action; c) the sponsoring state has not taken response action within a reasonable time after its notice. In other words, this annex attempts to avoid a repetitive carrying out of response actions. However, these provisions require that both the sponsoring state and other parties wishing to take action notify their action before taking these actions, which will take up precious response time. Whether this system is effective has yet to be tested. If the sponsoring state takes response action, and is willing to be assisted by another party or parties, the former should coordinate the response actions. That means, if the sponsoring state joins the remedy actions, it organise the whole process.

Fourthly, where there is no clear sponsoring State or there is more than one sponsoring State, any party may make the best endeavours to consult as appropriate and notify the Secretariat, if the notice is practicable. This approach is also inappropriate for an environmental emergency response. The provision does not regulate the time limits for consultation between sponsoring States, nor does it clarify how an emergency should be dealt with where there is no certain sponsoring State. Questions such as whether there is any institution able to respond and monitor an emergency, and who will take action if there is no sponsoring state, or if there is no result from the consultation among sponsoring states remain unanswered in Annex VI.

Fifthly, when a party takes response action, it is to consult and coordinate with three other kinds of States: a) other States also taking response actions; b) States conducting activities in the vicinity of the environmental emergency; and c) States impacted by the environmental emergency. The first case has been regulated in the third and fourth points above. Consultation and cooperation with the latter two categories of States is important, as they may either be impacted by the environmental emergency and the response, or these States may be associates of the responding State(s) who could provide necessary and prompt assistance.

The annex provides a system for response to an environmental emergency. Principally, the operators and sponsoring states are obliged to the take responsive action. However, if both fail to do so, or if there is some other obstacle, other parties could notify the Secretariat which will then determine whether that party is an appropriate responding party on a case by case basis. The complicated response system, particularly the notification and consultation requirements, may prevent other parties from taking prompt and effective action, and making it difficult to image the extent to which this system will be effective if a real emergency arises.

c) Liability

The liability provisions, detailed in Article 6, are the critical component of this annex. An operator who is liable respond and fails to do so has to reimburse the costs of the response action. It is not a punitive payment, but rather a compensation of what the responding actor has paid. Thus, operators simply pay the same amount as if they had responded themselves, an approach that does not seem likely to discourage an initial failure to respond.

If an operator does not respond to an environmental emergency, and there are no other response actions from alternative parties, the operator is to pay the costs of the response action into a fund specified in Article 12 of the annex. The fund, managed by the Secretariat, is to provide, *inter alia*, for the reimbursement of reasonable and justified costs incurred by a party or parties. Any party who responds to an environmental emergency may request reimbursement, with the decision to pay falling to the ATCMs. The idea of establishing a fund goes back to the liability fund proposal put forward in CRAMRA.[1139] Where no response action is taken, payment made into the fund is divided according to two levels. If a state operator is involved, it must pay the costs of the response action that should have been undertaken. If it is a non-state operator, it pays an amount of money that most closely reflects the costs of the response action that should have been taken.

Paragraph 3 of Article 6 also stipulates that "liability shall be strict", albeit without specific explanation, neither the elements of the liability nor the responsible person are explicitly defined. Conversely, CRAMRA provides a more detailed liability system than Article 6 of the Annex, imposing strict liability and that is categorised according to situation, including damage to the Antarctic environment and its ecosystems; loss of or impairment to an established use of the Antarctic and its ecosystems; loss of or damage to property of a third party or loss of life or personal injury of a third party; and reimbursement of reasonable costs by whomsoever incurred relating to necessary response action.[1140] Furthermore, under CRAMRA multiple operators are jointly and severally liable for joint activities. A party shall be liable if it fails to take an appropriate response action to an environmental emergency caused by its nationals.

d) Exemptions from and limits of liability

Article 8 of this annex provides exemptions for both the operators and the party states. An operator is exempted from liability for an environmental emergency if it was caused by: an act or omission necessary to protect human life or safety; an event constituting a natural disaster of an exceptional character; an act of terrorism; or an act of belligerency against the activities of the operator. When a party takes reasonable response action to an environmental emergency pursuant to Article 5(2) of this annex, which results in a new environmental emergency, the party will not be liable. Moreover, Article 9 of the Annex stipulates the maximum amount for an operator's liability.

1139 Article 8(7)(c)(iii), CRAMRA.
1140 Article 8(2), CRAMRA.

Unlike CRAMRA, Article 10 of this annex regulates that "a party state ... shall not be liable for the failure of an operator ... to take response action to the extent that the party took appropriate measures within its competence, including the adoption of laws and regulations, administrative actions and enforcement measures, to ensure compliance with this Annex." This provision excludes obligations of the party states', which makes the liability system much more ineffective. The loose liability, burdensome notification mechanism and indefinite responsibilities of party states under this Annex mean that responses may be delayed or completely lacking.

e) Conclusion

Annex VI to the Protocol, the first liability annex for protection of the Antarctic environment and its ecosystems, is far from an effective environmental instrument. It raises questions of whether the imposed liability is sufficient to restrict the operators' activities, whether the liability scheme works to achieve Antarctic environmental protection, and whether the liability system encourages timely response if the only result of delay is to require payment of the exact costs of those actions without additional punitive charges.

The annex only regulates that the operators – both state and non-state – carry the main obligation. However, it is overly tolerant toward party states, who could play a more significant role in responding to an environmental emergency than individual operators do. Protecting the Antarctic environment is a responsibility not only for operators but also for their sponsoring states. For the same environmental protection purposes, CRAMRA, by contrast, has established a detailed system of strict liability for operators; and also elaborated strict responsibility for sponsoring states who failed to exercise their supervisory obligations.[1141] In the annex, liability is limited to the costs of response actions, whereas CRAMRA stipulates several levels of response actions, including prevention, containment, clean up and removal measures.[1142]

Finally, the Annex leaves one considerable problem unsolved: who holds responsibility when no one responds to an environmental emergency, or when it is clear that no action will be taken within a reasonable time. Without an authority in charge of Antarctic environmental protection, it will be very difficult to achieve the purpose of preserving the Antarctic environment and its ecosystems. In all likelihood, there is only slight potential for the creation of such an organ, as parties to the ATS are not expected to voluntarily accept a mechanism controlling their actions and limiting their rights and freedom in the Antarctic. This conclusion has been proved by the fact that even such a loose liability system has not yet been accepted; making clear the level of challenge to stipulate a stricter and more fundamental regulation of Antarctic environmental protection.

1141 Article 8 and Article 27, CRAMRA.
1142 Article 8, CRAMRA.

V. Evaluation of the Protocol

The Protocol has several key differences to any preceding mechanism. It differs from the Whaling Convention, which is an international Convention separate from the Antarctic Treaty System, in that it is subject to the ATS. Unlike the Agreed Measures, which is an internal Measure of the Antarctic Treaty without clear clauses on implementation, the Protocol establishes a system of bodies and procedures to enforce its goals. Compared with CCAMLR, which deals with one ecosystem, the Protocol manages all life in the Antarctic. The Protocol appears to be an improvement in the quest to provide a comprehensive approach to the protection of the Antarctic environment.[1143] It provides a single governing instrument (with annexes) for the Antarctic environment, meaning the Antarctic environment is no longer governed by separate ad-hoc arrangements which are not always consistent in their approaches.[1144]

However, the improvement seems to be more one of form than of substance. Substantive improvements in environmental protection have been "significantly undermined" by the rapid negotiation of the Protocol "to satisfy the demands of environmental organizations and public pressures in some countries".[1145]

Firstly, critics have pointed out that the Environmental Impact Assessment regulated in the Protocol and its Annex I fall short of the highly protective measures expected for the Antarctic environment.[1146] Critics claim that the crucial concept of a "minor or transitory impact" was deliberately left undefined for each nation to determine its own standards of "minor"; thus the Protocol does not offer a reliable approach to environmental protection.[1147] It also lacks a mechanism for an independent review of the initial judgment of the operator, as this approach was equated with an all-powerful "Environmental Committee",[1148] deemed contrary to Antarctic traditions.[1149] There is some irony in the fact that it was the opposition to CRAMRA's

1143 Redgwell (note 1076), at 605. Redgwell claims: "The Protocol ... to a large extent is a consolidation of existing measures within the ATS."

1144 Overholt, Deborah H., "Environmental Protection in the Antarctic: Past, Present and Future", 28 CAN. Y.B. INT'L L. 227 (1990), at 248–54. The author identifies five major defects in the regime of Antarctic environmental protection that existed prior to the Protocol: the lack of enforcement mechanisms, sanctions or other formal procedures to address violations; the non-binding nature of certain measures under the Antarctic Treaty System, even as to some parties to the Antarctic Treaty; the absence of an overall conservation strategy; the failure in practice to require environmental impact assessments; and the lack of a forum to resolve disputes among parties to the Treaty.

1145 Vicuña (note 1062), at 3.

1146 Lyons (note 1127), at 111.

1147 Lyons (note 1127), at 114–115.

1148 Vicuña (note 1062), at 5.

1149 Vicuña, Francisco Orrego, *The Protection of the Antarctic Environment: International Antarctic Regime Project Publication Seires, No. 5,* (The Fridtjof Nansen Institute, 1992), at 20.

enforcement mechanism approved by a number of Consultative Parties that led to the independent review mechanism being left out of the Protocol.[1150] In fact, CRAMRA made considerable efforts to protect the Antarctic environment by prescribing standards and procedures to limit human activities. Additionally, the Protocol does not specifically allocate authority to determine whether a particular activity is allowed or prohibited. It does provide that the decision is to be based upon the Comprehensive Environmental Evaluation and other relevant considerations; but is very unclear as to by whom.[1151] Moreover, it requires an environmental impact assessment for almost any activity,[1152] which could interfere with the development of Antarctic science[1153] and contradict the purpose of "a natural reserve, devoted to peace and science".[1154]

Secondly, Annex II of the Protocol updates the Agreed Measures and subsequent recommendations adopted by the ATCMs.[1155] It introduces only a few new obligations of substance, such as a prohibition on taking or harmfully interfering with native fauna and flora,[1156] and prohibiting the introduction of non-native species, parasites and diseases.[1157] Additionally, it broadens exceptions relating to cases of emergency[1158] and extends protections to the flora and fauna of the marine environment.[1159] However, most of this updating is rather formal,[1160] as Annex II to the Protocol does not resolve the jurisdictional problems related to the use of permits.[1161] Additionally, the exclusion of activities committed under the Whaling Convention, the Seals Convention and CCALMR contradicts the aim of providing "comprehensive protection of the Antarctic environment" through an integrated system of environmental protection.[1162]

1150 Articles 39–54 of CRAMRA, application, permit and environmental impact assessment requirements for exploration and development. Vicuña (note 1062), at 5.

1151 Article 4, Annex I to the Protocol.

1152 Lyons (note 1127), at 114.

1153 Laws, Richard, "Science as an Antarctic Resource", in Cook, Grahame (ed.), *The Future of Antarctica: Exploitation Versus Preservation*, (New York: Manchester University Press, 1990), 8–24 at 20–21.

1154 Article 2 of the Protocol.

1155 Blay (note 962), at 392.

1156 Article 3, Annex II to the Protocol.

1157 Article 4, Annex II to the Protocol.

1158 Article 2(1), Annex II to the Protocol, which reads: "This Annex shall not apply in cases of emergency relating to the safety of human life or of ships, aircraft, or equipment and facilities of high value, or the protection of the environment."

1159 Article 1(d), Annex II to the Protocol: "'Native invertebrate' means any terrestrial or freshwater invertebrate, at any stage in its life cycle, indigenous to the Antarctic Treaty area".

1160 Vicuña (note 1062), at 6.

1161 Article 1(f) and (e), Annex II to the Protocol defines that a "permit" is issued by "an appropriate authority", which includes "any person or agency authorized by a Party to issue permits under this Annex".

1162 Article 2, the Protocol.

Thirdly, the Protocol establishes a new organization for the ATS – the Committee for Environmental Protection. Although the Committee has the function of monitoring the implementation of the Protocol and reporting on progress at ATCMs, its position as an expert body assisting the ATCPs means it does not have any power to enforce the Protocol. Unlike the Commission for the CCAMLR, the Committee for the Protocol was not designed as a permanent body with its own headquarters and staff. This weak institutional structure produces another inevitable shortcoming – implementation and enforcement are dependent upon individual states. Article 8 of the Protocol adopts a dual approach, leaving a loophole by which party members can exercise jurisdiction on a territorial and/or personal basis.[1163] This approach might cause jurisdictional disputes not covered by the mandatory dispute settlement provisions of the Protocol. In contrast, CRAMRA, whose leading principle is the preservation of "the status *quo ante* of the Antarctic environment in spite of mineral resource activities",[1164] provided explicit regulation of critical activities and improved institutional machinery for environmental protection including vesting the Consultative Meetings with law-making functions, which in contrast, are absent under the Protocol.[1165] It may be that the Protocol does not adopt CRAMRA-like provisions in respect of all human activities to avoid eroding ATCP national sovereignty.[1166] But this lack means that the Protocol relies heavily upon domestic measures, which poses some substantial hurdles for effective implementation. Additionally, the Protocol's vague regulation of human activities other than mineral resource activities devalues environmental protection in the face of economic and sovereignty concerns connected with mining.[1167]

C. Conclusion

The Protocol was born in an enthusiastic era of environmental protection, which cloaks this international instrument with much political significance. Since the "Question of Antarctica" was first submitted to the UNGA, the Antarctic has been spotlighted on the international stage. The most commonly voiced challenge is a criticism of ATS legitimacy and the authority of the ATCMs, and there has been a call for the application of the common heritage principle in Antarctica. Although ATCPs maintained their position of legitimacy in the management of Antarctica, the differences between them undermined the hitherto united front they had presented to the

---

1163 See Triggs, G.D., "the Antarctic Treaty Regime: Some Jurisdictional Problems", in Triggs, G.D. (ed.), *The Antarctic Treaty Regime: Law, Environment and Resources* (Cambridge: Cambridge University Press, 1987), 88–109.

1164 Wolfrum, "The Exploitation of Antarctic Mineral Resources: Risks and Stakes", in Verhoeven, J. et al. (eds.), *The Antarctic Environment and International Law*, (Londen: Graham and Trotman Limited, 1992), 27–32 at 29.

1165 Wolfrum (note 1101), at 89

1166 Wolfrum (note 1101), at 94.

1167 Redgwell (note 1076), at 634.

international community. This resulted in a strong desire among the ATCPs to return to the consensus which had characterized the ATS since its inception in 1959.[1168]

The Protocol does generally contribute to international environmental law. Firstly, it sets an example both in the Antarctic region and on the global stage[1169] as it "stands as the first instrument of international agreement to provide a comprehensive system of rules and procedures binding upon and applicable to all human activity in an area beyond the limits of national jurisdiction."[1170] Secondly, the Protocol supports some new components of international environmental law, such as the "polluter pays" principle and environmental impact assessments.

Besides Annex I (Environmental Impact Assessment) and Annex VI (Liability Arising from Environmental Emergency), the Protocol contains other four annexes. Annex II, namely Conservation of Flora and Fauna, improves the 1964 Agreed Measures for the Conservation of Antarctic Fauna and Flora, and aims to protect the diversity of native fauna and flora and avoid introduction of non-native species, parasites and diseases. Annex III is related to waste disposal and management, and requires the amount of wastes produced or disposed of should be reduced for the good of the environment in Antarctica. However, it fails to address the origin of waste and its prevention.[1171] Annex IV addresses marine pollution, and incorporates the provisions of MARPOL 73/78,[1172] which makes Antarctica a special area under MARPOL 73/78 for oil and garbage disposal.[1173] Annex V regulates area protection and management, and establishes two areas, Antarctic Specially Protected Areas (ASPAs) and Antarctic Specially Managed Areas (ASMAs), to replace the current five forms of designation.[1174] The ASPAs aim to protect "outstanding environmental, scientific, historic, aesthetic or wilderness values, any combination of those values, or ongoing or planned scientific research"[1175], and they may not be entered without specific permission.[1176] The ASMAs are areas "where activities are being conducted or may in the future be conducted" and are designated as such to "assist in the planning and coordination of activities, avoid possible conflicts, improve cooperation between parties or minimize environmental impacts".[1177] Activities in ASMAs

---

1168 Redgwell (note 1071), at 141.
1169 Redgwell (note 1076), at 633.
1170 Francioni (note 1072), at 51.
1171 Redgwell (note 1076), at 624.
1172 "MARPOL 73/78" is the International Convention for the Prevention of Pollution form Ships, 1973, as amended by the Protocol of 1978 relating thereto and by any other amendment in force thereafter.
1173 Redgwell (note 1076), at 628.
1174 These five forms of designation are: Specially Protected Areas (SPAs), Sites of Special Scientific Interest (SSISs), Sites of Historic Interest (SHIs), Specially Reserved Areas (SRAs) and Multiple-Use Planning Areas (MPAs).
1175 Article 3(1), Annex V to the Protocol.
1176 Article 3(4), Annex V to the Protocol.
1177 Article 4(1), Annex V to the Protocol.

should be subject to a code of conduct regulating various topics such as access to and movement within or over the area; activities which are or may be conducted within the area; and the installation, modification, or removal of structures.[1178]

However, like other international environmental instruments, the Protocol also shares some general weaknesses: it adopts general standards, creates a system using non-uniform national standards, non-independent institutional structure, individual national implementation and enforcement; and a weak monitoring system.[1179] The lack of an independent institution in particular impedes the exercise of the procedures and principles enshrined in the Protocol. Another weakness of the Protocol is its subordination to other components of the ATS.[1180] On one hand, the Protocol recognizes that the Antarctic Treaty and other instruments in the ATS will not be impaired by the Protocol; on the other hand, it seems untenable that these existing rights arising from the ATS not be subject to limitation. The Final Act of the XI SATCM reiterated that "nothing in the Protocol shall derogate from the rights and obligations of parties under the Convention on the Conservation of Antarctic Marine Living Resources, the Convention for the Conservation of Antarctic Seals and the International Convention for the Regulation of Whaling."

As compared to CRAMRA, the Protocol does not provide a strict liability system to control human activities in the Antarctic Treaty area. A clear line of demarcation is lacking between scientific action and mineral resource activities, making it difficult to distinguish genuine scientific activities from further exploration activities under the Protocol. CRAMRA, however, explicitly regulates that prospecting activities include "dredging and excavations ... or drilling ... not exceeding 25 meters", and subjecting strict limitations to any activities beyond these limits. In fact, many of the high-standard principles and mechanisms for the protection of the Antarctic environment and its ecosystems in CRAMRA have already significantly influenced the content of the Protocol.[1181] For example, Article 3, the heart of the Protocol, is primarily based on Article 4 of CRAMRA; Article 13, regarding compliance with this Protocol, could be found its original form in Article 7 of CRAMRA, and approaches to dispute settlement also originate from an identical system in CRAMRA.

Although the content of CRAMRA addresses preserving "the status *quo ante* of the Antarctic environment in spite of mineral resource activities", the fact that its name relates to mineral activities resulted in CRAMRA being subjected to tremendous pressures both internally, from the ATCPs, and externally, from the international community. In contrast, the Protocol, carrying the name of an environmental protection treaty, has obtained much broader support, notwithstanding that its effectiveness in environmental protection is doubtful.

1178 Article 5(3)(j), Annex V to the Protocol.
1179 Redgwell (note 1076), at 633.
1180 Francioni (note 1072), at 71.
1181 Wolfrum (note 1101), at 89.

Part 5: Exploitation of Antarctic Ice

Freshwater scarcity has become threat[1182] in light of anticipated rapid population growth in the foreseeable future[1183] and the effects of global warming[1184]. More than three-quarters of all freshwater resources are locked up in ice formations, mainly in the polar regions.[1185] More than 98 percent of the Antarctic continent, an area of 13,800,000 square kilometers, is covered by an ice sheet, containing an estimated water volume of 25–30 million cubic kilometers.[1186] It is thus not surprising that political and commercial interests are expected to someday drive States and entrepreneurs to exploit the ice resources found in the Antarctic. However, international law lacks the necessary rules to set out the legal status of ice and the jurisdictional competence over its use.[1187] With regards to Antarctic ice, some unavoidable questions have arisen again, such as the legal status of the ice, the appropriate international legal system and the appropriate international law for addressing the expected exploitation. This part of the dissertation seeks to make a modest contribution to the legal discussion by undertaking a three-fold analysis.

1182 Sauder, R. A., *The Lost Frontier: Water Diversion in the Growth and Destruction of Owens Valley Agriculture*, (Tucson: University of Arizona Press, 1994); Gleick, P. H., "Water and Conflict: Fresh Water Resources and International Security", 18 International Security 79 (1993); Gleick, P.H. / Yolles, P. / Hatami, H., "Water, War and Peace in the Middle East", 36 Environment 6 (1994); Kukk, C.L. / Deese, D.A., "At the Water's Edge: Regional Conflict and Cooperation over Fresh Water", 1 UCLA Journal of International Law and Foreign Affairs 21 (1996); OECD, *Water and Violent Conflict*, (OECD Issues Brief, 2005).

1183 DESA, *World Population 2300*, DESA Population Division, Doc. ST/ESA/SER.A/236 (2004).

1184 Parry, M.L. et al. (eds.), *Climate Change 2007: Impacts, Adaptation and Vulnerability, Contribution of Working Group II to the Fourth Assessment Report of the Intergovernmental Panel on Climate Change*, (Cambridge: Cambridge University Press, 2007), at 173–210. According to the IPCC, "[o]bservational records and climate projections provide abundant evidence that freshwater resources are vulnerable and have the potential to be strongly impacted by climate change, with wide-ranging consequences on human societies and ecosystems", in IPCC (ed.), *Technical Paper on Climate Change and Water*, Doc. IPCC-XXVIII/Doc.13 (8.IV.2008) (April 10, 2008), at 4.

1185 Goetz, P. (ed.), *The New Encyclopaedia Britannica: Macropaedia Vol. 20*, (15th Ed., Encyclopaedia Britannica, 1985), at 795. However, the World Bank estimates that polar region ice contains almost 70 percent of all freshwater resources, see World Bank, *World Development Report 2010: Development and Climate Change, 2010*, (Washington D.C.: The World Bank, 2010), at 139.

1186 Goetz (note 1185), at 797.

1187 Joyner, Christopher C., "Ice-Covered Regions in International Law", 31Natural Resources Journal 213 (1991), at 214.

Firstly, a brief assessment is made of the legal status of various forms of Antarctic ice. Secondly, an analysis is conducted which asks what legal system would govern Antarctic ice, UNCLOS or the Antarctic legal system. Thirdly, appropriate rules for the exploitation of Antarctic ice will be distilled through a further examination of the appropriate legal system.

A. Legal status of Antarctic Ice

Different forms of ice exist in Antarctica, including glacier ice (also known as sheet ice), sea ice (also called pack ice), shelf ice, icebergs and ice islands. As regards **glacier ice**, it is formed on Antarctic land by recrystallizing snow and is permanent in character. Analogous to land, the status of glacier ice is dependent on the status of the land on which it is located. Therefore, the question returns to familiar ground – the legal status of Antarctic – as analyzed in Part I. Based on the earlier exposition, the sovereign claims to Antarctica by seven claimant States lack a clear legal foundation and do not command the approval of international society. Therefore, the legal status of the glacier ice on Antarctic terra firma cannot be linked to a sovereign jurisdiction.[1188]

Sea ice, normally nonpermanent, forms by the freezing of seawater and perhaps from accumulated snow. Such ice drifts are unattached to land. Their area expands to more than 20 million square kilometers in winter and shrinks to about 4 million square kilometers in summer.[1189] International law gives it the same legal status to temporary sea ice as the surrounding seawater. In other words, the legal status of sea ice depends on the status of the ocean on which the ice is located. In regards to the Antarctic area, the absence of coastal States in legal terms means the sea ice is not under any State's jurisdiction and should be regarded as water surrounding the Antarctic. Moreover, Article VI of the Antarctic Treaty regulates that the Treaty applies to the area south of 60° South Latitude, including "the whole landmass, extensive sea adjacent to the continent, as well as all ice shelves". Therefore, the sea ice should be under the control of the Antarctic Treaty and other ATS instruments.

Shelf ice is a horizontal ice form extending beyond the margin of land and into the sea. It is fed by the accumulation of snow and the movement of glacier ice from the land, and covers more than one third of the Antarctic coastline.[1190] When parts of ice shelves break off and move into the sea, they are treated as icebergs. Since most ice shelves are usually attached to the land or seafloor, it is hard to distinguish where glacier ice ends and ice shelves begin. Therefore, ice shelves fixed to the Antarctic's

1188 Joyner (note 1187), at 222.
1189 Joyner (note 1187), at 223.
1190 Joyner (note 1187), at 225.

continental shelf carry the same legal status as glacier ice, while ice floating on the ocean surface, should be treated the same as seawater.[1191]

Icebergs are chunks of ice that break off from an ice shelf and drift into the sea. Antarctic icebergs are formed by the vast continental land-ice sheet, which extends hundreds of miles and breaks off into the sea.[1192] Exploiting icebergs is the most viable method for appropriating ice resources as a means for the acquisition of fresh-water. Antarctica holds almost 93 percent of the world's total iceberg mass;[1193] however, discussions in international law about the assignment of rights and duties for Antarctic iceberg appropriation have not yet reached a stage necessary to allow for sufficient governance of exploitation.

To some extent, the legal status of Antarctic icebergs depends on the legal status of ice shelves and their locations. The legal status of the ice shelf will partly decide the legal status of the relevant iceberg depending on its character. If an ice shelf is considered part of the high seas, icebergs related to it could also be treated the same. If an ice shelf is seen as analogous to the continent, the icebergs from it could be regarded as parts of land. Moreover, the legal status of Antarctic icebergs is affected by the regimes governing the Antarctic.[1194] The Antarctic Treaty plays an important role in defining the legal status of Antarctic icebergs. It can be anticipated that exploitation of Antarctic ice resources will once again give rise to sovereignty arguments between claimant and non-claimant States. As discussed above, sovereignty claims in Antarctica have been frozen by Article IV of the Treaty. With a strict interpretation of this article, no sovereign claim could involve icebergs, pack ice, fast ice or ice lands.[1195] Moreover, no coastal states exist in Antarctica, and there is no approval by the international community as to the claims on Antarctica. Waters around Antarctica should be treated as special high seas under the ATS regime. Hence, while Antarctic icebergs floating in the Southern Ocean should be treated as a water resource of the high seas and thus accessible by any State[1196], the appropriation of Antarctic icebergs should be subject to the ATS.

Ice islands, a unique type of iceberg, are an interesting classification of ice-formation. Scholars have tried to distinguish them based on different standards,

---

1191 Through analyzing the "shelf-as-high-seas approach" and the "shelf-as-land approach", Joyner concludes that: "[p]ortions of the high seas solidly frozen to the seafloor beneath the ice cap, as well as ice shelves and ice tongues fixed to Antarctica's continental shelf, should legally be regarded as having the status of land. Ice shelves not firmly fixed to the seafloor, but which are merely floating extensions of the glacier sheet on the ocean surface, should be accorded high seas status." For more detailed exposition please see Joyner (note 1187), at 225–228.

1192 Joyner (note 1187), at 232

1193 Goetz (note 1185), at 811.

1194 Joyner (note 1187), at 233.

1195 Joyner (note 1187), at 233.

1196 See Lundpuist, T.R., "The Iceberg Cometh?: International Law Relating to Antarctic Iceberg Exploitation", 17 Nature Resource Journal 1(1977), at 23.

stability and size.[1197] Moreover, being stable ice, they tend to be subject to the relevant land regime. Conversely, unstable ice islands are analogous to the high seas. Therefore, Antarctic ice islands have the same legal status as land only when they are stable. Joyner distinguishes ice islands from icebergs on the basis of size, and suggests new international regimes should be established to regulate the special legal status of ice islands.[1198] However, both of these classifications may cause problems, as neither the size nor movement of a given mass of ice remains constant.[1199] It is unreasonable and impossible to fix the legal status of Antarctic ice islands based on either of the classifications above. Nevertheless, it is unquestionable that any kind of ice-formation in Antarctica should be subject to the ATS within its scope of application.

In conclusion, given that ice formations are solid and thus fundamentally different from liquid water, it would not be surprising if similar legal norms failed to neatly apply in both cases. Thus, it might be inappropriate to extend the ocean law principles to ice and ice-covered polar regions. But as ice-covered areas are different from real land, legal norms for terra firma cannot be extended to ice-covered areas.[1200] It is therefore necessary to develop new international law to regulate ice and its appropriation.

B. Relevant Regimes for the Exploitation of Antarctic Ice

I. UNCLOS

The only reference to ice in UNCLOS is in Article 234, which regulates that coastal States can adopt laws regarding ice-covered areas within the scope of their EEZ. Apparently, this sole reference is not intended for Antarctica, but for those areas under domestic jurisdiction. Sovereign claims in Antarctica have not been endorsed by the international community, and claimant States cannot be treated as "coastal States" under UNCLOS. Therefore, it is hard to say that Article 234 could apply

---

1197 Francesca Trombetta-Panigadi claims that ice islands are floating freely in open water and can melt or break up into bergs and floes. Trombetta-Panigadi, Francesca, "The Exploitation of Antarctic Icebergs in International Law", in Francioni, F. / Scovazzi, T. (eds.), *International Law for Antarctica*, (The Hague: Kluwer Law International, 1996), 225–257 at 230.

1198 Joyner tell distinguishes islands from icebergs on the basis of size. He defines ice islands as being a type of iceberg which has a tabular structure, is 20 to 200 feet thick and covers an area up to 300 hundred square miles. Joyner (note 1187), at 237–242.

1199 An example of an iceberg is the Trolltunga iceberg, which was tracked by satellite. It drifted following the Weddell Sea coast for two years, and then grounded on a shoal near Berkner Island for a period of five years. After the stable period, Trolltunga moved again, and broke into smaller pieces before melting.

1200 Joyner (note 1187), at 242.

to the Antarctic ice. In fact, the Third UN Law of the Sea Conference deliberately avoided the specific problems associated with Antarctica,[1201] as the Antarctic issue was an especially tough obstacle which might have impeded the process of the Law of the Sea Conference. Moreover, the ATS has been evolving and improving gradually, and it has proven itself capable of governing the Antarctic effectively and maintaining its peaceful state.

II. Antarctic Treaty System

1. Forbidden or not?

Article 1(6) of CRAMRA defines mineral resources as "all non-living natural non-renewable resources, including fossil fuels, metallic and non-metallic minerals." If ice is considered to be "non-renewable", it is possible that ice falls under the category of mineral resources. However, the Final Act of the Fourth Special Antarctic Treaty Consultative Meeting (SATCM) on Antarctic Resources rejects that Article 1(6) includes ice.[1202] Moreover, the Final Act of the SATCM XI states that "[t]he Meeting noted that the harvesting of ice was not considered to be an Antarctic mineral resource activity".[1203] Hence, it is clear that ice is not regarded as a mineral, and thus the regulation prohibiting mineral resource activity, i.e. Article 7 of the Protocol, would not apply if the harvesting of ice were to become viable.

Article VI of the Antarctic Treaty provides that the ocean surrounding the Antarctic classified as high seas, which has also been reserved by Article IV(1) of the Convention for the Conservation of Antarctic Marine Living Resource (CCAMLR) and Article 4 of the Protocol. These provisions intend to respect States' right of access to natural oceanic resources in Antarctica that include ice resources. That UNCLOS does not prohibit the harvesting of ice provides another indication for the possible exploitation of Antarctic ice.

2. Restrictions

Although the ATS does not forbid the appropriation of Antarctic ice resources, it does set high standards limiting such activities. The Final Act of the XI SATCM explicitly indicates that if the possibility of harvesting of ice becomes reality in the

1201 Mangone specifically states that "Antarctica was deliberately excluded from any deliberations of the Third UN Law of the Sea Conference".

1202 Final Report of the Fourth Special Antarctic Treaty Consultative Meeting on Antarctic Mineral Resources, adopted at Wellington, 2 June 1988, http://www.ats.aq/documents/SATCM4_12/fr/SATCM4_12_fr001_e.pdf.

1203 The Final Act of the XI Antarctic Treaty Special Consultative Meeting was adopted at Madrid, October, 1991. At: http://www.ats.aq/documents/SATCM11_4/fr/SATCM11_4_fr002_e.pdf.

future, the provisions of the Protocol, other than Article 7, would apply.[1204] Until the ATCPs conclude specific norms for the exploitation of Antarctic ice, the Protocol will continue to play a critical role in governing such activities.

Moreover, the exploitation of Antarctic ice has been addressed in the Antarctic Treaty Consultative Meeting (ATCM) in 1989. The ATCM adopted Recommendation XV-21 dealing with the "Exploitation of Icebergs", which essentially called for the environmental consideration of such potential exploitation. Additionally, the precautionary principle, as a widely accepted international law principle, also limits ice-exploitation until adequate information proves that the environment would not be damaged.

III. Common Heritage of Mankind

In order to participate in Antarctic affairs, some developing States within the UN advanced the idea that Antarctica should be under the "common heritage of mankind" regime, adopted in Part XI of the UNCLOS regarding the "Area" – i.e. the seabed under the high seas. According to the common heritage of mankind principle, it is the international community, rather than individual State, which can exploit resources located in the relevant area for the benefit of mankind as a whole through an international authority.[1205] Several developing States complained that the two-tier governing structure of the ATS constructed Antarctica as a paradise for Antarctic Treaty members, especially those developed members.

The desire to put Antarctica under international management had been motivated by negotiations on the exploitation of Antarctic mineral resources. The non-party developing States expressed their dissatisfaction through the UN forum and required that any exploitation of the natural resources of Antarctica should ensure, *inter alia*, international management and an equitable sharing of the benefits of such activities.[1206] However, the common heritage regime cannot apply to Antarctica for three reasons. Firstly, even though not accepted by the international community, there are seven territorial claims over Antarctica. These claims effectively rule Antarctic out of those areas "beyond the limits of national jurisdiction". Secondly, Antarctica, unlike the deep seabed, has already been subject to an effective legal system. Thirdly, the common heritage regime demands yet another international authority governing Antarctic resources, which would cause unnecessary conflicts between the new authority and the ATCM, and might even lead to disorder in the region. Thus, the common heritage of mankind theory cannot apply to Antarctica.

Consequently, although international law, for now, provides no clear guidelines for determining the legal status of different Antarctic ice formations, which could

---

1204 Same as above note. http://www.ats.aq/documents/SATCM11_4/fr/SATCM11_4_fr002_e.pdf

1205 For more detailed discussion of exploitation, please see Part II Section C.

1206 See UN Doc. A/RES/38/77, 15 December 1983; UN Doc. A/RES/39/52, 17 December 1984; UN Doc. A/Res 40/156 (A) and (B), 16 December 1985.

be seen as either land or the high seas, this issue would not affect the appropriation of ice resources. Any potentially exploitative activities would be governed by the ATS. Moreover, relevant customary international laws should be respected and duly considered. However, as with the mineral exploitation approach, it is also impossible to adopt the common heritage regime for the exploitation of Antarctic ice.

C. Exploitation of Antarctic Ice

Although the ATS does not prohibit the exploitation of Antarctic ice, a comprehensive legally binding document on these activities has not been concluded by the ATCM. However, some documents of the ATS address the issue to varying degrees, such as Recommendation XV-21, adopted by the ATCM in 1989, the Convention for the Conservation of Antarctic Marine Living Resources (CCAMLR) and the Madrid Protocol. Moreover, the precautionary principle, as customary international law, also effects the exploitation of Antarctic ice.

I. Recommendation XV-21[1207]

At the XVth Antarctic Treaty Consultative Meeting, the issue of the utilization of ice was discussed and Recommendation XV-21, titled "Use of Antarctic ice", was adopted. Although it is not binding, this Recommendation laid out some basic principles for potential exploitation. Firstly, Antarctica shall be used for peaceful purposes only,[1208] and exploitation of Antarctic ice should be under the framework provided by the Antarctic Treaty.[1209] Secondly, such activities should be controlled in order to protect the Antarctic environment and its ecosystems.[1210] Third, since there was not sufficient scientific information on the environmental impacts of such activities, which might produce a number of environmental or other related issues,[1211] no commercial exploitation may occur, in any case, prior to an examination of these issues.[1212] The Recommendation, which eventually entered into effect on 29 January 2004, 15 years after its adoption, essentially laid down basic environmental protection requirements for the potential exploitation of Antarctic ice. However, no further follow-up action has been taken at subsequent meetings.[1213]

1207 Recommendation XV-21, adopted in 1989 at the XVth ATCM, at http://10.10.4.6/download/9667508/11060615/2/pdf/181/191/1331951251637_703/ATCM15_fr002_e.pdf.

1208 Recommendation XV-21, preamble, recital 3.

1209 Recommendation XV-21, preamble, recital 8.

1210 Recommendation XV-21, preamble, recital 5.

1211 Recommendation XV-21, preamble, recital 6–7.

1212 Recommendation XV-21, preamble, recital 9.

1213 Viñuales, Jorge E., Iced Freshwater Resources: A Legal Exploitation, 20 Yearbook of International Environmental Law186 (2009), at 195.

II. CCAMLR

It would appear that the CCAMLR does not apply to ice harvesting activities, since ice cannot be designated as a living resource. However, in conjunction with the purpose of the CCAMLR to conserve Antarctic marine living resources,[1214] "[a]ny harvesting and associated activities" shall be subject to the convention.[1215] The convention proclaims that it is important to safeguard "the environment and protect the integrity of the ecosystem of the seas surrounding Antarctica".[1216] Moreover, the convention provides a principle governing any harvesting or associated activities in its applicable area. The principle requires preventing or minimizing change or the risk of change "in the marine ecosystem which are not potentially reversible over two or three decades, taking into account ... the effects of associated activities on the marine ecosystem and ... the effects of environmental changes, with the aim of making possible the sustained conservation of Antarctic marine living resources."[1217] In other words, for the sake of conservation of Antarctic marine living resources, any activities that might cause irreversible changes in the marine ecosystem should be avoided. Hence, in the absence of adequate knowledge regarding the environmental effects of exploiting Antarctic ice, the provisions of the CCAMLR should be respected, especially when such activities could affect marine living resources.

Moreover, the CCAMLR has established the Commission for the Conservation of Antarctic Marine Living Resources.[1218] The Commission is authorized to "identify conservation needs and analyze the effectiveness of conservation measures",[1219] and to take necessary conservation measures to fulfill convention objectives, including "measures concerning the effects of harvesting and associated activities on components of the marine ecosystem other than the harvested populations".[1220] Based on these provisions, the Commission is entitled to take measures to manage the exploitation of Antarctic ice when such activities would affect the objectives of the convention. Additionally, conservation measures adopted by the Commission are binding upon all contracting parties.[1221]

1214 Article II, 1, the CCAMLR.
1215 Article II, 3, the CCAMLR.
1216 The Preamble of the CCAMLR, paragraph 1.
1217 Article II, 3(c), the CCAMLR.
1218 Article VII, the CCAMLR.
1219 Article IX, 1(e), the CCAMLR.
1220 Article IX, 2(i), the CCAMLR.
1221 Article IX, paragraph 6, the CCAMLR reads: "[c]onservation measures adopted by the Commission in accordance with this Convention shall be implemented by Members of the Commission", and the same paragraph (b) gives conservation measures binding effect. Article VII, paragraph 2, the CCAMLR, regulates that "each Contracting Party which participated in the meeting at which this Convention was adopted shall be a Member of the Commission". Thus, all contracting parties should obey conservation measures unless a member enters a reservation according to Article IX, paragraph 6, (c) and (d).

In summary, the exploitation of Antarctic ice should be limited by the requirements of conserving marine living resources and preserving the marine ecosystem. If ice appropriation has the potential damage the Antarctic ecological environment, the Commission could adopt binding conservation measures to prevent those activities.

III. The Madrid Protocol

Although the Protocol does not directly address the exploitation of Antarctic ice, the special meeting where the Madrid Protocol was adopted noted that "if the harvesting of ice were to become possible in the future, it was understood that the provisions of the Protocol, other than Article 7, would apply."[1222] Hence, potential ice exploitation activities should be limited by provisions and principles provided by the Madrid Protocol.

1. Applicable Principles

In order to achieve the purpose of the Madrid Protocol, all activities in the Antarctic area, including both the planning and conducting of activities, should fundamentally consider the protection the Antarctic environment and dependent and associated ecosystems.[1223] It is not surprising that the exploitation of Antarctic ice must satisfy the environmental principles elaborated by the Madrid Protocol. These principles include the following points:

Firstly, exploitation activities should limit or avoid adverse impacts on the Antarctic environment and ecosystems, such as climate or weather patterns, air or water quality, and distribution or productivity of species of fauna and flora.[1224] Thus, any plan for the exploitation of Antarctic ice should carefully consider its harmful effects on the Antarctic ecological environment. Such judgments should be determined according to the standard laid out in the second point.

The second point states specifically that it is necessary to obtain sufficient information to assess and judge possible impacts on the Antarctic environment and ecosystems which might be caused by activities planned for and conducted in Antarctic Treaty area.[1225] For the exploitation of Antarctic ice, the necessary information includes the scope and the cumulative impacts of the activity and its effects on any other activity in the Antarctic Treaty area, e.g. fishing and scientific research. Additionally, the exploitation should be able to be operated safely, able to provide

1222 Final Act of the Eleventh Antarctic Treaty Special Consultative Meeting, at: http://www.ats.aq/documents/keydocs/vol_1/vol1_3_AT_Final_Act_Eleventh_SATCM_e.pdf.

1223 Article 2, "Objective and Designation", and Article 3, "Environmental Principles", the Madrid Protocol.

1224 Article 3, paragraph 2(a) and (b), the Madrid Protocol.

1225 Article 3, paragraph 2(c), the Madrid Protocol.

early warning of any adverse effects of the activity, and of a nature allowing for prompt and effective response to accidents.

Thirdly, scientific research in Antarctica has priority over any other activities in the Antarctic Treaty area.[1226] Thus, the exploitation of Antarctic ice must be limited if it might affect or impede the scientific use and preservation of Antarctica.

Consequently, avoiding any potential to cause adverse effects on any aspect of Antarctica is extremely when considering the exploitation of Antarctic ice. Moreover, any such activities may only be conducted with sufficient information and technological capacity to assess, judge and monitor such activities so as to preserve Antarctica. Even though comprehensive information and appropriate technology might be available in the future to prove that the exploitation of Antarctic ice would not damage the Antarctic environment and its ecosystems, the exploitation activities would still not be allowed if they affect scientific research in Antarctica. Although the Madrid Protocol does not explicitly address the issues surrounding the exploitation of Antarctic ice, the application of these principles to any potential activities is unquestionable.

2. Prior Environmental Assessment

According to Article 8 and Annex I of the Madrid Protocol, an environmental assessment should be completed for any proposed human activities, both governmental and non-governmental, in the Antarctic Treaty area. Moreover, any change in an activity which "arises from an increase or decrease in the intensity of an existing activity, from the addition the decommissioning of a facility, or otherwise" should also follow the assessment procedures set out in Annex I.[1227] Hence, theoretically, the first test would be to determine whether the exploitation of Antarctic ice satisfies these environmental standards.

As described above in Part 6.2.3.1, the environmental impacts of human activities are divided into three categories: those having less than a minor or transitory impact; those of a minor or transitory impact; or those having more than a minor or transitory impact.[1228] If the exploitation of Antarctic ice is determined as falling under the last two categories, the activity cannot proceed[1229] until an Initial Environmental Evaluation (IEE) has been prepared.[1230] If the IEE indicates that the exploitation activity has a more than minor or transitory impact, it is necessary to prepare a Comprehensive Environmental Evaluation (CEE), which requires many more details than an IEE[1231] and must be circulated to all parties and forwarded to

1226 Article 3, paragraph 3, the Madrid Protocol.
1227 Article 8, paragraph 3, the Madrid Protocol.
1228 Article 8, paragraph 1, the Madrid Protocol.
1229 Article 1, paragraph 2, Annex I of the Madrid Protocol.
1230 Article 2, Annex I of the Madrid Protocol.
1231 See Part V, Subsection E (II) about the concrete explication of contents of a CEE.

the Committee for Environmental Protection.[1232] The activities cannot proceed until the Antarctic Treaty Consultative Meeting (ATCM) has considered the draft CEE.[1233]

It is difficult to offer a verdict on whether the exploitation of Antarctic ice could satisfy the required environmental evaluation and be allowed to proceed. However, based on some scientific research, it has been shown that ice serves to maintain the salinity and the temperature of sea water – in which krill lives – with the consequence that some huge impediments will have to be overcome in order to exploit Antarctic ice.[1234] In summary, although the ATS does not prohibit the exploitation of Antarctic ice, such activities are limited as part of the scheme to protect the Antarctic environment and its ecosystems.

3. *Third parties*

With regard to States that are not a party to any instrument of the ATS being restricted by these environmental requirements, some might assert that in accordance with the Vienna Convention on the Law of Treaties, a treaty cannot bind a State that is not a party.[1235] However, such a contention is not valid in the Antarctic Treaty area.

Firstly, through its decades of development, the Antarctic regime has become a part of customary international law.[1236] The Antarctic Treaty has had more than 50 years of history and has been accepted worldwide. With the gradual improvement of the ATS, it has been proven that the existing Antarctic regime is the "most appropriate framework for fostering international efforts to guarantee the protection of the environment"[1237] and that the Antarctic regime has been playing an effective role in keeping Antarctica in an orderly status.[1238] Moreover, the international

---

1232 Article 3, Annex I of the Madrid Protocol.

1233 Article 4, Annex I of the Madrid Protocol.

1234 Incident report by Quilleré-Majzoub, F., "Glaces polaires et icebergs: Quid juris gentium?" 52 A.F.D.I. 432 (2006), at 447, referring to a report of the CCAMLR Scientific Committee, Report of the Working Group on Ecosystem Monitoring and Management, Kochi, India, August 1998, at paragraph 7.13 [WGEMM Report].

1235 Article 34, the Vienna Convention on the Law of Treaties, United Nations, 1155 Treaty Series 331.

1236 Charney, Jonathan I., "The Antarctic System and Customary International Law", in Francioni, F. / Scovazzi, T. (eds.), *International Law for Antarctica*, (The Hague: Kluwer Law International, 1996), 51–101 at 81–95.

1237 The preamble of Recommendation XV-21, Part II of the Recommendations adopted at the XVth Antarctic Treaty Consultative Meeting, p. 113. at http://10.10.4.6/download/9667508/11060615/2/pdf/181/191/1331951251637_703/ATCM15_fr002_e.pdf.

1238 This idea is stated in the preamble of Recommendation XV-21, named "Use of Antarctic Ice", which reads: "the structure established under the Antarctic Treaty has proved effective in promoting international peace, in keeping with the purposes and principles of the United Nations Charter." Part II of the Recommendations adopted at the XVth Antarctic Treaty Consultative Meeting, at 113:

community has recognized and approved of the contributions made by the ATS, and the UN has welcomed such contributions several times[1239]. Hence, from a legal standpoint, it cannot be denied that any human activities in the Antarctic Treaty area must be undertaken under the Antarctic regime.

Secondly, according to the theory of objective regimes, the Antarctic regime can be imposed on a non-party.[1240] The theory of objective regimes was introduced by the International Court of Justice in the *Reparation for Injuries Suffered in the Service of the United Nations* case, in which the Court illustrated that "fifty States, representing the vast majority of the members of the international community, had the power, in conformity with international law, to bring into being an entity possessing objective international personality, and not merely personality recognized by them alone, together with capacity to bring international claims."[1241] Moreover, Article X of the Antarctic Treaty,[1242] which employs an approach similar to the one used in Article 2(6) of the Charter of the United Nations,[1243] imposes the principles and purpose of the Treaty on non-parties. Therefore, based on the two reasons above, the conclusion to be drawn is that third parties must comply with the environmental requirements in the ATS.[1244]

---

http://10.10.4.6/download/9667508/11060615/2/pdf/181/191/1331951251637_703/ATCM15_fr002_e.pdf.

1239 In 1992, the contribution of the ATS was affirmed at the UN Conference on Environment and Development in Agenda 21, Chapter 17, and welcomed in the 1992 UNGA Resolution, Question of Antarctica, Resolution 47/57 (9 December, 1992), at http://daccess-dds-ny.un.org/doc/RESOLUTION/GEN/NR0/024/24/IMG/NR002424.pdf?OpenElement ; in 1993 UNGA Resolution, Question of Antarctica, Resolution 48/80 (16 December, 1993), at http://daccess-dds-ny.un.org/doc/RESOLUTION/GEN/NR0/711/64/IMG/NR071164.pdf?OpenElement ; and in 1994 UNGA Resolution, Question of Antarctica, Resolution 49/80 (15 December, 1993), at http://daccess-dds-ny.un.org/doc/UNDOC/GEN/N95/760/69/PDF/N9576069.pdf?OpenElement.

1240 Viñuales (note 1213), at 198.

1241 Reparation for Injuries Suffered in the Service of the United Nations, Advisory Opinion: I.C.J. Report (1949), 174 at 185. http://www.icj-cij.org/docket/files/4/1835.pdf.

1242 Article X of the Antarctic Treaty reads: "Each of the contracting Parties undertakes to exert appropriate efforts, consistent with the Charter of the United Nations, to the end that no one engages in any activity in Antarctic contrary to the principles or purposes of the present Treaty."

1243 Article 2(6) of the Charter of the United Nations reads: "The Organization shall ensure that states which are not Members of the United Nations act in accordance with these Principles so far as may be necessary for the maintenance of international peace and security."

1244 Viñuales (note 1213), at 198.

IV. The Precautionary Approach

As analyzed in former part, generally speaking, the precautionary approach adopts a four part structure, which includes a) under which level of damage the approach should be applied; b) the scientific criterion as being uncertainty of the relationship between a given activity and a given effect; c) the requirement for decision-makers to respond to the given activity; and d) the party who wants to carry out an activity must prove that the activity is not harmful to environment.[1245] For the exploitation of Antarctic ice, the precautionary principle will apply unless damage can be identified as having less than a minor or transitory impact.[1246] According to scientific criteria, sufficient scientific information is not yet available as to the impact of such exploitation on the Antarctic environment and its ecosystems. However, some scientists have undertaken initial research on the role of melting Antarctic ice in Antarctic climate change and its impact on the delicate ecosystem. According to this research, "warming on the Antarctic Peninsula is having documented impacts on terrestrial flora, seasonal snow cover, lake ecology, penguin distribution, ice-shelf distribution, glacier thickness and sea-ice duration."[1247] The melting of Antarctic ice sheet is associated with severe consequences as well: "abrupt changes are possible in ice sheets affecting sea level and ocean circulation."[1248] Additionally, these findings have been buttressed in a study on Antarctic birds: "... the large (12–20%) reduction in sea ice extent since the 1950s has been associated with a decline in abundances of krill and other marine organisms, which are major food resources for most Antarctic seabirds. Their decline may be driving the observed delays in bird arrival and breeding... The consequences of delayed phenology could be serious for these top predators if they become less synchronized with the phenology of their food supplies..."[1249] Although the result of exploitation of Antarctic ice is still uncertain, these studies have shown that there is a "reasonable scientific possibility"[1250] that

1245 Wolfrum, Rüdiger, "Precautionary Principle", in Beurier, Jean-Pierre / Kiss, Alexander / Mahmoudi, Said (ed.), *New Technologies and Law of the Marine Environment,* (Kluwer International Law, 2000), at 209; Manson, Neil A., "Formulating the Precautionary Principle",24 *Environmental Ethics* 263 (2002), at 265; Wang, Runyu, "The Precautionary Principle in Maritime Affairs", 10 WMU Journal of Maritime Affairs, October 143 (2011), at 149–154.

1246 Article 8 and Annex I Article 1 paragraph 2, the Madrid Protocol.

1247 Vaughan, David G. et al., "Recent Rapid Regional Climate Warming on the Antarctic Peninsula," 60 Climatic Change 243 (2003), at 267.

1248 Alley, R. B., et al., "Abrupt Climate Change", 299 Science5615 (2003).

1249 Barbraud, Christophe / Weimerskirch, Henri, "Antarctic birds breed later in response to climate change," 103 Proceedings of the National Academy of Sciences 6248 (2006), at 6248–6249.

1250 Hickey, James E / Walker, Vern R, "Refining the Precautionary principle in International Environmental Law", 14 Virginia Environmental Law Journal 423 (1994–1995), at 449–450. Hickey and Walker state that "an agreement might require a reasonable scientific 'probability' that the activity can produce the kind of adverse

the harvesting of ice will cause adverse effects on the Antarctic environment and associated ecosystems.

Therefore, in accord with the principle, policy-makers should take some steps to mitigate any damage that may arise when the exploitation of Antarctic ice becomes possible. For now, only the assessment procedures set out in Annex I of the Madrid Protocol must be adopted and applied in the planning processes leading to decisions about exploitation. Nonetheless, the precautionary principle provides a theoretical foundation that demands prudential consideration of the impacts of harvesting ice in Antarctica before any relevant decisions are taken.

D. Conclusion

In Antarctica, it is still uncertain what legal status is to be attributed to various ice structures. They are normally regarded as having either the same legal status as Antarctic land or the status of high seas. No matter which kind of legal status the Antarctic ice has, it should only be managed under the ATS when it is located in the Antarctic Treaty area.

Regarding the issues of exploitation of Antarctic ice, the ATS explicitly indicates that ice is not a kind of mineral resource, and the prohibition of mineral activities, covered in the Madrid Protocol, does therefore not apply. However, the harvesting of ice in Antarctica is not a free activity without control as it is subject to restrictions designed to protect the Antarctic environment and related ecosystems. Pursuant to the ATS, preservation requirements for marine living resources can be identified in the CCAMLR, and pursuant to the Madrid Protocol, an environmental impact assessment would be necessary before making any decision on exploitation. Moreover, even third parties to the ATS have to obey these requirements. Additionally, as a widely accepted international law principle, the precautionary principle provides another legal foundation for limiting such exploitation so as to preserve Antarctica. Therefore, the exploitation of Antarctic ice would become permissible only when these environmental requirements are met. Nevertheless, in order to achieve the goal of preserving Antarctica, the ATCPs must conclude more specific regulations making such limits clear and complete.

---

effect that is to be avoided, but only a reasonable scientific 'possibility' that the exposure contribution from the contemplated activity will produce the adverse effect."

Summary of Study

Antarctica is the coldest continent in the world and is also the potentially richest continent in the world in respect to mineral resources. It is an ice-covered desert without human inhabitants, but it is also a controversial continent which has been drawing more and more international attention. The energy crisis seen as potentially impeding mankind's development may prove to be one of the severest problems. In March 2011, the nuclear catastrophe in Japan[1251] rang the warning-bell once again, clearly showing that nuclear energy is not perfect and may result in serious disasters and environmental pollution. Greenpeace reports that the radiation from Japan's nuclear crisis is accumulating in marine life off Japan's coast above legal limits for food contamination.[1252] One potential result of this incident is the reconsideration of nuclear energy as safe energy, with the result that the world returns its focus to traditional energy sources, such as mineral resources. This will result in Antarctica, the only undeveloped continent in the world, remaining the focus of contest within the international community, much as it has been since its initial discovery and exploration.

The struggle for Antarctica stems from its controversial history. No one knows who was the first to find it or when it was first found. This unclear history provides ample opportunity for different states to assert rights over Antarctica. The seven states claiming territorial sovereignty over Antarctica can offer historical, factual and legal evidence to support their respective claims. However, the rest of the world rejects these claims, even among the claimant states there is no clear acceptance, with an overlap in the territories claimed by three of the seven states, Argentina, Chile and the UK. These territorial claims, based on discovery, occupation, sector principles, inherited rights and contiguity, are untenable from the perspective of international law. Nevertheless, the reality is that there are seven claimants who have asserted sovereignty in Antarctica, a situation that may be further complicated if additional states announce similar claims at a later date. In the cold war climate of the late 1950s, there was a common interest in postponing difficult issues surrounding Antarctica, such as the issue of territorial sovereignty, and at the same time there was a common fear that if the relevant states who were interested in Antarctica failed to agree among themselves, "the matter would be taken out of their

1251 On 11 March, 2011, a devastating earthquake struck Japan which cut off the electricity for the nuclear reactor at Fukushima. An hour later a tsunami destroyed the back-up diesel generators. The cooling system consequently failed which caused the first of several explosions and radiation leaks.

1252 For more information, see: "Greenpeace: Japan Nuclear Plant Radiation Accumulating in Marine Life", May 26th, 2011, at: http://news.blogs.cnn.com/2011/05/26/greenpeace-japan-nuclear-plant-radiation-accumulating-in-marine-life/?iref=allsearch.

hands".[1253] Therefore, the USA convened the negotiation of the Antarctic Treaty, which was accomplished in just over eighteen months.

The Antarctic Treaty is a unique success and has preserved the peaceful situation in Antarctica. It recognizes that Antarctica should be a demilitarized zone allowing for freedom of scientific investigation. It freezes the territorial claims and establishes a functionary organ to manage the matters of Antarctica, the Antarctic Treaty Consultative Meetings. The ATCMs can issue decisions, measures and regulations. Measures approved by all the ATCPs constitute the components of the Antarctic Treaty System, and become binding on parties to the Antarctic Treaty. Alongside the series of measures making up the ATS, the ATCMs also concluded two separate conventions, namely the Conservation of Antarctic Seals and the Conservation of Antarctic Marine Living Resources, as well as the Protocol on Environmental Protection to the Antarctic Treaty. Although the Convention for the Regulation of Antarctic Mineral Resource Activities has failed to enter into force, it is still a part of the ATS. In sum, the Antarctic Treaty establishes the legal framework for activities within its area, and the Antarctic has become a unique international area beyond national jurisdictions.

The development of the ATS has interacted with numerous international legal regimes, UNCLOS being the most important. UNCLOS delimits offshore maritime zones, which implies the presence of coastal states, raising the question of whether the delimitation of maritime zones could apply to water area around Antarctica. According to UNCLOS, a coastal state could claim an EEZ and thus own a continental shelf adjacent to its land territory. Coastal states exercise jurisdiction over their respective water area and continental shelf. However, UNCLOS cannot simply be applied to Antarctica for two reasons. Firstly, the territorial claims in Antarctica have not been accepted by the international community. This means that the legal status of Antarctica is not the same as other national territories in other continents: Antarctica cannot be described as falling under a national jurisdiction and therefore maritime zones around Antarctica cannot be legally delimited. Secondly, the unique geophysical factors of Antarctica prevent the application of UNCLOS, as it is difficult to determine the baseline required. Although a baseline is available, the icebergs and ice-packs in the region challenge the norms of UNCLOS. Therefore, delimitating various maritime zones is inappropriate for water areas around Antarctica. The water areas within Antarctic Treaty area could be treated as special high seas, within which the rights of high seas are respected, and activities are made subject to the ATS.

Moreover, norms of UNCLOS regarding the continental shelf cannot apply to the Antarctic continental margin. This is due to the fact that coastal sovereignty is untenable from the perspective of international law, and because claiming continental shelves violates Article IV of Antarctic Treaty that rules out new territorial

1253 Maddox, John, "Antarctic Mining Regime at Risk", 307 Nature 105 (January 1984), at 105.

claims. If the Antarctic continental margins are not under coastal states' national jurisdictions, pursuant to UNCLOS, deep seabeds beyond national jurisdictions are the common heritage of mankind, which is under the control of the International Seabed Authority. However, Antarctica is not common heritage. Firstly, in contrast to *res communes* there are seven territorial claims to the Antarctic continent. Secondly, there is a legal system managing Antarctic matters. Thirdly, applying the common heritage principle to the Antarctic will cause conflicts between the ISA and the ATCPs. Therefore, continental margins within the Antarctic Treaty area are neither under national jurisdiction nor under the authority of ISA, but rather fall under the Antarctic Treaty system. However, the ATCPs are capable of establishing a convention to control mineral exploitation in the Antarctic continent and its margins, as demonstrated by CRAMRA.

The ISA, which is authorized to manage mineral exploration in common areas, provides some experience for a future Antarctic minerals regime. Most importantly, ISA is a compromise between developed and developing states. In order to obtain the support of developed states, the original norms in UNCLOS and its Annex have been modified by the Implementation Agreement, which reduces the privileges of developing states and protects the rights of developed states. Also important is that an Antarctic mineral exploration regime should be concluded between the claimants group and non-claimants group, with CRAMRA being the first attempt to find a balance between the two groups. It has however failed in this attempt, as the biggest claimant was dissatisfied with the benefits sharing system. A universally acceptable solution may require greater consideration of claimants' privileges, without however disregarding the right of developing states to benefit, particularly with regard to the right of participation for non-aligned states.

Formal negotiations of CRAMRA started in 1982 and finished in 1988, remarkably swift for a complicated question considering both claimants and non-claimants interests. The adoption of CRAMRA is a huge triumph for the ATS as a whole and for the protection of the Antarctic environment. CRAMRA appears to be a mineral exploration convention, however, its purpose is to prohibit unregulated mineral resource activities in Antarctica, and its provisions are extremely strict with the aim of environmental protection. It provides environmental protections mechanisms and principles as embodied in the 1991 Environmental Protocol. In attempting to balance claimant and non-claimant state interests, CRAMRA provides for a Commission as a general functionary organ for all ATCPs to determine general and basic matters relating to mineral resources activities. The Regulatory Committees are aimed at coordinating the special requirements of the relevant claimants whenever proposed mineral resource activities are to be carried out in areas under their claim. CRAMRA also accommodates external players, providing an opportunity for the UN and other international organizations to be observers at the meetings of the Commission, at Special Meeting of Parties and at the Advisory Commission. However, this design does not fully satisfy the international community's need to fully participate in the Antarctic minerals regime.

Antarctic mineral exploration has not only caused legal and political controversy between various interest-groups – between claimants and non-claimants, the ATCPs and non-aligned parties, developing states and developed states, etc. – but has also provoked environmentalists. A surge in criticism of CRAMRA was observed in both the UN forums and the NGO movements, bringing huge pressure to bear on CRAMRA and the ATCPs. Although a few parties to the Antarctic Treaty had signed CRAMRA, Australia and France unequivocally rejected it, which prevented its entry into force. Although both Australia and France declared that their rejections of CRAMRA were based on environmental protection considerations, their political aims and their consideration of commercial benefits was clearly an underlying subtext. Although CRAMRA failed to enter into force, it does provide a good example of how to deal with resource exploitation in an area subject to controversy, beyond national jurisdictions, especially given the high standard for environmental protection that significantly influenced not only the 1991 Protocol, but also the ISA Regulation on Polymetalic Nodules.

The 1991 Protocol prohibits any mineral resources activity during its life; nevertheless, it leaves the opportunity for re-initiating mineral resources activities. Moreover, the increasingly pressing international demand for mineral resources will drive people to rethink this remote, unexplored ice-covered continent. It is therefore not beyond credulity to imagine that a new round of discussion on Antarctic mineral exploration will be held in the foreseeable future. The new Antarctic minerals regime is primarily based on the CRAMRA framework, and it in turn yields three important insights. Firstly, protection of the Antarctic environment and its ecosystem is an important consideration during mineral resources activities. Secondly, negotiation procedures should be as open as possible to the international community: while it is not necessary to grant veto rights to non-aligned states, it is important to provide an opportunity to participate in the negotiations as observers. Thirdly, to obtain claimant support, the new regime should consider assigning more rights or new mechanisms that provide special management rights for mineral resources activities in their sectors.

Antarctica, the most unique continent in the world, provides interested States a stage on which they may struggle for political and commercial benefits. Although no one knows what the ultimate Antarctic minerals regime will look like, at least one thing is beyond doubt: CRAMRA could provide not only precious legal experience but political experience as well.

Literature

"Beeby Draft II", reprinted in Greenpeace International, *The Future of the Antarctic: Background for a Second U.N. Debate*, (Greenpeace International, 1984), Appendix 8.

"Beeby's Package Deal", *ECO*, Vol. 33, No. 1, p. 1 (23 September-4 October, 1985).

"Memorandum of the Soviet Government on the Antarctic Regime Question", in 10 USSR Information Bulletin (23 June 1950), at 380.

"The Antarctic Minerals Regime: The Beeby Draft", reprinted in Greenpeace International, *Greenpeace International, The Future of the Antarctic: Background for a Second U.N. Debate*, (Greenpeace International, 1983), Appendix 8.

Akehurst, Michael, *A Modern Introduction to International Law*, (3rd ed., London: George Allen and Unwin Ltd., 1977).

Alexander, Frank C. (Jr.), "Legal Aspects: Exploitation of Antarctic Resources – A Recommended Approach to the Antarctic Resource Problem," *Miami Law Review*, Vol. 33, pp. 371–425 (1978–1979).

Alley, R. B. / Marotzke, J. / Nordhaus, W. D. / Overpeck, J. T. / Peteet, D. M. / Pielke, R. A. (Jr.) / Pierrehumbert, R. T. / Rhines, P. B. / Stocker, T. F. / Talley, L. D. / Wallace, J. M., "Abrupt Climate Change", *Science*, Vol. 299, No. 5615, pp. 2005–2010 (2003).

Alley, Roderic, "New Zealand and Antarctica", *International Journal*, Vol. 39, No. 4, Polar Politics pp. 911–931 (Autumn, 1984).

Anand, Ram P., "Freedom of the Seas: Past, Present and Future", in Gutiérrez-Girardot, Rafael / Ridder, helmut / Sarin, Manohar Lal / Schiller, Theo (eds.), *New Directions in International Law Essays in Honour of Wolfgang Abendroth: Festschrift zu seinem 75. Geburtstag*, (Frankfurt/New York: Campus Verlag, 1982), pp. 215–233.

Auburn, F.M., "A Sometime World of Men: Legal Rights in the Ross Dependency", *American Society of International Law*, Vol. 65, No 3, pp. 578–582 (July 1971).

Auburn, F.M., "The Ross Dependency – An Undeclared Condominium", *Auckland University law Review*, Vol. I, No. 3, pp. 89–106 (1970).

Auburn, F.M., "The White Desert", *International and Comparative Law Quarterly*, Vol. 19, pp. 229–256, (1970).

Auburn, F.M., *Antarctica law and politics*, (London: C. Hurst & Company, 1982).

Balch, Edwin Swift, *Antarctica*, (Philadelphia: Press of Allen, Lane & Scott, 1902).

Balch, Thomas Willing, "Arctic and Antarctic Regions and the Law of Nations", *American Journal of International Law*, Vol. 4, pp. 265–275 (1910).

Barbraud, Christophe / Weimerskirch, Henri, "Antarctic Birds Breed Later in Response to Climate Change," *Proceedings of the National Academy of Sciences*, Vol. 103, pp. 6248–6251, (2006).

Bargagli, Roberto, *Antarctic Ecosystems Environmental Contamination, Climate Change, and Human Impact*, (Berlin Heidelberg: Springer-Verlag, 2005).

Beck, Peter J., "British Antarctic Policy in the Early 20th Century", *Polar Record*, Vol. 21, No. 134, pp. 475–483 (1983).

Beck, Peter J., "Convention on the Regulation of Antarctic Mineral Resource Activities: A Major Addition to the Antarctic Treaty System", *Polar Record*, Vol. 25, pp. 19–32 (1989).

Beck, Peter J., "The Antarctic Resource Conventions Implemented: Consequences for the Sovereignty Issues", in Jørgensen-Dahl, Arnfinn, / Østreng, Willy, (eds.), *The Antarctic Treaty System in World Politics* (Palgrave Macmillan, July 1991) Chapter 15.

Beck, Peter J., "The Future of the Antarctic Region: Is This Great White Continent Turning Green?", presented at the 33rd Annual Convention of the International Studies Association, Atlanta, Georgia, USA, April 1992.

Beck, Peter J., "The United Nations and Antarctica, 2005: the End of the 'Question of Antarctica'?", *Polar Record*, Vol. 42, (2006), pp. 217–227.

Beck, Peter J., *The International Politics of Antarctica*, (New York: St. Martin's Press, 1986).

Bednarek-Ochyra, H. / Bana, J./ Ochyra, R / Smith, R.I. Lewis, *The Liverwort Flora of Antarctica*, (Cracow: Polish Academy of Sciences, Institute of Botany, 2000).

Beeby Draft I, in Greenpeace International, *The Future of the Antarctic: Background for a U.N. Debate, Appendix 8 (1983)*; ECO, Vol XXIII, No. 1, Special Consultative Meeting on Antarctic Mineral Resources, Bonn, West Germany, (July 11–22, 1983).

Beeby, C.D., "An Overview of the Problems Which Should Be Addressed in the Preparation of a Regime Governing the Mineral Resources of Antarctica", in Vicuña, Francisco Orrego (ed.), *Antarctic Resources Policy: Scientific, Legal and Political Issues*, (Cambridge: Cambridge University Press, 1983), pp. 191–198.

Behrendt, J.C., "Recent Geophysical and Geological Research in Antarctica Relate to the Assessment of Petroleum Resource and Potential Environmental Hazards to Their Development", in Splettstoesser, J.F. / Dreschhoff, G.A.M. (eds.), *Mineral Resource Potential of Antarctica, Antarctic Research Series Vol. 51*, (Washington D.C.: American Geophysical Union, 1990), pp. 163–174.

Behrendt, John C, "Are There Petroleum Resources in Antarctica?", *AAPG Bulletin*, Vol. 68, Issue 4, p. 453 (1984).

Bentley, Charles R., "Glacial and Subglacial Geography of Antarctica", in Wexler, H./ Rubin, M. J./ Caskey, J. E. (eds.), *Matthew Fontaine Maury Memorial Symposium on Antarctic Research*, (Richmond, Virginia: William Byrd Press, 1962), pp. 18.

Bergin, Patrick T, "Antarctica, the Antarctic Treaty regime, and Legal and Geopolitical Implications of Natural Resource Exploration And Exploitation", *Florida International Law Journal*, Vol. 4, No. 1, pp. 1–40 (1988).

Berguno, J., "The Antarctic Park: The Issue of Environmental Protection", in Verhoeven, J. / Sands, P. / Bruce, M. (eds.), *The Antarctic Environment and International Law* (London: Graham and Trotman Limited, 1992), pp. 103–107.

Berkman, Paul Arthur, *Science into Policy: Global Lessons from Antarctica*, (California, Academic Press, 2002).

Bernard H. Oxman / Richardson, Elliot L. / Aguilar, Andres M. / Kronmiller, Theodore / Schneider, Jan / Highet, Keith, "Law of the Sea", *Proceedings of the Annual Meeting (American Society of International Law)*, Vol. 77, pp. 150–168 (April 14–16, 1983).

Bernhardt, J. Peter A, "Sovereignty in Antarctica", *California Western International Law Journal*, Vol. 5, pp. 297–349 (1974–1975).

Bertand, Kenneth J., *Americans in Antarctica, 1775–1948*, (New York: American Geographical Society, 1971).

Bilder, Richard B., "Control of Criminal Conduct in Antarctica", *Virginia Law Review*, Vol. 52 No. 2, pp. 231–285, (1966).

Blay, S.K.N., "New Trends in the Protocol of the Antarctic Environment: the 1991 Madrid Protocol", *The American Journal of International Law*, Vol. 86, No. 2, pp. 377–399 (April 1992).

Blewett, N. / Duffy, M. / Crean, S., "Australia Enhances its Power Over Marine Conservation and Resources", (News Release by Minister for Foreign Affairs and Trade, Canberra, September 24, 1991).

Boczek, Boleslaw A., "The Soviet Union and the Antarctic Regime", *The American Journal of International Law*, Vol. 78, pp. 834–858 (1984).

Boetzinger, Joachim, *Völkerrecht der Antarktis*, (Dissertation, Göttingen: 1958).

Bonner, W.N., "Recent Developments in Antarctic Conservation", in Triggs, Gillian Doreen, (ed.), *The Antarctic Treaty Regime: Law, Environment and Resources*, (Cambridge: Cambridge University Press, 1987), pp. 143–149.

Botsford, Louis W. / Micheli, Fiorenza / Hastings, Alan, "Principles for the Design of Marine Reserves", *Ecological Applications*, Vol. 13, Issue 1, pp. 25–31 (Supplement, 2003).

Brennan, Keith, "Criteria for Access to the Resources of Antarctica: Alternatives, Procedure and Experience Applicable", in Vicuña, Francisco Orrego edited, *Antarctic Resources Policy: Scientific, Legal and Political Issues*, (Cambridge: Cambridge University Press, 1983), pp. 217–230.

Brierly, James Leslie, *The Law of Nations*, (Oxford: The Clarendon Press, 1928).

Brinie, Patricia, "Effect of Article VI of the Antarctic Treaty on Scientific Research", in Wolfrum, Rüdiger (ed.), *Antarctic Challenge III, Conflicting Interests, Cooperation,*

Environmental Protection, Economic Development: Proceedings of An Interdisciplinary Symposium, (Berlin: Duncker & Humblot, 1988), pp. 105–120.

Brown, D.E. (ed.), *The Areas within National Jurisdiction, Volume 1 of Seabed Energy and Mineral Resources and the Law of the Sea,* (London, Graham & Trotman, 1984).

Brown, Lesley (ed.), *The New Shorter Oxford English Dictionary on Historical Principles,* (Oxford: Clarendon Press, 1993).

Brownlie, Ian, *Principles of Public International Law,* (4th ed., Oxford: Oxford University Press, 1990).

Brundtland, Gro Harlem, *Report of the World Commission on Environment and Development, Our Common Future,* (Oxford: Oxford University Press, 1987).

Burke, William T., " Legal Aspects of the IWC Decision on the Southern Ocean Sanctuary", in *Ocean Development & International Law,* Vol. 28, No. 3, pp. 313–327 (1996).

Burke, William T., "Memorandum of Opinion on the Legality of the Designation of the Southern Ocean Sanctuary", in *Ocean Development & International Law,* Vol. 27, pp. 315–326 (1996).

Bush, W.M., *Antarctica and International Law: A collection of the Inter-state and National Documents II,* (Dobbs Ferry, New York: Oceana Publications, 1982).

Cameron, James /Abouchar, Juli, "The Precautionary Principle: A Fundamental Principle of Law and Policy for the Protection of the Global Environment", *Boston College International and Comparative Law Review,* Vol. 14, p. 1–27 (1991).

Campbell, William, "International Law Notes-Maritime Legislation Amendment Act", *Public Law Review,* Vol. 5, pp. 141–143 (1994).

Charney, Jonathan I., "The Antarctic System and Customary International Law", in Francioni, Francesco / Scovazzi, Tullio (eds.), *International Law for Antarctica,* (The Hague: Kluwer Law International, 1996), pp. 51–101.

Child, Jack, *Antarctica and South American Geopolitics: Frozen Lebensraum,* (New York: Praeger Publishers, 1988).

Churchill, Robin Rolf / Lowe, Alan Vaughan, *The Law of the Sea,* (2nd ed., Manchester University Press, 1988).

Coates, D.A. / Stricker, G.D. / Landis, E.R., "Coal geology, Coal Quality and Coal Resources in Permian Rocks of the Beacon Supergroup, Transantarctic Mountains, Antarctica", in Splettstoesser, J.F. and Dreschhoff, G.A.M. (eds.), *Mineral Resource Potential of Antarctica, Antarctic Research Series Vol. 51* (Washington D.C.: American Geophysical Union, 1990), pp. 133–162.

Colson, David A., "The Antarctic Treaty System: The Mineral Issue", *Law and Policy in International Business,* Vol. 12, pp. 841–902 (1980).

Colson, David A., "The United States Position on Antarctica", *Cornell International Law Journal,* Vol. 19, pp. 291–310 (1986).

Commission for the Conservation of Antarctic Marine Living Resources, *CCAMLR's Management of the Antarctic*, (Habart, Australia, Published by CCAMLR, http://www.ccamlr.org/pu/E/e_pubs/am/man-ant/e-management.pdf, 2001).

Conforti, Benedetto, "Territorial Claims in Antarctica: A Modern Way to Deal with an Old Problem", *Cornell International Law Journal*, Vol. 19, pp. 249–258 (1986).

Cook, F. A., "Captain Fabian Gottlieb Von Bellingshausen, 1819–1821. The Discovery of Alexander I., Peter I., and Other Islands", *Bulletin of the American Geographical Society*, Vol. 33, No. 1, pp. 36–41 (1901).

Cordonnery, Laurence, "Environment Protection in Antarctica: Drawing Lessons from the CCAMLR Model for the Implementation of the Madrid Protocol", *Ocean Development and International Law*, Vol. 29, pp. 125–146 (1998).

Costa, Jean Fitch Da., *Souveraineté sur l'Antarctique*, (Paris: Librairie Generale De Droit Et De Jurisprudence, 1958).

Couratier, Josyane, "Regime for the Convention of Antarctica's Living Resources", in Vicuña, Francisco Orrego edited, *Antarctic Resources Policy: Scientific, Legal and Political Issues*, (Cambridge: Cambridge University Press, 1983), pp. 139–149.

Crary, A. P. / Gould, L. M. / Hulburt, E. O. / Odishaw, Hugh/ Smith, Waldo E (eds.), *Antarctica in the International Geophysical Year: Based on a symposium on the Antarctic*, (Baltimore, Maryland: Waverly Press, 1956).

Crockett, R.N. / Clarkson, P.D., "The Exploitation of Antarctic Minerals", *Environment International*, Vol. 13, Issue 1, pp. 121–132 (1987).

Cumpston, John S., "Introduction the Antarctic", in Simpson, Frank A. (ed.), *The Antarctic Today: A Mid-Century Survey by the New Zealand Antarctic Society*, (A. H. & A.W. Reed and New Zealand Antarctic Society, 1952), p. 17 et suq.

David, T.W. Edgeworth / Priestley, Raymond E., *British Antarctic Expedition 1907–1909 Reports on the Scientific Investigations: Geology, Vol. 1 Glaciology, Physiography, Stratigraphy and Tectonic Geology of South Victoria Land*, (London: Heinemann, 1914).

Davies, William E., "Antarctic Stratigraphy and Structure", in Crary, A. P. / Gould, L. M. / Hulburt, E. O. / Odishaw, Hugh/ Smith, Waldo E (eds.), *Antarctica in the International Geophysical Year: Based on a symposium on the Antarctic*, (Baltimore, Maryland: Waverly Press, 1956), pp. 44–51.

Davis, B.W., "Antarctica as a Global Protected Area: Perceptions and Reality", *Australian Geographer*, Vol. 23, No. 1, pp. 39–43 (1992).

de la Mare, W.K., "CCAMLR's Management Task and the Definition of its Operational Objectives", in Kock, Karl-Hermann (ed.), *Understanding CCAMLR's Approach to Management*, (http://www.ccamlr.org/en/system/files/am-all.pdf, 2000) pp. 8–10.

de Vattel, Emer, *Le droit des gens ou principes de la loi naturelle appliqués à la conduite et aux affaires des nations et des souverains, Klassiker des Völkerrecht*, Band.3, (Tübingen: Mohr, 1959).

Drewry, D. J. / Jordan, S.R., "The bedrock surface of Antarctica", in, *Antarctica: Glaciological and Geophysical Folio*, (Cambridge, England, UK: Scott Polar Research Institute, 1983).

Dufek, George J, *Operation Deepfreeze*, (New York: Harcourt, Brace and Company, 1957).

Dunnett, Harding McGregor, *Shackleton's Boat – The Story of the James Caird*, (Neville & Harding, 1996).

Dupuy, René-Jean, "Le statut de l'Antarctique", *Annuaire français de droit international*, Vol. 4, pp. 196–229 (1958).

Dyke, Jon M. Van / Brooks, Robert A., "Uninhabited Islands: Their Impact on the Ownership of the Oceans' Resources", *Ocean Development and International Law*, Vol. 12, Issue 3 &4, pp. 265–300 (1983).

Edwards, J. A. / Smith, R.I. Lewis, "Photosynthesis and Respiration of Colobanthus Quitensis and Deschampsia Antarctica from the Maritime Antarctic", *British Antarctic Survey Bulletin*, Vol. 81, pp. 43–63 (1988).

Eklund, Carl R., "Antarctic Fauna and Some of its Problems", in Crary, A. P./ Gould, L. M. / Hulburt, E. O. / Odishaw, Hugh/ Smith, Waldo E (eds.)., *Antarctica in the International Geophysical Year: Based on a symposium on the Antarctic*, (Baltimore, Maryland: Waverly Press, 1956), pp. 117–123.

Evans, G, / Duffy, M., "Australia Extends Territorial Sea", *Australian Foreign Affairs and Trade*, Vol. 61, p. 816 (November 1990).

Evteev, Sveneld, "Antarctica and its Place in the Contemporary Environmental Movement", in Jørgensen-Dahl, Arnfinn / Østreng Willy, (eds.), *The Antarctic Treaty System in World Politics* (Palgrave Macmillan, July 1991), pp. 147–152.

Fairbridge, Rhodes W., "The Geology of the Antarctic", in Simpson, Frank A. (ed.), *The Antarctic Today: A Mid-Century Survey by the New Zealand Antarctic Society*, (A. H. & A.W. Reed and New Zealand Antarctic Society, 1952), pp. 56–101.

Fauchille, Paul, *Traité de Droit International Public Vol. 1* (Paris: A. Rousseau and Co., 1925).

Fifield, Richard, *International Research in the Antarctic*, (Scientific Committee on Antarctic Research and the ICSU Press by Oxford University Press, 1987).

Fisher, F. C., "The Arbitration of the Guatemalan-Honduran Boundary Dispute", *The American Journal of International Law*, Vol. 27, No. 3, pp. 403–427 (July 1933).

Fox, Hazel, "The Relevance of Antarctica to the Lawyer", in Triggs, Gillian Doreen (ed.), *The Antarctic Treaty regime: Law, Environment and Resources*, (Cambridge: Cambridge University Press, 1987), pp. 77–87.

Frakes, Jennifer, "The Common Heritage of Mankind Principle and the Deep Seabed, Outer Space, and Antarctica: Will Developed and Developing Nations Reach a Compromise?", in *Wisconsin International Law Journal*, Vol. 21, pp. 409–434 (2003).

Frakes, L.A. / Moreton, D.L.E., "Manganese Nodule Provinces of the Southern Ocean", in Splettstoesser, J.F. / Dreschhoff, G.A.M. (eds.), *Mineral Resource Potential of Antarctica, Antarctic Research Series 51* (Washington D.C.: American Geophysical Union, , 1990), pp. 217–221.

Francioni, Francesco, "Legal Aspect of Mineral Exploitation in Antarctica", *Cornell International Law Journal*, Vol. 19, pp. 163–188 (1986).

Francioni, Francesco, "Madrid Protocol on the Protection of the Antarctic Environment", *Texas International Law Journal*, Vol. 28, pp. 47–72 (1993).

Francioni, Francesco, "Resource Sharing in Antarctica: for Whose Benefit?", *European Journal of International Law*, Vol. 1, No. 1, pp. 258–268 (1990).

Gardiner, Stephen M., "A Core Precautionary Principle", *The Journal of Political Philosophy*, Vol. 14, Number 1, pp. 33–60 (2006).

Glasby, G. P. (ed.), *Antarctic Sector of the Pacific, Elsevier Oceanography Series 51*, (Amsterdam: Elsevier Science Publishers B.V., 1990).

Glasby, G.P. (ed.), *Antarctic Sector of the Pacific*, (New York: Elsevier Science Publisher, 1990).

Gleick, Peter H. / Yolles, P. / Hatami, H., "Water, War and Peace in the Middle East", *Environment*, Vol. 36, Issue 3, pp. 6–42 (1994).

Gleick, Peter H., "Water and Conflict: Fresh Water Resources and International Security", *International Security*, Vol. 18, No. 1, pp. 79112 (1993).

Goetz, P. (ed.), *The New Encyclopaedia Britannica: Macropaedia Vol. 20*,(15th Ed. Encyclopaedia Britannica, 1985).

Gould, Laurence McKinley, *The Polar Regions in Their Relation to Human Affairs*, (New York: American Geographical Society, 1958).

Gould, Lawrence McKinley, *Antarctica in World Affairs*, Vol. 121, (Foreign Policy Association, 1958).

Greenpeace, "Japan Nuclear Plant Radiation Accumulating in Marine Life", May 26th, 2011, at: http://news.blogs.cnn.com/2011/05/26/greenpeace-japan-nuclear-plant-radiation-accumulating-in-marine-life/?iref=allsearch.

Grotius, Hugo, *The Freedom of the Seas* or *The Right Which Belongs to the Dutch to Take Part in the East Indian Trade*, Magoffin, Ralph Van Deman, translated, (New York: Oxford University Press, 1916).

Guggenheim, Paul, *Lehrbuch des Völkerrechts*, Band 1, (Basel: Verlag für Recht und Gesellschaft, 1948).

Gulland, John A., "The Antarctic Treaty System as a Resource Management Mechanism", in Triggs, Gillian Doreen, (ed.), *The Antarctic Treaty regime: Law, Environment and Resources,* (Cambridge: Cambridge University Press, 1987), pp. 116–127.

Hackworth, Green Haywood, *Digest of International Law Vol. I.* (Washington, DC: Government Printing Office, 1940).

Hall, William Edward, *A Treatise on International Law,* Higgins, A. Pearce (ed.) (8th ed., Oxford: Clarendon Press, 1924).

Hambro, Edvard, "Some Notes on the Future of the Antarctic Treaty Collaboration", in *The American Journal of International Law,* Vol. 68, No. 2, pp. 217–226, (April 1974).

Hanessian, John, "The Antarctic Treaty 1959", *International and Comparative Law Quarterly,* Vol. 9 Issue 3, pp. 436–480 (July 1960).

Hanevold, Truls, "Inspections in Antarctica", *Cooperation and Conflict,* Vol. 6, pp. 103–114 (1971).

Hansom, James D. / Gordon, John E., *Antarctic Environment and Resources: A Geographical Perspective,* (New York: Addison Wesley Longman, 1998).

Harris, Colin M. / Meadows, Janice, "Environmental Management in Antarctica – Instruments and Institutions", *Marine Pollution Bulletin,* Vol. 25, Issues 9–12, pp. 239–249, (1992).

Harry, Ralph L., "Antarctic Regime and the Law of the Sea Convention: An Australian View", *Virginia Journal of International Law,* Vol. 21, No. 4, p. 727–746 (1981).

Hatherton, T, "Preferred Futures: A Scientific View", in New Zealand Institute of International Affairs, *The Antarctic: Preferred Futures, Constraints and Choices: Proceedings of a Seminar Held by the New Zealand Institute of International Affairs at Wellington on 17 and 18 June 1983, (Wellington: The Institute, 1983),* pp. 63–68.

Hawke, Bob, "Speech on the Theme 'Antarctica's Future: Continuity or Change' at the 16th national conference of the Australian Institute of International Affairs in Hobart on November 18", 1989, *Australian Foreign Affairs and Trade – Monthly Record,* Vol. 60, No. 11, p. 627–628 (November 1989).

Hayasif, Moritaka, "The Antarctica Question in the United Nations", in *Cornell International Law Journal,* Vol. 19, pp. 275–290 (1986).

Hayton, Robert D., "The 'American' Antarctic", *The American Journal of International Law,* Vol. 50, pp. 583–610 (1956).

Hayton, Robert D., "The Antarctic Settlement of 1959", *The American Journal of International Law,* Vol. 54, No. 2, pp. 348–371, (April 1960).

Headland, Robert K., *Chronological List of Antarctic Expeditions and Related Historical Events,* (Cambridge: Cambridge University Press, 1989).

Heap, John A. / Holdgate, Martin W., "The Antarctic Treaty System as an Environmental Mechanism: An Approach to Environmental Issues", Polar Research

Board (ed.) *The Antarctic Treaty System as An Assessment – An Approach to Environmental Issues,* (US National Research Council, 1986) pp. 195–210.

Heap, John A., "Has CCAMLR Worked? Management Policies and Ecological Needs", in Jørgensen-Dahl, A. / Østreng, W. (eds.), *The Antarctic Treaty System in World Politics,* (Basingstocke: Macmillan Academic and Professional, 1991), pp. 43–53.

Hedberg, Hollis Dow, "Ocean Boundaries for the Law of the Sea", *Marine Technology Society Journal,* Vol. 10, No. 5, pp. 6–11 (June 1976).

Heim, Barbara Ellen, "Exploring the Last Frontiers for Mineral Resources: A Comparison of International Law Regarding the Deep Seabed, Outer Space, and Antarctica", *Vanderbilt journal of Transnational Law,* Vol. 23, pp. 819–850 (1990–1991).

Hendry, I. D., "The Antarctic Minerals Act 1989", *International and Comparative Law Quarterly,* Vol. 39, No. 1, pp. 183–190 (January 1990).

Herber, Bernard P., "The Common Heritage Principle: Antarctica and the Developing Nations", *American Journal of Economics and Sociology,* Vol. 50, No. 4, pp. 391–406 (October 1991).

Hickey, James E / Walker, Vern R, "Refining the Precautionary principle in International Environmental Law", *Virginia Environmental Law Journal,* Vol. 14, pp. 423–454 (1994–1995).

Hill, Norman, *Claims to Territory in International Law and Relations,* (London, New York, Toronto: Oxford University Press, 1945).

Honnold, Edward, "Thaw in International Law? Rights in Antarctica under the Law of Common Spaces", *The Yale Law Journal,* Vol. 87, No. 4, pp. 804–859 (March 1978).

Hutchinson, D.N., "The Seaward Limit to Continental shelf Jurisdiction in Customary International Law", *The British Year Book of International Law,* Vol. 56, Issue I, pp. 111–188 (1985).

Infante, María Teresa, "The Continental Shelf of Antarctica: Legal Implications for a Regime on Mineral Resources", in Vicuña, Francisco Orrego (ed.), *Antarctic Resources Policy: Scientific, Legal and Political Issues,* (Cambridge: Cambridge University Press, 1983), pp. 253–264.

Intergovernmental Panel on Climate Change, *Technical Paper on Climate Change and Water,* Doc. IPCC-XXVIII/Doc.13 (8.IV.2008) (April 10, 2008).

International Panel on Climate Change, *Climate Change 2001: The Scientific Basis,* (Cambridge: Cambridge University Press, 2001).

International Whaling Commission, "Chairman's Report of the Forty-Fourth Meeting, Appendix 4. Resolution on a Sanctuary in the Southern Hemisphere", in *Report of the International Whaling Commission,* pp. 41–48 (1993).

International Whaling Commission, "Report of the Intercessional Working Group to Review Data and Results from Special Permit Research on Minke Whales in the Antarctic", in the *Reports of the International Whaling Commission*, Vol. 48, pp. 377–412 (1998).

International Whaling Commission, "Report of the Scientific Committee", in *Report of the International Whaling Commission*, Vol. 6, pp. 17–24 (April 1955).

Isachsen, Fridtjov, "The New Norwegian Dependency in the Antarctic", *Le Nord*, No. 1, pp. 67–78 (1939).

Jankovié, Branimir M., *Public International Law*, (Dobbs Ferry, New York: Transnational Publisher, 1984).

Jessup, Philip C. / Taubenfeld, Howard J., *Controls for Outer Space and the Antarctic Analogy*, (New York: Columbia University Press, 1959).

Joyner, Christopher C, "The Antarctic Minerals Negotiating Process", *The American Journal of International Law*, Vol. 81, No. 4, pp. 888–905 (October 1987).

Joyner, Christopher C, "The Evolving Minerals Regime for Antarctica", *Ocean Development & International Law*, Vol. 19, Issue 1, pp. 73–95 (1988).

Joyner, Christopher C. / Lipperman, Peter J., "Conflicting Jurisdictions in the Southern Ocean: the Case of an Antarctic Minerals Regime", *Virginia Journal of International Law*, Vol. 27, No. 1, pp. 1–38 (Fall 1986).

Joyner, Christopher C. and Ewing, Blair G. Jr, "Antarctica and the Latin American States: the Interplay of Law, Geopolitics and Environmental Priorities", *Georgetown International Environmental Law Review*, Vol. 4 pp. 1–46 (1991–1992).

Joyner, Christopher C., "Antarctica and the Law of the Sea: Rethinking the Current Legal Dilemmas", *San Diego Law Review*, Vol. 18, pp. 415–442 (1981).

Joyner, Christopher C., "Ice-Covered Regions in International Law", in *Natural Resources Journal* , Vol. 31, pp. 213–242 (1991).

Joyner, Christopher C., "Legal Implications of the Concept of the Common Heritage of Mankind", *International & Comparative Law Quarterly*, Vol. 35, pp. 190–199 (1986).

Joyner, Christopher C., "Protection of the Antarctic Environment: Rethinking the Problems and Prospects", *Cornell International Law Journal*, Vol. 19, pp. 259–273 (1986).

Joyner, Christopher C., "Review Article: the Evolving Antarctic Legal Regime", *The American Journal of International Law*, Vol. 83, No. 3, pp. 605–626 (1989).

Joyner, Christopher C., "The Evolving Antarctic Minerals Regime", *Ocean Development and International Law*, Vol. 19, Issue 1, pp. 73–95 (1988).

Joyner, Christopher C., *Antarctica and the Law of the Sea*, (Dordrecht/ Boston/ London: Martinus Nijhoff Publishers, 1992).

Joyner, Christopher C., *Governing the Frozen Commons -- The Antarctic Regime and Environmental Protection*, (University of South Carolina Press, 1998).

Kammerer, Jörn-Axel, "The Protocol on Environmental Protection to the Antarctic Treaty", *Law and State*, Vol. 45, pp. 68–80 (1992).

Keller, Arthur Schopenhauer / Lissitzyn, Oliver James / Mann, Frederick Justin, *Creation of Rights of Sovereignty Through Symbolic Acts, 1400–1800*, (New York: Columbia University Press, 1938).

Keys, J. R., "Chapter 4: Ice", in Glasby, G.P. (ed.), *Antarctic Sector of the Pacific*, (New York: Elsevier Science Publisher, 1990) pp. 95–124.

Kimball, Lee A., "Special Report on: the Antarctic Mineral Convention", in Spletts-toesser, J.F. / Dreschhoff, G.A.M., (eds.), *Mineral Rsources Potentional of Antarctica* (Washington, D.C.: American Geophysical Union, 1990).

Kimball, Lee A., *Report on Antarctica*, (Washington, D.C., World Resources Institute, January 1991), (copy at SPRI).

Kindt, John warren / Parriott, Todd Jay, "Ice-Covered Areas: The Competing Interests of Conservation and Resource Exploitation", *San Diego Law Review*, Vol. 21, pp. 941–984 (1983–1984).

Kock, Karl-Hermann, (ed.), *Understanding CCAMLR's Approach to Management*, (http://www.ccamlr.org/pu/E/e_pubs/am/am-all.pdf , 2000).

Kock, Karl-Hermann, "Fishing and Conservation in Southern Water", *Polar Record*, Vol. 30, No. 172, pp. 3–22 (1994).

Kosack, Hans-Peter, " Wie gross ist das Südpolargebiet?", *Petermann's Geographische Mittheilungen*, vol. 94, pp. 202–206 (1950).

Kukk, Christopher L., / Deese, David A., "At the Water's Edge: Regional Conflict and Cooperation over Fresh Water", *UCLA Journal of International Law and Foreign Affairs*, Vol. 1, pp. 21–431 (1996).

Lagoni, Rainer, "Antarctica's Mineral Resources in International law", *Zeitschrift für ausländisches öffentliches Recht und Völkerrecht* , Vol. 39, pp. 1–37 (1979).

Lakhtine, W., "Rights over the Arctic", *American Journal of International Law*, Vol. 24, pp. 703–717 (1930).

Larminie, F.G., "Mineral Resources: Commercial Prospects for Antarctic Minerals", in Triggs, G.D. (ed.), *The Antarctic Treaty Regime: Law, Environment and Resource*, (Cambridge: Cambridge University Press, 1987), pp. 176–181.

Laws, Richard, "Science as an Antarctic Resource", in Cook, Grahame (ed.), *The Future of Antarctica: Exploitation Versus Preservation*, (New York: Manchester University Press, 1990), pp. 8–24.

Lindley, Mark Frank, *The Acquisition and Government of Backward Territory in International Law: Being a Treatise on the Law and Practice Relating to Colonial Expansion*, (Longmans, Green and Company, 1926).

Lodge, Michael W., "International Seabed Authority's Regulations on Prospecting and Exploration for Polymetallic Nodules in the Area", *Journal of Energy & Natural Resource Law*, Vol. 20, No. 3, pp. 270–295 (2002).

Lodge, Michael W., "The International Seabed Authority and Article 82 of the UN Convention on the Law of the Sea", *The International Journal of Marine and Coastal Law*, Vol. 21, No. 3, pp. 323–333 (2006).

Lowe, Vaughan, *International Law*, (Oxford: Oxford University Press, 2007).

Luard, Evan, *The Control of The Seabed – A New International Issue*, (London: Heinemann, 1974).

Lucas, P.H.C., "Finding Ways and Means to Conserve Antarctica", in McNeely, Jeffrey A. / Miller, Kenton (eds.), *National Parks, Conservation, and Development: The Role of Protected Areas in Sustaining Society: Proceedings of the World Congress on National Parks, Bali, Indonesia, 11–22 October 1982*, (Smithsonian Institution Press, 1984) pp. 369–374.

Lundpuist, Thomas R., "The Iceberg Cometh?: International Law Relating to Antarctic Iceberg Exploitation", *Nature Resource Journal*, Vol. 17, No. 1, pp. 1–41 (1977).

Lyons, David, "Environmental Impact Assessment in Antarctica under the Protocol on Environmental Protection", *Polar Record*, Vol. 29, Issue 169, pp. 111–120 (1993).

Maddox, John, "Antarctic Mining Regime at Risk", *Nature*, Vol. 307, pp. 105–106 (January 1984).

Malanczuk, Peter, *Akehurst's modern introduction to International Law*, (7th ed., London and New York: Routledge, 1997).

Manson, Neil A., "Formulating the Precautionary Principle", *Environmental Ethics*, Vol. 24, pp. 263–274 (2002).

May, John, *The Greenpeace Book of Antarctica: A New View of the Seventh Continent*, (Child & Associates, 1988).

Mikhailov, S., "Imperialisticheskaya Bor'ba za Antarctiku", *Voprosy Ekonomiky*, No. 10, pp. 66–73 (October 29, 1949).

Mitchell, Barbara / Kimball, Lee, "Conflict over the Cold Continent", Foreign Policy, No. 35, pp. 124–141 (Summer, 1979).

Mitchell, Barbara, *Frozen Stakes: The Future of Antarctic Mineral*, (London: International Institution for Environment and Development, 1983).

Moelwyn-Hughes, Ronw ed., *International Law*, (6th ed., London & Toronto: J.M. Dent & Sons Ltd., 1927).

Molodtsov, S.V., *Sovremennoe mezhdunarodno-pravovoe polozhenie Antarktiki*, (Moscow: Gosyurizdat, 1954).

Müller, Scharnhorst, "The Impact of UNCLOS III on the Antarctic Regime", in Rüdiger Wolfrum (ed.), *Antarctic Challenge II*, (Berlin: Duncker & Humblot, 1984), pp. 169–176.

Murray, John, "The Renewal of Antarctic Exploration", *The Geographical Journal*, Vol. 3, No. 1, pp. 1–27 (January 1894).

Myhre, Jeffrey D., *Antarctic Treaty System, Politics, Law and Diplomacy*, (Boulder: Westview Press, 1986).

Newman, E. Paul, "The Antarctica Mineral Resources Convention: Developments from the October 1986 Tokyo Meeting of the Antarctic Treaty Consultative Parties", *Denver Journal of International Law and Policy*, Vol. 15, pp. 421–434 (1986–1987).

Nordenskjöld, Otto/ Mecking, Ludwig, *The Geography of the Polar Regions*, (New York: The Commonwealth Press, 1928).

O'Connell, Daniel Patrick, *International Law*, (2nd ed., London: Stevens, 1970).

OECD, *Water and Violent Conflict*, (OECD Issues Brief, 2005).

Opeskin, Brian R., / Rothwell, Donald R., "Australia's Territorial Sea: International and Federal Implications of Its Extension to 12 Miles", *Ocean Development and International Law*, Vol. 22, Issue 4, pp. 395–431(1991).

Oppenheim, L., *International Law*, (7th ed., London, New York, Toronto: Longmans, 1948).

Overholt, Deborah H., "Environmental Protection in the Antarctic: Past, Present and Future", *The Canadian Yearbook of International Law*, Vol. 28, pp. 227–262 (1990).

Øvstedal, D.O. / Smith, R.I. Lewis, *Lichens of Antarctica and South Georgia a Guide to their Identification and Ecology*, (Cambridge: Cambridge University Press, 2001).

Parkinson, Claire L., "Southern Ocean Sea-Ice Distribution and Extents", *Philosophical Transactions: Biological Sciences*, Vol. 338, No. 1285, Antarctica and Environmental Change (Nov. 30, 1992), pp. 243–250.

Parriott, Todd Jay, "Territorial Claims in Antarctica: Will the United States Be Left Out the Cold?", *Stanford Journal of International Law*, Vol. 22, pp. 68–121 (1986).

Parry, M.L. / Canziani, O.F. / Palutikof, J.P. / van der Linden, P.J. / Hanson, C.E. (eds.), *Climate Change 2007: Impacts, Adaptation and Vulnerability, Contribution of Working Group II to the Fourth Assessment Report of the Intergovernmental Panel on Climate Change*, (Cambridge: Cambridge University Press, 2007).

Peterson, M. J., "Antarctica: The Last Great Land Rush on Earth", *International Organization*, Vol. 34, No. 3, pp. 377–403 (Summer, 1980).

Peterson, M.J., *Managing the Frozen South-The Creation and Evolution of the Antarctic Treaty System (Studies in International Political Economy) (Vol. 20)*, (Berkeley, Los Angeles, London: University of California Press, 1988).

Pharand, A.D., *The Law of the Sea of the Arctic: With Special Reference to Canada*, (Ottawa: University of Ottawa Press, 1973).

Phillpot, H.R., "Physical Geography: Climate", in Bonner, W.N. / Walton, D.W.H. (eds.), *Key Environments: Antarctica*, (Oxford: Pergamon Press, 1985), pp. 23–38.

Plott, Barry Merrill, *The Development of United States Antarctic Policy*, (Fletcher School of Law and Diplomacy, 1969).

Powell, D.L., "Scientific and Economic Considerations Relating to the Conservation of Marine Living Resources in Antarctica", in Vicuña, Francisco Orrego (ed.), *Antarctic Resources Policy: Scientific, Legal, and Political Issues*, (Cambridge: Cambridge University Press, 1983), pp. 111–118.

Powell, Darry, "Antarctic Marine Living Resources and CCAMLR", in Herr, R.A. / Hall, H.R. / Haward, M.G. (eds.), *Antarctica's Future Continuity or Change?* (Hobart: Australian Institute of International Affairs, 1990), pp. 61–70.

Quigg, Philip W., *A Pole Apart: The Emerging Issue of Antarctica: A Twenty Century Fund Report*, (New York: New Press McGraw-Hill Book Company, 1983).

Quilleré-Majzoub, Fabienne, "Glaces polaires et icebergs: quid juris gentium?" *Annuaire Français de droit international*, Vol. 52, Issues 52, pp. 432–454 (2006).

Redgwell, Catherine, "Antarctica: Wilderness Park or Eldorado Postponed?", *Environmental Politics*, Vol. 1, Issue 1, pp. 137–143 (1992).

Redgwell, Catherine, "Antarctica", *International and Comparative Law Quarterly*, Vol. 39, pp. 474–481 (1990).

Redgwell, Catherine, "Environmental Protection in Antarctica: the 1991 Protocol", *International and Comparative Law Quarterly*, Vol. 43, Issue 03, pp. 599–634 (July 1994).

Reeves, J.S., "Antarctic Sector", *American Journal of International Law*, Vol. 33, pp. 519–521, (1939).

Robin, G. de Q., "Changing the Sea Level", in Bolin, B. / Döös, B.R./ Jäger J. / Warrick, R. A. (eds.), *SCPOE 29 The Greenhouse Effect, Climate Change, and Ecosystems*, (Chichester: John Wiley & Sons, 1986), pp. 323–359.

Rose, G. / McElroy, C.T., *Coal Potential of Antarctica, Resource Report 2.* (Bureau of Mineral Resources, Geology and Geophysics, Australian Government Publishing Service, Canberra, 1987).

Rowley, Peter D. / Williams, Paul L. / Pride, Douglas E., "Mineral Occurrences of Antarctica", in Behrendt, John C (ed.), *Petroleum and Mineral Resources of Antarctica*, (USA Geol Surv,1983), pp. 25–49.

Rubin, M.J., "Thaddeus Bellingshausen's Scientific Programme in the Southern Ocean 1818–1821", *Polar Record*, Vol. 21, Issue 132, pp. 215–229 (1982).

Sauder, Robert A., *The Lost Frontier: Water Diversion in the Growth and Destruction of Owens Valley Agriculture*, (Tucson: University of Arizona Press, 1994).

Schenk, Dieter, *Kontiguität als Erwerbstitel im Völkerrecht*, (Ebelsbach: Gremer, 1978).

Schmitz, Ernst / Friede, Wilhelm, "Souveränitätsrechte in der Antarktis , *Zeitschrift für ausländisches öffentliches Recht und Völkerrecht*, Vol. 9, pp. 219–263 (1939–1940).

Scullz, Tucker / Kimball, Lee A., "Antarctica: Is There Life after Minerals? The Minerals Treaty and Beyond", *Marine Policy*, Vol. 13, Issue 2, pp. 87–98 (1989).

Seppelt, R.D. / Green, T.G.A., "A Bryophyte Flora for Southern Victoria Land, Antarctica", *New Zealand Journal of Botany*, Vol. 36, Issue 4, pp. 617–635 (1998).

Shapley, Deborah, *The Seventh Continent: Antarctica in a Resource Age*, (Washington D.C.: Johns Hopkins University Press, 1985).

Shusterich, Kurt M, "The Antarctic Treaty System: History, Substance, and Speculation", *International Journal*, Vol. 39, pp. 800–827 (1984).

Simpson, Frank A. (ed.), *The Antarctic Today: A Mid-Century Survey by the New Zealand Antarctic Society*, (Wellington: A. H. & A.W. Reed and New Zealand Antarctic Society, 1952).

Simsarian, James, "The Acquisition of Legal Title to Terra Nullius", *Political Science Quarterly*, Vol. 53, No. 1, pp. 111–128 (March 1938).

Sinclair, Ian, *The Vienna Convention on the Law of Treaties*, (2en. Ed., Manchester University Press, 1984).

Siple, Paul A., "Antarctic Geography", Crary, A. P. / Gould, L. M. / Hulburt, E. O. / Odishaw, Hugh / Smith, Waldo E. (eds.), *Antarctica in the International Geophysical Year – Based on a symposium on the Antarctic*, (Baltimore, Maryland: Waverly Press, 1956), pp. 13.

Skotnicki, M. L. / Selkirk, P. M., "Plant Biodiversity in an Extreme Environment: Genetic Studies of Origins, Diversity and Evolution in the Antarctic", in Bergstrom, D. M. / Convey, P. /Huiskes, A.H.L. (eds.), *Trends in Antarctic Terrestrial and Limnetic Ecosystems*, (Dordrecht: Springer, 2006), pp. 161–175.

Smedal, Gustav, *Acquisition of Sovereignty over Polar Areas*, (J. Dybwad Publisher, 1931).

Smith, R.I. Lewis, "Terrestrial Plant Biology of the Sub-Antarctic and Antarctic", in Laws, R.M. (ed.), *Antarctic Ecology*, (London: Academic Press, 1984), pp. 61–162.

Smith, R.I. Lewis, "The Enigma of Colobanthus Quitensis and Deschampsia Antarctica in Antarctica", in Huiskes, A.H.L. / Gieskes, W.W.C. / Rozema, J. / Schorno, R.M.L. / van der Vies, S.M. / Wolf, W.J. (eds.), *Antarctic Biology in a Global Context*, (Leiden: Backhuys, 2003), pp. 234–239.

Sollli, Finn, "The Development of the Antarctic Treaty: Trends and Issues", in Wolfrum, Rüdiger (ed.), *Antarctic Challenge II*, (Berlin: Duncker & Humblot, 1984), pp. 17–37.

Spivak, J, "Now the Energy Crisis Spurs Idea of Seeking Oil at the South Pole", *Wall Street Journal*, Feb. 21, 1974, P.1.

Splettstoesser, J.F., "Antarctic Geology and Mineral Resources", *Geology Today*, Vol. 1, Issue 2, pp. 41–45 (1985).

St John, B. "Antarctica: Geology and Hydrocarbon Potential", in Halbouty, M.T. (ed.), *Future Petroleum Resources of the World, American Association of Petroleum Geologists Memoir*, (Tulsa, Oklahoma 1986), pp. 55–100.

Stäblein, Gerhand, "Historische Aspekte der deutschen geowissenschaftlichen Polarforschung", *Polarforschung*, Vol. 52, No. 2, pp. 219–225, (1981).

Statement by Mr Stevenson, US Representative to the Second Committee, 41st Meeting, II Off. Rec. 291, para. 20 (1974).

Sullivan, Walter, "Antarctica in a Two-Power World", *Foreign Affairs*, Vol. 36, No. 1, pp. 154–166 (October 1957).

Sullivan, Walter, *Quest of A Continent*, (New York, McGraw-Hill, 1957).

Taubenfeld, Howard J., "A Treaty for Antarctica", *International Conciliation*, Vol. 33, No. 531, pp. 245–322 (1959–1961).

Toma, Peter A., "Soviet Attitude towards the Acquisition of Territorial Sovereignty in the Antarctic", *American Journal of International Law*, Vol. 50, No. 3 pp. 611–626, (July 1956).

Triggs, Gillian D., *International Law and Australian Sovereignty in Antarctica*, (Sydney: Legal Books Ltd., 1986).

Triggs, Gillian Doreen, "The Antarctic Treaty System: Some Jurisdictional Problems", in Triggs, Gillian Doreen (ed.), *The Antarctic Treaty Regime: Law, Environment and Resources*, (Cambridge: Cambridge University Press, 1987), pp. 88–109.

Trolle-Anderson, Rolph, "The Antarctic Scene: Legal and Political Facts", in Triggs, Gillian D. (ed.), *The Antarctic Treaty Regime: Law, Environment and Resources*, (Cambridge: Cambridge University Press, 1987), pp. 57–64.

Trombetta-Panigadi, Francesca, "The Exploitation of Antarctic Icebergs in International Law", in Francioni, Francesco / Scovazzi ,Tullio (eds.), *International Law for Antarctica*, (The Hague: Kluwer Law International, 1996), pp. 225–257.

Tronchetti, Fabio, *The Exploitation of Nature Resources of the Moon and other Celestial Bodies: a Proposal for a Legal Regime*, (Leiden/Boston: Martinus Nijhoff Publishers, 2009).

U.S. Antarctic Projects Officer, *An Introduction to Antarctica*, (Washington D.C.: U.S. Antarctic Projects Officer, 1961).

U.S. Policy with Regard to Mineral Exploration and Exploitation in the Antarctic: Hearing before the Subcommittee on Oceans and International Environment of the Senate Committee on Foreign Relations, 94th Congress, 1st Session 68 (1975).

United Nations, *Memorandum on the Regime of the High Seas*, (reputed to have been prepared by the French jurist Gidel), UN General Assembly Doc. No. A/CN. 4/32, 14 July 1950.

United Nations, *United Nations Conferences on the Law of the Sea Third Conference, Official Records*, Vol. 11–17, (New York, William S. Hein & Co., Inc., 2000).

United Nations, *Year Book of the International Law Commission Volume II: Documents of the Second Part of the Seventeenth Session and of the Eighteenth Session Including the Reports of the Commission to the General Assembly*, (New York: United Nations, 1967).

United Nations Division for Ocean Affairs and the Law of the Sea, Office of Legal Affairs, *The Law of the Sea*, http://www.un.org/Depts/los/index.htm.

Vaughan, David G. / Marshall, Gareth J. / Connolley, William M. / Parkinson, Claire / Mulvaney, Robert / Hodgson, Dominic A. / King, John C. / Pudsey, Carol J. / Turner, John, "Recent Rapid Regional Climate Warming on the Antarctic Peninsula," *Climatic Change*, Vol. 60, pp. 243–274 (2003).

Verdross, Alfred, *Völkerrecht*, (4th ed., Wien: Springer, 1959).

Vicuña, Francisco Orrego, "The Protocol on Environmental Protection to the Antarctic Treaty: Questions of Effectiveness", in *The Georgetown International Environmental Law Review*, Vol. 7, pp. 1–13, (1994–1995).

Vicuña, Francisco Orrego, *The Protection of the Antarctic Environment, International Antarctic Regime Project Publication Series*, No. 5 (The Fridtjof Nansen Institute, 1992).

Vigni, Patrizia, "The Interaction between the Antarctic Treaty System and the Other Relevant Conventions Applicable to the Antarctic Area: A Practical Approach versus Theoretical Doctrines", in Frowein, Jochen A./ Wolfrum, Rüdiger (eds.), *Max Planck Yearbook of United Nations Law*, Vol. 4 , (The Hague, London, Boston, Kluwer Law International,2000), pp. 481–542.

Viñuales, Jorge E., "Iced Freshwater Resources: A Legal Exploitation", in *Yearbook of International Environmental Law*, Vol. 20, Issue 1, pp. 186–206 (2009).

von der Heydte, Friedrich August Freiherr, "Discovery, Symbolic Acquisition and Virtual Effectiveness in International Law", *American Journal of International Law*, Vol. 29, pp. 448–471 (1935).

von Glahn, Gerhard, *Law Among Nations: An Introduction to Public International Law*, (2nd ed., London, Macmillan Company, 1970).

von Münch, Ingo, "Völkerrechtsfragen der Antarktis", *Archiv des Völkerrechts*, Vol. 7, pp. 225–262 (1958–1959).

Waldock, C.H.M., Disputed sovereignty in the Falkland Islands Dependencies, *British Yearbook of International Law*, Vol. 25, pp. 310–353 (1948).

Wang, Runyu, "The Precautionary Principle in Maritime Affairs", *WMU Journal of Maritime Affairs*, Volume 10, Number2, pp. 143–165 (2011).

Watts, Arthur D., "Lessons from Resource Negotiations", in Wolfrum, Rüdiger (ed.), *Antarctic Challenge III, Conflicting Interests, Cooperation, Environmental Protection, Economic Development: Proceedings of An Interdisciplinary Symposium*, (Berlin: Duncker & Humblot, 1988), pp. 319–332.

Watts, Arthur D., "The Convention on the Regulation of Antarctic Mineral Resource Activities 1988", *International & Comparative Law Quarterly*, Vol. 39, pp. 169–182 (1990).

Watts, Arthur D., *International Law and the Antarctic Treaty System* – (Hersch Lauterpacht Memorial Lecture Series; Vol. 11), (Cambridge: Grotius Publications Limited, 1992).

Westermeyer, William E., *The Politics of Mineral Resource Development in Antarctica: Alternative Regimes for the Future,* (Boulder: Westview Press, 1984).

Westermeyer, William E., "Alternative Regimes for Future Mineral Resource Development in Antarctica", *Ocean Management,* Vol. 8, Issue 3, pp. 197–232 (September 1983).

Wexler, H./ Rubin, M. J./ Caskey, J. E. (eds.), *Matthew Fontaine Maury Memorial Symposium on Antarctic Research,* (Richmond, Virginia: William Byrd Press, 1962).

White, Mary Victoria, "The Common Heritage of Mankind: An Assessment", in *Case Western Reserve Journal of International Law,* Vol. 14, pp. 509–542 (1984).

Whiteman, Marjorie M., *Digest of International Law Vol. II,* (Washington, DC: Government Printing Office, 1963).

Wilson, Robert E., "National Interests and Claims in the Antarctic", *Arctic,* Vol. 17, No. 1, pp. 15–31 (1964).

Wolfrum, Rüdiger (ed.), *Antarctic Challenge II, Conflicting Interests, Cooperation, Environmental Protection, Economic Development: Proceedings of an Interdisciplinary Symposium,* (Berlin: Duncker & Humblot, 1984).

Wolfrum, Rüdiger (ed.), *Antarctic Challenge III, Conflicting Interests, Cooperation, Environmental Protection, Economic Development: Proceedings of an Interdisciplinary Symposium,* (Berlin: Duncker & Humblot, 1988).

Wolfrum, Rüdiger (ed.), *Antarctic Challenge: Conflicting Interests, Cooperation, Environmental Protection, Economic Development,* (Berlin: Duncker & Jumblot, 1984).

Wolfrum, Rüdiger, "Precautionary Principle", in Beurier, Jean-Pierre / Kiss, Alexander / Mahmoudi, Said (eds.), *New Technologies and Law of the Marine Environment,* (The Hague et al.: Kluwer International Law, 2000), pp. 203–213.

Wolfrum, Rüdiger, "The Exploitation of Antarctic Mineral Resources: Risks and Stakes", in Verhoeven, J. / Sands, P. / Bruce, M. (eds.), *The Antarctic Environment and International Law* (London: Graham and Trotman Limited, 1992), pp. 27–32.

Wolfrum, Rüdiger, "The Principle of the Common Heritage of Mankind", *ZaöRV* Vol. 43, pp. 312–337 (1983).

Wolfrum, Rüdiger, "The Use of Antarctica Non-Living Resources: The Search for A Trustee?", in Wolfrum, Rüdiger (ed.), *Antarctic Challenge: Conflicting Interests, Cooperation, Environmental Protection, Economic Development,* (Berlin: Duncker & Jumblot, 1984) pp. 143–163.

Wolfrum, Rüdiger, Die Internationalisierung staatsfreier Räume, (Springer-Verlag, Berlin Heidelberg New York Tokyo, 1984).

Wolfrum, Rüdiger, *The Convention on the Regulation of Antarctic Mineral Resource Activities: An Attempt to Break New Ground,* (Heidelberg: Springer-Verlag, 1991).

World Bank, *World Development Report 2010: Development and Climate Change*, (Washington D.C.: The World Bank, 2010).

Zacharias, Mark A. / Gerber, Leah R. / Hyrenbach, K. David, "Review of the Southern Ocean Sanctuary: Marine Protected Areas in the context of the International Whaling Commission Sanctuary Programme", *Journal of Cetacean Research and Management*, Vol. 8, Issue 1, pp. 1–12 (2006).

Zegers, Fernando, "The Canberra Convention: Objectives and Political Aspects of Its Negotiation", in Vicuña, Francisco Orrego (ed.), *Antarctic Resource Policy: Scientific, Legal and Political Issues*, (Cambridge: Cambridge University Press, 1983), pp. 149–156.

Table of Cases

Anglo Norwegian Fisheries Case	International Court of Justice, *United Kingdom v. Norway*, Reports of Judgments, Advisory Opinion and Orders Judgment of December 18th 1951, ICJ Report 1951, p. 117
Antarctica Cases	International Court of Justice, *United Kingdom v. Argentina; United Kingdom v. Chile*, ICJ Report 1956, p. 12, p. 15
Asylum Case	International Court of Justice, *Colombia v. Peru*, Reports of Judgments, Advisory Opinion and Orders Judgment of November 20th , ICJ Report 1950, p. 395
Case Concerning Military and Paramilitary Activities in and against Nicaragua	International Court of Justice, *Nicaragua v United States of America*, Merits of the Court of 27 June 1986, ICJ Reports 1986, p. 14
Case of S.S. "Lotus"	Permanent Court of International Justice, *France v Turkey*, Judgment of the Court of 7 September1927, Series A. No. 10
Case Relating to the Territorial Jurisdiction of the International Commission of the River Oder	Permanent Court of International Justice, *United Kingdom, Czechoslovakia, Denmark, France, Germany, Sweden v. Poland*, Judgment of 10 September 1929 (including the text of Declarations by Judge de Bustamante and by Judge Pessôa),Series A No. 23
Clipperton Island Case	*Mexico v France*, "Judicial Decisions Involving Questions of International Law: Arbitral Award on the Subject of the Difference Relative to the Sovereignty over Clipperton Island Source", Decision rendered at Rome, 28 January 1931, *The American Journal of International Law*, Vol. 26, No. 2 (April 1932), pp. 390–394
Free Zones of Upper Savoy and the District of Gex	Permanent Court of International Justice, *France v Switzerland*, Judegement of the Court of 7 June 1932, Series A./B. No. 46
Legal Status of Eastern Greenland	Permanent Court of International Justice, *Denmark v Norway*, Judgement of the Court of 4 April 1933, Series A./B. No. 53

North Sea Continental shelf Cases	International Court Justice, *Federal Republic of Germany v. Denmark; Federal Republic of Germany v. Netherlands,* ICJ Report 1969 p. 3
Reparation for Injuries Suffered in the Service of the Nations	International Court of Justice, Advisory Opinion of the Court of 11 April 1949, ICJ Report 1949, p. 174
Tasmanian Dam Case	High Court of Australia, *Commonwealth of Australia v. State of Tasmania,* Judgment of 1 July 1983, Commonwealth Law Reports, Vol. 158, p. 1 (1983)
The Island of Palmas Case (or Miangas)	Permanent Court of Arbitration, *United States of America v the Netherland,* Arbitral Awards of 4 April 1928, Reports of International Arbitral Awards, vol. 2, p. 829
The Legal Status of Eastern Greenland Case	Permanent Court of International Justice, *Norway v Denmark,* Judgement of the Court of 5 April 1933, Series A./B. No. 53

International Treaties

1969 Civil Liability Convention	1969 International Convention on Civil Liability for Oil Pollution Damage, adopted on 29 November 1969, entered into force on 19 June 1975; being replaced by 1992 Protocol: adopted on 27 November 1992, entered into force on 30 May 1996, International Legal Materials, vol. 9, pp. 45 et seq.
1958 Convention on the High Seas	Convention on the High Seas, adopted on April 29 1958, entered into force on 30 September 1962, United Nations Treaty Series, vol. 450, pp. 11 et seq.
1969 Vienna Convention on the Law of Treaties	1969 Vienna Convention on the Law of Treaties, adopted on 23 May 1969, entered into force on 27 January 1980, Treaty Series, vol. 1155, pp. 331 et seq.
1973/1978 MARPOL Convention	International Convention for the Prevention of Pollution form Ships, adopted on 2 November 1973 as amended before entry into force by the Protocol of 1978 relating thereto (adopted 17 February 1978, entered into force 2 October 1983), International Legal Materials, vol. 12, pp. 1319 et seq.
1979 the Moon Treaty	1979 Agreement Governing the Activities of States on the Moon and Other Celestial Bodies, adopted on 5 December 1979, entered into force on 11 July 1984, International Legal Materials, vol. 18, pp. 1434 et seq. (1979)
1982 UNCLOS	1982 United Nations Convention on the Law of the Sea, adopted on 10 December 1982, entered into force on 16 November 1994, International Legal Materials, vol. 21, pp. 1261 et seq. (1982)
Convention on the continental shelf	Convention on the continental shelf, adopted on 29 April 1958, entered into force on 10 June 1964, United Nations Treaty Series, vol. 499, pp. 311 et seq.
ICJ Statute	1945 Statute of the International Court of Justice, adopted on 26 June 1945, entered into force on 24 October 1945, the Supplement to volume 39 of the American Society of International Law, pp. 215 et seq. (1945)
International Whaling Convention	1946 International Convention for the Regulation of Whaling, adopted on 2 December 1946, entered into force on 10 November 1948, United Nations Treaty Series, vol. 161, pp. 72 et seq.

Outer Space Treaty	1967 Treaty on Principles Governing the Activities of States in the Exploration and Use of Outer Space, including the Moon and Other Celestial Bodies, adopted on 27 January 1967, entered into force on 10 October1967, International Legal Materials, vol. 6, pp. 386 et seq. (1967)
CLC Protocol 1992	Protocol of 1992 to amend the International Convention on Civil Liability for Oil Pollution Damage of 29 November 1969, adopted on 27 November 1992, entered into force on 30 May 1996, United Nations Treaty Series, vol. 1956, pp. 255 et seq.
Stockholm Declaration	Declaration of the United Nations Conference on the Human Environment, adopted on June 16, 1972, by the United Nations conference on the Human Enivornment, UN Doc. A/Conf.48/14/Rev. 1(1973); International Legal Materials, vol. 11, pp. 1416 et seq. (1972)
Universal Declaration on the Human Genome and Human Rights	Universal Declaration on the Human Genome and Human Rights, published on 11 November 1997, Bulletin of Medical Ethics, vol. 126, pp. 9–11 (1997)
World Charter of Nature	World Charter of Nature, adopted on 28 October 1982, International Legal Materials, vol. 22, pp. 455 et seq. (1983)
World Heritage Convention	Convention Concerning the Protection of the World Cultural and Natural Heritage, adopted on 16 November 1972, entered into force on 15 December 1975, 1037, pp. 151 et seq.
World Population in 2300	World Population in 2300, concluded by Department of Economic and Social Affairs Population Division, Doc.ST/ ESA/SER.A/ 236 (2004), ESA/P/ WP.187/ Rev.1, released on 24 March 2004, in Population and Development Review, vol. 30, Issue 1, pp. 181–187 (March 2004)

UN Documents

A/37/PV.10	Verbatim Record of the 10th Meeting, United Nations General Assembly, 37th Session, 25 October 1982
A/CONF.62/C.2/ L.98/Add.2	Calculation of Areas Illustrated beyond 200 Miles, United Nations General Assembly, The Third Nations Conference on the Law of the Sea,3 May 1978
A/CONF.62/WP.10	Third United Nations Conference on the Law of the Sea: Informal Composite Negotiating Text, the 62nd Session, 23 May to 15 July 1977
A/RES/1348 (XIII)	Resolution of Question of the Peaceful use of Outer Space, United Nations General Assembly, the 13th Session, 13 December 1958
A/RES/38/132,	Draft Code of Offences against the Peace and Security of Mankind, United Nations General Assembly, the 38th Session, 19 December 1983
A/RES/38/193	Conversion of the United Nations Industrial Development Organization into a Specialized Agency, United Nations General Assembly, the 38th Session 20 December 1983
A/RES/38/77	Resolution on the Question of Antarctica, United Nations General Assembly, the 38th Session, 15 December 1983
A/RES/39/52	Resolution on the Cessation of all Explosions of Nuclear Weapons, United Nations General Assembly, the 39th Session, 17 December 1984
A/RES/40/156 (A) and (B)	Resolution on the Question of Antarctica, United Nations General Assembly, the 40th Session, 16 December 1985.
A/RES/41/88	Resolution on the Question of Antarctica, United Nations General Assembly, the 41st Session, 4 December 1986
A/RES/42/46	Resolution on the Question of Antarctica, United Nations General Assembly, the 42nd Session, 30 November 1987
A/RES/43/83	Resolution on the Question of Antarctica, United Nations General Assembly, the 43rd Session, 7 December 1988
A/RES/44/124(B)	Resolution on the Question of Antarctica, United Nations General Assembly, the 44th Session, 15 December 1989
A/Res/60/47	Resolution of Question of Antarctica, United Nations General Assembly, the 60th Session, 8 December 2005

A/C.1/38/PV.44	Statement of the Representative of Pakistan, United Nations General Assembly, 44th meeting of the 38th Session, 29 November 1983
A/C.1/44/L.69	Draft Resolution of Question of Antarctica, submitted by Antiqua and Barbuda, Bangladesh, Brunei Darussalam, Cameroon, Ghana, Indonesia, Kenya, Kuwait, Lesotho, Malaysia, Mali, Mexico, Nepal, Nigeria, Oman, Pakistan, Rwanda, Sir Lanka, Sudan, Uganda, Zaire, Zambia and Zimbabwe, the 44th Session, 20 November 1989.
A/C.1/60/PV.23	Action on all draft Resolutions Submitted under all Disarmament and International Security Agenda Items, United Nations General Assembly, the 23rd meeting of the 60th Session, 1 November 2005.
A/RES/S-6/3201	Resolution of Declaration on the Establishment of a New International Economic Order, United Nations General Assembly, the Sixth Special Session , Supp. (No. 1), 1 May 1974

ATS Documents

Antarctic Treaty	Antarctic Treaty, adopted in Washington on 1 December 1959, entered into force on 23 June 1961, International Legal Materials, vol. 19, pp. 860 et seq. (1980)
Convention for the Conservation of Antarctic Marine Living Resources	Convention for the Conservation of Antarctic Marine Living Resources, adopted on 20 May 1980, entered into force on 7 April 1982, International Legal Materials, vol. 19, pp. 841 et seq. (1980)
Convention for the Conservation of Antarctic Seals	Convention for the Conservation of Antarctic Seals, adopted on 1 June 1972, entered into force on 11 March 1978, International Legal Materials, vol. 11, pp. 251 et seq. (1972)
Convention on the Regulation of Antarctic Mineral Resource Activities	Convention on the Regulation of Antarctic Mineral Resource Activities, adopted on 2 June 1988, not entry into force, International Legal Materials, vol. 27, pp. 868 et seq. (1988)
Final Act of the Fourth Special Antarctic Treaty Consultative Meeting on Antarctic Mineral Resources	Final Act of the Fourth Special Antarctic Treaty Consultative Meeting on Antarctic Mineral Resources, adopted on June 2 1988, International Legal Materials, vol. 27, pp. 859 et seq. (1988)
Final Report of the XV ATCM	Final Report of the XV ATCM, adopted by ATCM at the 15th Meeting in Paris on 9–20 October 1989
Interim Report of the XI SATCM,	Interim Report of the XI SATCM, adopted by ATCM at the 11th Meeting in Vina del Mar on 19 November to 6 December 1990
Madrid Protocol	Protocol on Environmental Protection to the Antarctic Treaty, adopted on 4 October 1991, entered into force on 14 January 1998, International Legal Materials, vol. 30, pp. 1455 et seq. (1991)
Measures for the Protection of Living Resources	Measures for the Protection of Living Resources, adopted in 1961 in Canberra, at the First ATCM annexed to Recommendation ATCM I-VIII, effective on 10 October 2003
Recommendation IX–1	Recommendation on Antarctic Mineral Resources, adopted at the 9th ATMC (19 September to 7 October 1977), London, effective on 8 September 1983

Recommendation VII-6,	Recommendation on Antarctic Resources – Effects of Mineral Exploration, adopted at the 7th ATCM (30 October to 10 November 1972), Wellington
Recommendation XV-3	Recommendation on Human Impact on the Antarctic Environment: Waste Disposal, the 15th ATCM (9 to 20 October 1989), Paris
Recommendation XV-4	Recommendation on Human Impact on the Antarctic Environment: Prevention, Control and Response to Marine Pollution, the 15th ATCM (9 to 20 October 1989), Paris, the 15th ATCM (9 to 20 October 1989), Paris
Recommendation XV-5	Recommendation on the Antarctic Environment: Environmental Monitoring in Antarctica, the 15th ATCM (9 to 20 October 1989), Paris, not yet effective
Recommendation XV-6	Recommendation on the Antarctic Protected Area System: New Sites of Special Scientific Interest, the 15th ATCM (9 to 20 October 1989), Paris
Report of the IX ATCM	the 9th ATMC (19 September to 7 October 1977), London
XV ATCM/WP/14	"Comprehensive Measures for the Protection of the Antarctic Environment and Dependent and Associated Ecosystem: Proposed Outline of a Draft Recommendation by the Swedish Delegation", at the 15th ATCM, Annex C to the Final Report of the 15th ATCM (Paris, 9–20 October 1989), pp. 262–264
XV ATCM/WP/4	"Working Paper on the Protection of the Antarctic Environment Submitted by New Zealand", at the 15th ATCM, Annex C to the Final Report of the 15th ATCM (Paris, 9–20 October 1989), pp. 232–245
XV ATCM/WP/8	"Comprehensive Measures for the Protection of the Antarctic Environment and Dependent and Associated Ecosystems, submitted by the United States" at the 15th ATCM, Annex C to the Final Report of the 15th ATCM (Paris, 9–20 October 1989), pp. 256–261

www.ingramcontent.com/pod-product-compliance
Ingram Content Group UK Ltd.
Pitfield, Milton Keynes, MK11 3LW, UK
UKHW021922120926
8556IPUK00042BA/1121

9 783631 724255